Art and Technology

An ICUS Book

Science and Values series

Humankind has pursued truth about the extended, material world in recent centuries through scientific methods. Using the knowledge gained through these methods, man has achieved a remarkable degree of mastery over the world around him. Yet all this material progress holds limited promise for solving the deep and widespread societal problems which trouble humanity today. To ease this malaise, there is a need for a clear, universal value-system to guide mankind in personal life, in sociopolitical relations, economic relations and in the application of the power made available through science in the service of mankind.

Books in the Science and Values series seek to create an academic environment in which value considerations can be brought into scientific discourse and application. In particular, books in this series offer different perspectives regarding the dilemma of a "value-free" scientific approach while the world faces global-level environmental and societal crises. It is hoped that this series will help bring the value perspective into the application of man's knowledge.

Books within this series represent an attempt to direct scholarly discourse toward the essential interface of Science and Values precisely because this interface is so important yet so difficult to clarify. Each book differs in its relation to the overarching theme and none is expected to be able to consider all of the aspects elaborated in this statement. Nonetheless, in their cumulative impact the Science and Values books should make significant contributions toward harmonizing these two disparate domains.

Art and Technology

Edited by René Berger and Lloyd Eby

An ICUS Book

Paragon House Publishers
New York

Published in the United States by
PARAGON HOUSE PUBLISHERS
2 Hammarskjold Plaza
New York, New York 10017

An International Conference on the Unity of the Sciences Book

Library of Congress Cataloging-in-Publication Data
Main entry under title:

Art and Technology.

Papers of Committee III of the Twelfth International Conference on the Unity of the Sciences, held in Chicago, Nov. 24–27, 1983.
"An ICUS book."
Bibliography:
Includes Index.
1. Technology and The Arts—Congresses. I. Berger, René, 1915– . II. Eby. Lloyd, 1943– .
III. International Conference on the Unity of Sciences (12th 1983: Chicago, Ill.). Committee III.
NX180.T4A78 1985 700'.1'05 85-21552
ISBN 0-89226-029-7

Contents

Preface

This book has been created from the papers of Committee III of the Twelfth International Conference on the Unity of the Sciences (ICUS XII), which was convened in Chicago on November 24 to 27, 1983, under the overall theme of "Absolute Values and the New Cultural Revolution." The Committee, which had as its theme "Art and Technology," was conceived, brought together and chaired by René Berger. Professor Berger gives the rationale and the organizing issues for the Committee in his introduction to this volume. A list of contributors, along with some information about them, occurs at the end of this volume.

Differences of language caused difficulties both for the conference in Chicago and for this book. As everyone who has any experience with a language other than his own knows, each language exerts its own force, compelling thought and expression in its own characteristic ways. This is as true of English as any other language, and there are even subtle differences between American English and British English. With but one or two exceptions—who are British—all of the writers of the major papers here are persons whose first language is other than English. The papers and discussion have therefore been at least partially "translated" into American idiom, spelling and punctuation, so even those British exceptions have been affected by this language problem. As Marshall McLuhan reminded us in *The Gutenberg Galaxy: The Making of Typographic Man*, it is not just the new electronic technology which influences the messages we send and receive; we have also been profoundly molded by that older example of new technology—the printing technology introduced by Gutenberg and his fellows.

A word about the division of labor between the two editors: René Berger brought together the participants for the Committee, worked with them on paper assignments, convened and chaired the Committee, worked with the authors to revise and gather together the manuscripts prior to publication, and oversaw the process of assembly of the materials for the book. Lloyd Eby served as the chairman's assistant during the conference, received the revised manuscripts, and edited and assembled these revised manuscripts into the book; he also prepared the incidental materials and information included in it. Since he resides in New York City, he also served as liaison with the publisher.

If anything, the growth of technology has accelerated in the two years since the meeting of Committee III. The recent election campaign in the USA was conducted almost completely *on* or *for* TV. Computers have become ubiquitous, at least in the United States, and more and more powerful and innovative programs are constantly being introduced for them. Small, lightweight, sophisticated color television cameras are available now for less than five hundred dollars, and video recorders are widespread. Everyone has a music Walkman, or even several, the TV Walkman is now available, and electronic means for the creation, reproduction, and dissemination of music have become very highly developed (laser disks and electronic simulators, for example) and widely available and used. Most communities have cable television. Many of the most innovative video techniques have been adopted by commercial broadcasting and especially by music videos such as MTV. Cellular car telephones have become readily available, and there has been a confident prediction that within ten years every individual will have his own permanent personal wireless telephone that he carries with him at all times. The success of NASA's "Buck Rogers backpack" in making it possible for humans to venture into space unconnected with any other vehicle, and of its Challenger spacecraft in plucking two defective satellites from orbit and returning them to Earth has led to renewed speculation and hope about extraterrestrial adventures, stations and colonies. What people *do* with all this technology, however, remains as problematic as ever.

It is noteworthy that, either during the discussions in Chicago or in their papers, *all* the participants in the Committee who are natives of the Far East expressed some objection to Western logic, Western technology or Western organizing principles. Whether these objections mean that there is something genuinely wrong with Western things or whether the objections merely represent certain cultural biases deserves consideration. None of the cultural and sociological questions raised here has finally been decided, nor the various questions about values. But *technology is now an inescapable and perhaps even overwhelming fact*, a fact which impinges on and even influences almost all our human activities and ways of thinking, and this fact is what occupies our consideration in these papers.

This book itself both expresses and is a symptom of human confusion and uncertainty in the face of this new technology. Many of these papers contain neologisms and unfamiliar terminology, and some contain expressions of bewilderment. Almost all of them express the need for new definitions and new ways of thinking in the face of these technological changes that now threaten to overwhelm us. We, perhaps, especially need a new definition, or at least a new understanding, of what it is nowadays for something to be an art, for the old definitions and understandings have begun to fail us now.

René Berger
Lloyd Eby

Lausanne & New York

Acknowledgements

Many people helped in the making of this book, more than we can hope to acknowledge. Dr. Morton A. Kaplan of the University of Chicago, the Conference Chairman of ICUS XII, provided us with inspiration for the making of this book from the papers of Committee III, as well as co-ordination with the Sponsor and the Publisher; without his urging this book might never have been produced. The Conference itself was sponsored by the International Cultural Foundation, Inc., and the various personnel of ICF have been most helpful. Stephen Henkin, Director of ICUS Publications, deserves special mention for his work of co-ordination and for working to secure the rights for the illustrations, and Glenn Carroll Strait has helped with the financial arrangements. Each of the authors of the papers contributed a great deal. Nick Rudell, Artistic Director of the Court Theatre at the University of Chicago, read the entire manuscript and gave helpful comments and suggestions that contributed to the final editing. Pauline Pilote Eby translated the Epilogue from the original French; she also typed some of the other manuscripts. Helen Driller, our production editor at Paragon House, and Jack Ehn of Ehn Graphics, the designer of this book, also contributed a great deal, for which we give them our profuse thanks.

R. B.
L. E.

Special Note: Chapter Seven, "Social Aspects of Computer Art," was originally intended to be of similar length to the other chapters, but the unexpected withdrawal of a paper just as this book was going to press left us with no alternative but to publish the book with only one paper in this chapter. The shortness of Chapter Seven does *not* mean that we regard this topic as less important or less interesting than the others.

THE EDITORS

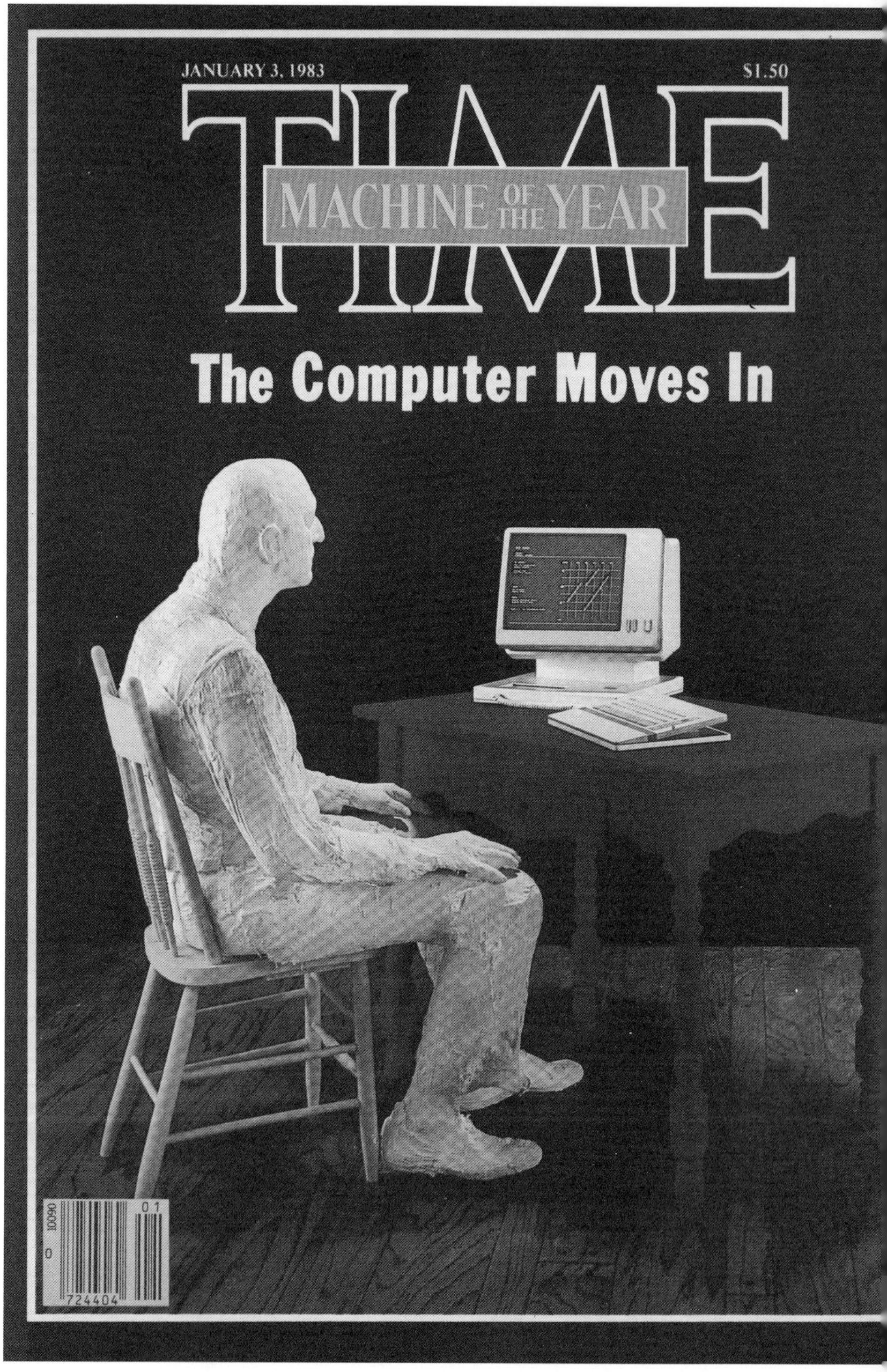

Intro. Time Magazine cover, January 8, 1983
(Copyright 1982 Time Inc., Reprinted by permission from Time).

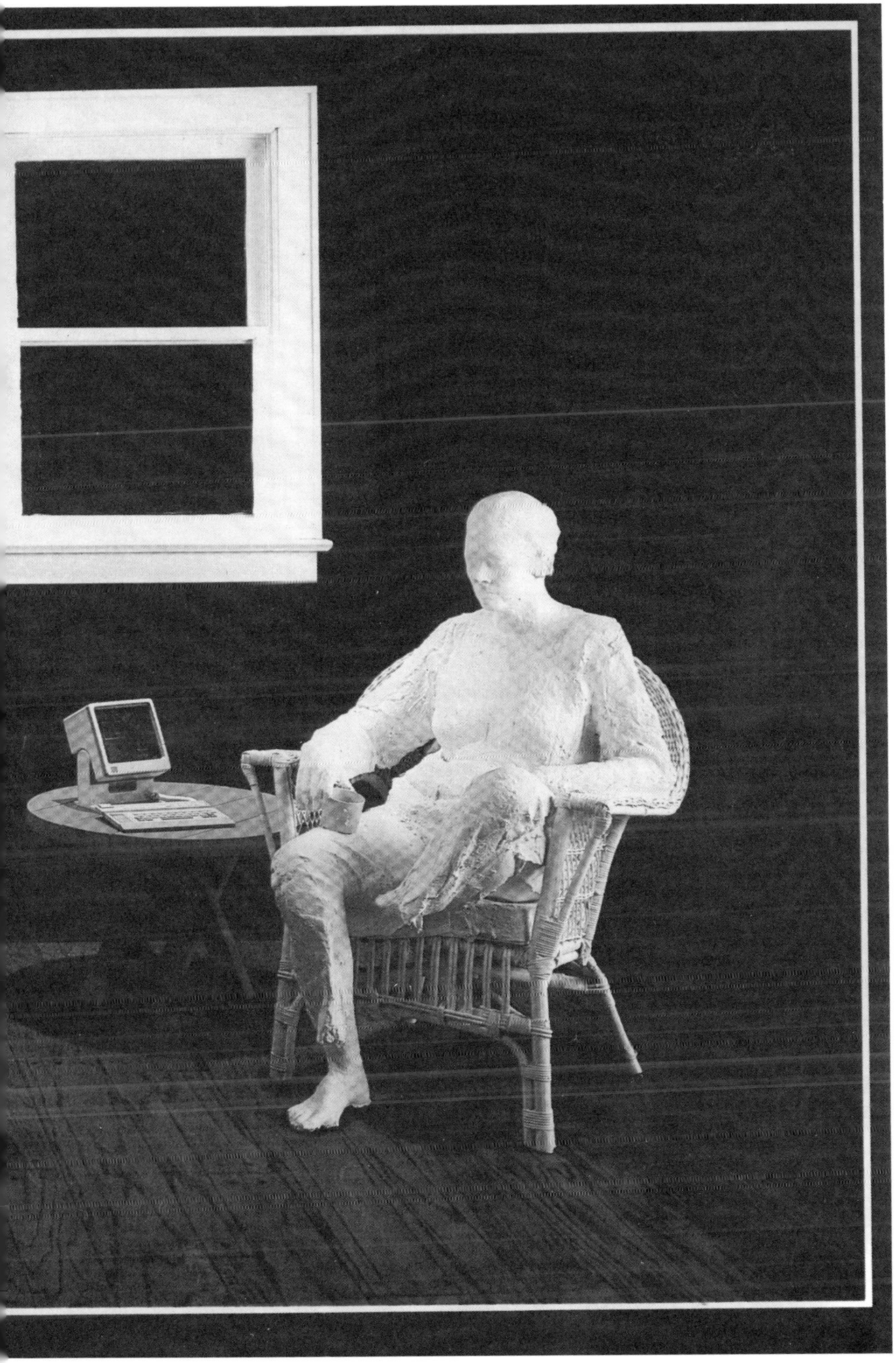

Introduction

On January 3rd, 1983 *Time Magazine* greeted the New Year with a revolutionary cover page: "For the past 55 years, Time's Man of the Year covers have depicted real individuals or, on four occasions, a symbolic representation of a group of people: G.I.Joe (1950), the Hungarian Patriot (1956), the Young (1966) and Middle Americans (1969)." The editor went on giving the reasons for his choice: "Several human candidates might have represented 1982, but none symbolized the past year more richly, or will be viewed by history as more significant, than a machine: the computer."[1]

Nothing like it had ever happened before: for the first time a magazine of *Time*'s importance, attuned to millions of readers all over the world, broke away from the historical portrait (*Time*'s Man of the Year) and the allegorical image (The Young . . .) and *endowed a machine with full symbolic power.*

Of course, man has long known that his fate is inextricably entwined with that of his tools: in fact, he has been aware of this state of affairs ever since the first silex was purposefully cut in prehistoric days. However, it had never been stated with such a profound conviction that man and tool share the same destiny. And yet, this realization had given rise to a menacing disquiet. Hence, *Time Magazine* felt a need to go to some length in order to justify its daring choice, even at the risk of repeating its arguments: "A new world beckons, created by a technological upheaval that is bringing computers to millions. Since no one person dominated this process, *Time*'s Man of the Year for 1982 is not a man, but the computer itself."[2]

Would it then be true that the computer is altering all our ways of doing things? Would it be true that a new world is taking shape before our very eyes, in our work places and our homes? A new world? Or is it a new ruler of the world?

Still, the computer does not sum up the entire technology—far from it. But it is, nonetheless, its most spectacular aspect. Why is it so? Undoubtedly, because the computer has moved far beyond all machines which preceded it, machines that operated without ever fraying man's feeling that he was capable of keeping them under control.[3] But what happens when the machine grows able to handle ever more varied, ever more complex operations, functions at a ceaselessly increasing speed and acquires a well-nigh unlimited memorizing capacity?[4-5] Furthermore, how are we to take the fifth generation computers now being developed in Japan and the United States which will be capable of processing the equivalent of thousands of billions of instructions per second in the form of logical inferences, an inference requiring perhaps a hundred to a thousand instructions per second?[6]

These few points hint at the prodigious fields opened wide by micro-informatics which, let us not forget, is a little more than ten years old. It is, therefore, quite clear why the public at large should feel somewhat dizzy. What is more, this feeling is shared by laymen and specialists alike, both being equally prone to tell themselves that "machines are mindless," thus implying that "man alone has a mind" because he is the only one to conceive the programs implemented by the machines. However trivial this man-worship may be, it is, nonetheless, most telling. Is not this urge to downgrade the machine an evidence of the fact that man and machine are being compared, and that the machine is being surreptitiously endowed with a self-awareness which must be repressed at any cost? Some scientists, open minded enough not to reduce the technological evolution to a cruel fratricide, do not hesitate to oppose this view. Herbert A. Simon, Nobel Prize winner for Economics (1978) and one of the pioneers of Informatics and Management Sciences, has this to say: "It would be an error to believe that computers mirror only logical thoughts. They mirror only their programs and it is not difficult to program them in such a way that they may be induced to commit errors in judgement or logic. The notion of intuition is usually brought into play when an expert, for instance a brilliant chess player, shows the capacity of instantaneously grasping all the implications of a given situation. But if we take a closer look at what happens in such a case, we realize that the key to intuition is but recognition: the expert recognizes the indices which enable him to benefit from a vast knowledge stored in the course of a lifetime of experience. Much in the same way, the computer equipped with a rich data bank would be perfectly able to act in an intuitive manner."[7] And the author bravely rejects the traditional objection that a computer can reflect only that which has been provided by its conceiver:" A computer may know all sorts of things ignored by its programmer: some thinking Robots already manage to defeat their conceiver in a chess game; some conceive mathematical equations which are more elegant than those worked out by mathematicians, while others arrive at diagnostics which surpass those of the doctors. If computers had, as we have, free access to information provided by informational circuits of their own choice, they would evidently have experiences unknown to their programmers. Furthermore, if we add their learning capacity, it is certain that computers could create for themselves a whole new range of competences and would then be able to "program themselves" in order to solve new problems . . ."[8]

If I evoke, be it in the broadest possible terms, the striking history of the computer, it is because in the last few years this machine has taken a firm grip on the whole of our activities. Confined initially to the complex problems of management and military operations, it has now made its way into the fields of education, teaching, learning, games, office and factory work, scientific research and pre-set homecare (washing machines, garden watering, window closing, etc.); telematics, mediatics, robotics, privatique, bureautique, communautique, créatique—a whole host of newly-coined English and French terms stem straight from the computer, as do CAD (Computer Aided Design) or CAP (Computer Aided Production), and many other acronyms heralding the victory of the machine. The media move ahead at full speed; printing has lost its time-honored smell of lead; reporters type their texts directly, which the paper's computer lays out straightaway; infography or transmission of drawings at great distances has become a daily occurrence—and so has the telemeeting which brings together on the screen interlocutors from different cities, countries or continents. The arts have been equally affected: architecture, painting, theatre, dance, music, as well as the "public" or mass-oriented arts: radio, TV, animated cartoons, advertising, etc. Being products of cultural industries, they make full use of the resources of technology.[9] And, for all the hostility or rejection it may encounter, the "informatization" of society relentlessly marches on.[10]

In little over half a century audiovisual devices have widened their grip on practically all our endeavours and their impact has reached its climax in the triumph of the mass media, mainly that of television. Now their very foundations are being shaken. Not only does the computer solve an ever-increasing multitude of problems, but it proves capable of transforming analog signals into digital signals.[11] Images, sound, music—all messages can be coded in sets of numbers. Furthermore, it is now possible to create artificially images devoid of any correlative in the preexisting reality, and such images are a true synthesis of a different, self-referring reality.[13, 14] It is, indeed, tempting to explore even further these possibilities which appear boundless. But in a foreword such as this it is perhaps more befitting to submit to a closer scrutiny the actual principle embedded in the very root of these events: what, in fact, is technology?

Certain words share the fate of certain flags; rather than being displayed in pomp and ceremony they are used for passion-unleashing purposes: Marxism, liberalism . . . Is it not the same with "technology"—a term constantly used and abused, and yet so hard to evaluate in all serenity?[15] At this stage a few clarifying remarks may not be out of place.

The term "technique," such as it is currently used, derives from the Greek *techne* and was originally equivalent to "fabricate, build, produce." It denotes the particular means, or sets of means, ensuring the attainment of a given goal. This notion of technique embraces at least three factors: 1) the most evident is the tool or instrument which is being used; 2) the least evident is the mental model at the root of given practical activity; 3) the rational process inherent in any technique makes it a decisive socializing factor.

It is, therefore, clear that any social evolution, from the invention of the first tool and the first language onwards, is a history of the techniques transmitted from generation to generation. Without going as far as to subscribe to the Marxist dogma that the state of the "productive forces"—in other words, of the techniques—determines the nature of human relations and therefore of social organizations, it is, indeed, hard not to consider the views of Walter B. Wriston, president of the largest banking concern in the world, whose every word is literally worth its weight in gold.[16] Wriston is president of Citicorp, which rules over an empire of nearly 2500 agencies spread across some hundred countries, and whose holdings in 1982 amounted to 130 billion dollars, equivalent to half the military budget of the USA. Wriston had this to say to the correspondent of the International Herald Tribune: "Whether you think it's good or bad, we've changed the environment." Leaving aside the concepts of "infrastructure" and "superstructure," this statement provides an unassailable confirmation of the fact that the techniques may not necessarily determine, but at least transform (and this is a notable reservation), the whole of our environment. This compels us to pay even greater attention to what more Wriston has to say: "The world is in the midst of a great transformation comparable to those of the Renaissance and the Industrial Revolution."[17] In the banking system this is manifest in the overall development of Informatics: its Home-Base system will enable everyone to handle all kinds of operations without leaving his home (purchases, sales, bookings, learning, telephone directories, surveillance, etc.).[18]

This example has been deliberately chosen to illustrate both the position and the evolution of the most powerful banking organization in the world and, therefore, by dint of analogy, the position and evolution of our economy-dominated society. It also better enables us to grasp the transition from "technique" to "technology," often implicit, yet dimly perceived. While the field of application of the former is somewhat limited, that of the latter embraces the full span of the "artificial world"[19] in which we live today—a world characterized precisely by an overall permanent mutation. This proposition will be further clarified if we distinguish at least three meanings assigned to the term "technology":

1. *this term has become synonymous with "technique" in its current usage in English speaking countries, and tending to gain ground also in French speaking countries as techniques become more sophisticated;*
2. *in a learned context this term denotes, according to its etymology, the theory or science of the production processes stemming from the Industrial Revolution and stresses the models based on mathematics;*
3. *a more insidious meaning of this term refers to W.B. Wriston's statement: "Whether you think it's good or bad . . ." What is meant by "technology" is a challenge to transform the environment—and therefore all of us —through the impact of current or future techniques.*

In view of all that has been said, the collective endeavor which we attempted under the title "Art and Technology" had to conform, on one hand, with the guidelines set for the conference in Chicago in November 1983 and, on the other hand, with the specific conditions for the publishing, not merely of its proceedings, but of a volume with clearly outlined and distinctive features.[20]

The brief for the research to be undertaken and the ensuing discussions and publishing read as follows: "to elucidate new relationships being established between the creative processes underway in various fields and the different means which the application of modern science is producing to achieve them. The concept of "art" is to be understood in a wide sense, including all the forms of art with a special stress on those which proceed from modern technology. What is meant by "technology" is especially related to the new techniques used by artists, such as movies, slides, video, registered music, records, tapes, computers, lasers, TV, etc. In this perspective, attention will be paid to two main uses of technology: (1) as a new medium in the *creative process*, such as videoart, computerart; (2) as a new medium in the process of *disseminating* art, such as TV, videocassettes, videodiscs, etc. This investigation should envisage all the factors which are shaping our sensitivity and bringing forth a new aesthetic for the masses, namely "techno-culture,"[21] e.g., posters, billboards, advertising; in brief, the urban environment in which we live. The thread linking the papers' topics is a social/cultural consideration of the phenomena underway in relation to the tremendous and accelerated change of our society. The purpose is not only to make interesting, even subtle statements, but to detect the dangers of a too rapid evolution in order to determine what can be done to prevent them."

In his paper *Science and Technology: Blessing or Curse to the Arts?* Nicholas Kurti underlines the relevance of present day science not only in relation to a more extensive and precise cognition of the arts, but also in relation to the techniques which have perfected existing procedures or originated new ones, such as Moog's celebrated synthesizer or the numerous uses to which computers are now being put. These techniques are also at the root of the prodigious movement of "vulgarization." Some of its consequences, such as the omnipresent background music known as Muzak[22] irritate the author, but he is nonetheless aware that these techniques have greatly favored the elaboration of methods of authentification and conservation unknown to this day. And Kurti concludes that science has the merit of contributing to an evolution on which we will be well advised to keep a vigilant eye.

Thus conceived, the paper leads, beyond the information provided and the stand taken by the author, to fundamental questions beneficial to further meditation: for instance, would it be possible—and this is but one of these questions—to define more precisely the nature of science, technology and the arts by elucidating the processes associated with them—cognitives in the first instance, transformational in the second, and ethical in the third.

Ervin Y. Galantay focuses his attention on one of the most ancient and complex collective works due to man: the city. Is it an organism or an artifact? asks the author, and answers by opting for the metaphor "ecological system." After a brief critical introduction he recalls the main methods of perceiving and understanding the city: agglomerate, symbolic, perspectivistic, statistically ordered, topological. The next chapter succinctly surveys the methods generated by new technologies with the help of computers excelling in quantification, formalization and simulation. The comparison between two achievements as characteristic as Brasilia and Ciudad Guayana leads to the conclusion that "intuition," "aspirations," "imagination," and "emotion" have not been, and cannot be, brushed aside. Thus, the most advanced devices which ceaselessly grow more and more sophisticated clash with human factors still considered irreducible. Technology may well unfold its mode of exponential development—it is nonetheless bound to encounter psychological thresholds which are, if not more, then at least just as important as economic and political conditions.

Nothing is more revealing, Freud assures us, than plays on words. And we can hardly doubt this assertion in re-reading the title which Pierre Rouve has given to his communication: *Reel to Real*. Right from the start, the author makes it clear that his intent is not to elaborate on the philosophy *of* technology, but to unearth the philosophy embedded *in* technology. A film maker of some repute and a rigorous theoretician inspired by Jakobson, he takes a highly individual turn by claiming that, in films, "the center of epistemological gravity falls not on a single photogram but on *two structurally conjoined photograms*." The dynamic kernel of the film or filmeme resides therefore in the relation, or rather in the tension, between two photograms. This leads the author, whose chain of arguments is supported at each successive stage by illustrations and diagrams, to the conclusion that "the Reel conclusively asserts its right of precedence over the Real." Hence, the subtitle of the paper: *The cinema as technological co-reality.*"

At this conjuncture emerges the temptation to query whether this model of co-reality evolved by the author for his analysis of the cinema may not be extended to technology as a whole. Thus, what reemerges is also the doubt whether reality is knowable, or at least accessible, in any other way but through the intermediary of a technique. And if all reality is but "mediated reality," then how heavy indeed is the weight of the media in our existence!

Unlike the cinema, music goes back to the dawn of mankind —and yet it does not remain outside the span of technology. Modern techniques, affirms Daniel Charles, considerably affect our musical memory which began with notation codes and continued with the phonograph, the radio, compact records, etc. But, beyond their conservation function, these techniques lend themselves, within the very framework of ambiguity inherent to all media, to the elaboration of new artistic forms. This is evidenced by pioneers like John Cage, whose HPSCHD goes as far back as 1969, or Pierre Boulez, a guiding mind that knows no respite, for whom the computer has become a full-time musician. The analysis of Charles draws further substance from examples of sound impossible to reproduce on a printed page.[23] Perhaps it would be most appropriate to explore, following the author's trend of thought, the paradox inherent in all media: on one hand, they are the main providers of mass-consumption products; on the other, they lend themselves to quests, or works, by and for the happy few.

Can we nowadays imagine a world without television? This would be utterly impossible for our children who treat the TV receiver as a member of the family. And yet this planetary state of affairs has been with us only a few decades! Its substance is examined by Robert Stéphane, who hastens to proclaim its new law: *"An event is not created by itself but by its audiovisual production."*[24] The author begins by surveying the potential and the limitations of the medium for the representation of the arts and their diffusion. Then, postulating that television is an art in its own right, he shows that it has its own logic, language and syntax—in a word, its own aesthetics. This transpires in the programming procedures or metaprograms, logos, jingles, studio design, anchormen and masters of ceremony styles, as well as in, at one end of the scale, advertising spots and video clips and, at the other end, in *Dallas* and *Dynasty*. We may, therefore, ask ourselves, in the ambit of these thoughts, if the arts which do not stem from television but which can no longer do without it—such as literature, dance, architecture, sculpture, poetry and the theatre—are not progressively losing their former "essence," ousted by an increasingly more pertinent "mass aesthetics."

The seasoned and methodical sociologist, Alfons Silbermann begins by reminding us that ". . . both art and technology should be regarded as *cultural items* in the sense of Tylor according to whom culture is, 'that complex whole which includes knowledge, belief, art, morals, law, custom and any other capabilities and habits acquired by man as a member of society.' "[25] On the other hand, all sociological phenomena are based on the *triangle relation*: person—society—culture (art/technology). The author corroborates this point by exploring the interaction between the mass and popular culture and stressing the present-day importance of the cultural industries.

Thus comics, in his view, have assumed the stature of a true mass medium. All social activities converge in them and lend themselves to their presentational mode: entertainment, information, learning, education, advertising. This confirms Silbermann's central thesis which is clearly conveyed by the title of his communication: *The Way Toward Visual Culture: Comics and Comic Films*, which brings to the forefront all the dialectical tensions provoked by such a mutation—for instance, the tensions between mass and individual, between monological and dialogical communication, between subjectivity and objectivity, etc. From this trail-blazing analysis of a phenomenon as yet hardly elucidated emerges the imperative necessity to devote similar rigorous studies in depth to the role of the cultural industries which are molding our behavior. Carried by the unavoidable tide of mass products, our traditional values are undergoing a substantial alteration or are simply annihilated to clear the way for new values. But what are they?

"The Computer moves in," the cover page of *Time Magazine* proclaimed in 1983. This provides Abraham Moles with the

opportunity of asking himself what exactly is meant by "Computer Art," an expression ". . . covering a wide range of aesthetic products which have been made in connection with, or directly by, a computing machine equipped with terminals which have made it possible to produce, from an assembly of elements, a directly acceptable product such as sounds (talking and singing computers), music (computer composition), shapes (sketching, designing, architectural projects), and, more recently, color and tactile patterns." However, the author does not hesitate to assert that this is an art doomed to extinction, a victim of its own success. And indeed technological society leads us to more and more frequently replace art with games; with the ultimate aim of installing, for all the motley arcades, new Computer Art Cultural Centers where everything would become possible by virtue of ad hoc installations. It is, in fact, something in the nature of Faust's "Verweile doch," which does not make any sense when the "Kitsch Computer" can, at any time, activate our pleasure-seeking urges. In ending, the author boldly declares that, ". . . in past times the artist was an 'intuitive programmer;' nowadays he is a graduate from a programming school just like the research engineer." Are not our time-honored notions of art, artists and works of art undergoing a radical transfiguration? Abraham Moles' essay has the merit of forcing us back in our safe retreats and compelling us to work out our personal answers.

My own contribution endeavors to shed some light on one of the least well-known creative aspects of the audiovisual. Parallel with the development of television—cable, satellite, etc.—the video market, opened only ten years ago, offers to the public at large the possibility of recording transmissions in one's own absence and, above all, a wide selection of pre-recorded cassettes whose number continues to grow. Thrillers, horror and pornographic movies figure prominently in this selection, not to mention a host of games. In such circumstances, the public is naturally prone to consider the video, in spite of television, merely as a reproduction and distribution device. It is, therefore, all to the credit of a couple of hundred brave artists to have proclaimed, some twenty years or so ago, that the video, too, could serve creative purposes. To prove it they created—and still do create—installations and environments which, unfortunately, can be seen only in a few museums or here and there at national and international festivals. Because of its scant social exposure and the indifference of collectors, video art is often misunderstood even by those who show some interest in it. It is, therefore, quite legitimate to query whether such a "marginal" art, challenging by its experiments the conventional audiovisual products, could ever emerge as a serious rival of the mighty commercial media.

Some readers may be disconcerted by Magoroh Maruyama's communication; some may take it, at best, with a condescending smile—but no one will remain indifferent to it. *Art and Technology in the Future* is an attempt to foresee some aspects of life such as it might be expected in "outer space communities." When the technical and economic problems have been settled—and the author is pretty sure that they will be—how will the needs of the first inhabitants be catered for? They will be provided with miniature earth-like landscapes and outsize laser-produced holographic images. A fundamental problem will still have to be solved: "in a highly controllable environment where temperature, humidity and other conditions can be set at will, and where everything is quite predictable, you may get a feeling that there is no difference between what you think and what actually happens." To dispel the menace of such solipsism, art will have to be brought in. The forms and modes of expression which it may take are associated by the author with the culture of the Mandenka or that of Japan, both laying a heavy stress on "contextualization," rather than with Western culture which Maruyama considers much too "conceptualized." Whatever the chosen path—and the author would prefer it to be a Far Eastern one—there will remain still another cardinal problem to solve: "The thinking required is far more than technological or economic. We will be in a position to invent new cultural patterns and then choose environmental designs to fit the desired cultural, aesthetic and philosophical goals." But who is to invent these new cultural patterns? Who is to decide on them? Who is to evaluate them? In whose name? On the strength of what? In view of what? The whole of our future looms large on the horizon across this cataract of questions.

What will be the interest of a publication such as this? To put it succinctly: it will help to pinpoint both the orientation and the meaning of the Third Industrial Revolution which is now in full sway.[26] All civilizations which have preceded us have been brought about by ethnic entities settled in clearly circumscribed localities and have flourished for shorter or longer periods: Egypt, Greece, Rome, China, India, pre-Columbian America. Each of them was imbued with a particular culture which assigned well-defined meanings to their beliefs, values and behavioral stances. The diversity of these cultures still fills us with amazement and admiration. But the fact is that our era shows a radical break with this past: technology—all technologies—tend to become universals. Airplanes may well display distinct national emblems, but they function nevertheless according to the same principle. If one of these airplanes were to abandon these principles and revert to obsolete artisan traditions, it would cease to fly and, if it did not crash on the ground, its pilot would turn at best into some sort of coachman. The same goes for all kinds of Know-How: none of them can wriggle out of the grip of a *generalized mutation*.

Though this mutation affects us all, it is by no means everybody's cup of tea. Not by any means! The investment required nowadays for a ten-year research and development program for a digital telecommunication network is evaluated at nearly one billion dollars.[27] Consequently, advanced technology or "high-tech" requires not only the cooperation of scientists and technicians but, above all, the blessing of financiers and politicians. They alone have the final say; they alone hold the power to create and to give the go ahead to any new initiative. And it is hardly necessary to add that neither finance nor politics has ever shown much interest in charitable deeds! The enterprises which they promote can feel secure only if, directly or indirectly, they make their way to profitable markets or, to put it in crude, realistic terms, to the *world at large*, be it in the field of telecommunication, informatics or armaments. Cultural industries—cinema, television, records, video cassettes, etc.—are very much in the same situation.

But if technology, inextricably linked to finance and politics, ultimately results, as we have seen, in a transformation of the environment, its immediate effect is to favor and reinforce the power of those who provide the means for its development: the multinationals in the United States, Japan and other countries with so called free enterprise systems, and the class at the helm of the State in the socialist-communist countries. In both cases, the operative values upheld by the system are *optimum yield and optimum efficiency*—and cannot be any different. It is precisely this exclusively technological world that Orwell and Huxley have denounced so vehemently; yet, we all slide towards it if we silence the urge which has spurred our species to transcend its animal nature and become human, if we cease to query the meaning of our existence and of our deeds. These are indeed the common features shared by all civilizations which have preceded us. Though technology is today the core and the spur of the Third Industrial Revolution, we refuse to submit to it as if it were our immutable destiny.

I am confirmed in this belief by the very cover page of *Time Magazine*, which provided me with my opening gambit: *The Computer Moves In*. Scrutinized at close quarters it reveals that neither the computer nor the man seated in front of it, neither the table nor the chair are real objects, or even semblances of real objects. The *Time* editor asked an artist, the American sculptor George Segal, to conceive and convey the upheaval in which informatics has hurled our world: "In search of a memorable depiction of the irresistible invasion of computers into American homes, Art Director Rudy Hoglund approached George Segal, the world-famous sculptor. Segal almost never accepts commercial commissions, but Hoglund thought that Segal's stark and dramatic settings in which the eye is drawn to the objects were perfectly suited to the first Machine of the Year. Segal enthusiastically agreed. Hoglund went to Richardson/Smith, a design firm in Columbus, to create the mock computers portrayed on the cover."[28]

Readers are thus faced not by a computer or its image, but by a *symbolic array* intended to make them think or, at least, react.[29] And indeed, what goes on between them and the cover page conceived and produced by Segal? Undoubtedly, the majority of readers will be startled. What could be the meaning of that computer perched on a red table with turned legs which all too obviously is not made for it? Is it an anachronism? Or a solution of continuity? And if we are to take it seriously, does it not show that "high-tech" is turning to our family setting after having invaded our professional environment?[30] This is stressed once again by the yellow wooden chair befitting some kitchen rather than a living room. In this uncommon array sits a man, or rather the plaster cast of a man with its unadulterated color and texture. Is he a nameless actor, halfway between ghost and puppet? His body crouches limply, his arms and hands rest inertly on his thighs and knees. Only his head strives to uproot itself from the shoulders as if to meet some unspoken challenge. In his own way,

Segal reverts to the primeval struggle between the species whose resurgence in present-day conditions is symbolized by his art. To survive (perhaps to prosper)—this is the fate promised to those who will succeed in adapting themselves to the changing environment. As for those who will not, does not the black background behind the computer stand for their failure, which will bury them in bleak nothingness?

It would be all too presumptuous to insist that this reading of mine is the one and only interpretation of *Time*'s cover. I do, however, believe that most of the magazine's readers will detect in Segal's work something of their own experience of men struggling, for the first time in history, with a machine which calculates, programs, reasons, deduces, draws, sings, and plays the piano or the trumpet.

This survey, however summary, hints at the fundamental function of art: to set us free from a destiny that would be strictly biological, economic or political. It is of the essence of destiny, whatever its form, to inflict on us conditions which we cannot bend or modify, let alone overcome. The might of art is to open wide the borderless expanse of reflection which charts also the span of our freedom. There, words and works offer themselves to our interpretation, which prevents their significance from being reduced to mere dessicated knowledge and keeps open the channels of dialogue and intercourse. In these days a fierce battle rages all around us and we are quite unaware of it; newly born but increasingly more sophisticated Technology confronts Art which, streaming from the deepest depth of millennia, still demands over and above the might of our machines that we should never abandon the quest for ultimate meanings, propulsed and propped up by aesthetic experiences.

Notes

1. Cf. *Time Magazine*, January 3, 1983, Vol.121, No. 1, p. 3. *A Letter from the Publisher.*
2. Cf. *Ibidem*, Index, Machine of the Year.
3. Micro-informatics, hardly a decade old, is at the root of the formidable spread of informatics. Micro-computers function by virtue of micro-processors (integrated circuits, chips) whose density and efficiency have increased in inverse proportion to the reduction of their production costs. Limited initially to vast enterprises, political or military organizations, micro-computers now hold sway over modest businesses and reach, as do television sets, the very heart of family life.

4. IBM and NTT respectively announce that they have developed an experimental circuit of IM-bits i.e. one million elementary informations items. Industrial production of 256 K-bits is guaranteed for 1984–1985. Cf. *Le Virage des 256 K* in *Science & Vie Micro*, No. 7, June 1984, p. 18.
5. Fujitsu has two models of a supercomputer, the VP-100 which the company said can perform 250 million operations and the VP-200 which it said can perform 500 million operations per second. Cf. Andrew Pollack, *Fujitsu to sell Supercomputer* in U.S. in *Herald Tribune*, June 28, 1984.
6. Tohru Moto-oka, *Les Ordinateurs de la cinquieme Generation* in *La Recherche* No. 154, April, 1984, p. 516–525. "The performances of the Fifth Generation Computers will be measured by the number of logical inferences which they will be able to realize per time unit, each inference representing the work necessary for the selection and application of a rule. The maximal capacity of such inference machines will be in the range of 100 Mlips to 1 Glips (logical inference per second). It is estimated that an inference realized by a present-day computer requires from 100 to 1000 instructions, thus one lips represents from 100 to 1000 instructions per second." (p. 520)
"For that purpose it will be necessary to develop a system of meta-rules (reasoning procedure achieved with the help of meta-rules, i.e. of rules governing the appropriate recourse to other rules) which will control the inference process of the different systems of problem solving and the cognitions processed by them. In another field, this meta-inference system will have to include also high level functions capable of simulating mental activities such as 'common sense' analogy or other forms of tacit reasoning." (p. 521).
7. *L'Ordinateur libere de ses Maitres*. Discussion between Herbert A. Simon and Didier Eribon, *Le Monde d'aujourd'hui*, 1–2 April 1984. Herbert A. Simon, *La Science des systemes. Science de l'artificiel*. EPI, Paris, 1974.
8. *Ibidem*.
9. I refer to the classic work by Gilbert Seldes *The Public Arts*. Simon & Schuster, New York, 1964.
10. According to Gilbert Seldes, among "the major marks of identification are these:

—The public arts are popular to the extent of being almost universally acceptable.
— They tend to be more and more professional, less and less to be practised privately.
— They are often produced by teams rather than by individuals. They are commissioned, the patron-sponsor-executive providing the pattern.
— They are by intention ephemeral, paying well initially but not increasing in value . . ." Cf. *Ibidem*, p. 298.

11. Broadly speaking, the term audiovisual is applied both to auditive techniques (radio, records) and to audio-verbal, visual or verbo-visual techniques (magnetophone, still or animated projections). In the strict sense of the term audiovisual techniques are films, television, videocassettes, videotelephone, cable or satellite television, videoclips, the totality of telematic services, videorecords, etc.
12. "Digital is to be distinguished from analog. An analog signal is continuous,

which means that if it goes from level A to level B, it must assume every value in between. In contrast, a digital signal has only two meaningful values, zero and one (Fig. I.1). In theory, none of the values in between is defined . . . It is this ability to reconstruct the original information that has allowed digital techniques to revolutionize electronics. It has clear advantages for any form of information storage, transmission or processing."
Myron W. Krueger, *Artificial Reality*. Addison, Wesley Publishing Company, Reading, Mass. p. 246, 247.

13. Cf. Francis Coupigny, *La Synthese des Images* in *La Recherche*, No. 144 special, May, 1983, p. 735–745.
14. Cf. Frank Brown, *La Musique et l'Ordinateur*. Paris, PUF, 1982, Coll. Wue-sais-je?
15. This is always the case, whether one swears unconditional allegiance, as do technocrats, or one behaves like a faithful watchdog, as politicians pretend to do, constantly clamoring for innovations while, in the same breath, warning us to beware of them. Is it not rather telling that even the *Encyclopedia Britannica* in its article on *Changes in Technology*, after relating the disenchantment of Aldous Huxley, examines at some length the "negative" attitude of a Jacques Ellul?
16. Cf. *International Herald Tribune*, June 1, 1983, p. 8.
17. Cf *Ibidem*.
18. It has been recently announced that John S. Reed, a fervent promoter of this evolution, will succeed Walter B. Wriston as head of Citicorp in September 1984: "The winner and new chairman is . . . After a four year competition Citicorp taps Reed for the top spot." In *Time Magazine*, May 2, 1984.
19. "Artificiality" is defined by H.A. Simon in his book *La Science des Systemes. Science de l'artificiel. Op. cit.*
20. Direct consequence of technology! This publication is couched in a form attuned to the logic of the printed page while, still aiming at the same target, a series of television transmissions would have based the essence of its demonstration on photographic and cinematographic documents and would have reduced the spoken word to a strict minimum.
21. On the concept of technoculture, see particularly the works of René Berger: *La Mutation des Signes*, Paris, Denoel, 1972; *La tele-fission. Alerte a la television*. Paris, Casterman, 1976; *L'effet des Changements technologiques. En mutation, l'art, la ville, l'imagine, la culture, NOUS* ! Lausanne, Ed. Pierre-Marcel Favre, 1983.
22. On July 9, 1984 the Herald Tribune paid tribute to Muzak's "Fifty Years of Mellowness": The object of Muzak is not to be really heard, but to serve as background music to relieve stress and give workers or shoppers a lift" . . . "Owned by Westinghouse Broadcasting & Cable Inc., Muzak is sold as 'environmental music.' It is sent by satellite to more than 11,000 U.S. businesses and syndicates in 17 other countries." The damning verdict on such music pronounced by learned music lovers has failed to stifle an event which has proved its staying power and spreads more and more widely infiltrating cars, supermarkets, airplanes, airports . . . This certainly gives food for further thought on Muzak's functions.

23. Yet another example of technological constraint. The printed page which can easily accommodate photographic reproductions is utterly incapable of directly evoking any music.
24. Edward B. Tylor, *Primitive culture*, 1871.
25. The impact of this assertion continues to grow with the mushrooming of audiovisual devices born in the wake of television: videotext, videotex, infographie—in brief the entire field labelled peritelevision or telematics.
26. The First Industrial Revolution, spawned in the mid 18th century, was fueled in Britain by steam power and coal; the Second, around 1900, by chemical and electrical industries in Germany. The Third, now under way, will probably be driven by the microchip, perhaps even with regenerative power of genes. Cf. Russ Hoyle, *The High-Tech Challenge* in *Time Magazine*, July 16, 1984.
27. ". . . the European Strategic Program for Research and Development in Information Technologies (ESPRIT) is believed to cost some $ 1.3 billion for five-year plan with the cooperation of twelve countries" Cf. *Ibidem*, p. 28.
28. Cf. *A Letter from the Publisher. Time Magazine*, January 3, 1983, p. 3.
29. I am considering the left part of the cover which was the only one to appear in the European edition.
30. Most probably, all computer owners have felt ill at ease much in the same way. Furniture items, particularly libraries, seem to recede in a dim distant past. Electronics: white magic or black magic?

Chapter 1

Science and Technology: Blessing or Curse to the Arts?

Science and Technology: Blessing or Curse to the Arts?

Nicholas Kurti

The effects and influences of science and technology on the arts may be divided roughly into two somewhat overlapping categories. A new technology, by becoming the germ of fundamental changes in the *creative process*, can lead to novel art forms. But new technology is also an important agent in the *process of dissemination*, this word being used in its widest sense. Thus, dissemination includes reproduction, re-creation, conservation and, more generally, new techniques for the study of works of art leading to their better historical and aesthetic understanding.

As the title indicates this is partly a critical essay. I shall try to indicate, wherever appropriate, whether and to what extent, the application of new technologies could be or is being abused, and could result in the lowering of active artistic experience.

Clearly this paper cannot cover the vast field of technological influences on the arts. A mere catalogue would be unsatisfactory, and to do the subject justice several lengthy essays would be needed. Professor Robert U. Ayres, in his very helpful and constructive comments on the text of this paper as presented at ICUS XII, has pointed out that a proper treatment of the subject might begin with a discussion of technologies based on new *materials*, new *tools* and novel *means of presentation*, followed by a review of technologies used in the *reproduction*, *analysis* and *dissemination* of art.

Professor Ayres' point is well taken, but my reason for choosing a different approach, namely, selecting *examples* of new technologies in the context of particular *art-forms*, was that it made it easier for me to weigh up the advantages and the disadvantages of their application. In choosing the art forms I found the following definition of art in the American Heritage Dictionary useful: "The conscious production or arrangement of sounds, colors, forms, movements or other elements that affects the sense of beauty." Artistic experience thus involves three senses: sight, hearing and touch, but not smell or taste — unless these are deemed to be included in "other elements."

I propose to adhere to this restrictive definition of art and exclude artistic experiences connected with smell and taste, and justify this on the following grounds. Although one talks loosely about the art of the perfumer or of culinary art, these activities have not yet been accepted as art forms in the same way as music, painting, sculpture, ballet—perhaps even figure-skating and cricket! I wonder, however, whether this discrimination is a subconscious compliment from the "scholar," i.e., someone engaged in the study of the humanities and the arts, to the scientist. The accepted art forms have one important thing in common: they are perceived by the senses which respond to *physical* stimuli capable of quantitative description. Thus it is possible, at least in principle, to represent in the language of physics, mathematics and computer science the stimuli that finally result in the perception of a certain rendering of, say, the first eight bars of Mozart's G minor symphony, or of the Mona Lisa. (I shall return to this idea when I discuss replicas of paintings.)

This, albeit a complicated process, cannot be used in the case of smell or taste, since these senses respond to chemical rather than physical stimuli and, in spite of great efforts to classify satisfactorily the stimuli producing certain subjectively perceived smells or tastes, there are no well-defined scientific disciplines covering the olfactory and gustatory senses analogously to those covering the auditory sense (acoustics), the visual sense (optics, spectroscopy) and the tactile sense (mechanics, heat). It is true that much is known about what atomic grouping in a molecule produces a generally agreed sensation, but this is a far cry from, e.g., optical stimuli which can be characterized by wave length, amplitude and polarization of all the components of the light that reaches the eye. I shall, therefore, restrict this discussion to the senses of touch, hearing and seeing.

Touch

The role of touch in artistic perception can be dealt with rapidly since it is insignificant. Although the arts concerned with modelling, e.g., ceramics, pottery, and more generally with the representation of solid objects are often called the "plastic arts," the tactile sense hardly ever gets involved. The familiar "Do not touch" notices in museums and exhibitions displaying sculptures and other plastic art objects militate powerfully against giving the sense of touch full rein in artistic appreciation. There are, admittedly, notable exceptions. For instance, I believe that Henry Moore encourages visitors to his exhibitions to touch the sculptures and, instead of just *seeing* them, to *feel* their shape and their texture. I doubt whether anyone has yet been brave enough to walk up to the Venus de Milo in the Louvre and stroke her superbly shaped shoulders. It would also be interesting to know how many museums and galleries make special arrangements for blind or nearly blind people to enjoy sculptures, etc., by touching them.

A general relaxation of the "Do not touch" rule would be desirable. It may even help to stamp out the tendency in our civilization of deliberately atrophying the sense of touch. It may encourage parents to let their children explore objects with their hands rather than forbid them to do so.

There are two objections to a general abolition by museums of the "Do not touch" rule. (1) Perspiration may cause chemical deterioration of the objects. This could be prevented by the provision of very thin gloves like those used by surgeons. While some details of the texture might be lost, appreciation of the shape of the object would be unimpaired. (2) The gradual abrasion of the object through the stroking motion of many thousands of hands week after week, year after year. It would be interesting to have some quantitative facts on this.

Hearing

Of all the art forms music, which relies on the auditory sense, has been influenced most by technology and science. The reason is simple. Sound, whether it is music or noise—and anything between them can be analyzed, broken up into its components, and described quantitatively with high accuracy. The contrary holds too. Once we know the components of any sound or sequence of sounds we can, at least in principle, synthesize it. We have both the scientific basis to do this and—thanks to the technological developments, mainly of the last three or four decades—the practical ability to enlist acoustics and electronics to serve music.

Musical reproduction. This has made enormous strides during one generation, and music of superlative quality has been brought to an ever-increasing audience. Nor has the rate of improvement seemed to have slowed down. Light mechanical pick-ups are soon to be replaced by laser pick-ups, stereophonic reproduction is becoming commonplace, the quality of home-recording on magnetic tape is advancing, and one is tempted to assert that, in this case, technology is pure blessing. But there are some blemishes which, unless one watches out, could turn into sores.

Because of the ease of the transmission of music we are increasingly being exposed to background music and are usually powerless to stop it. In hotels, restaurants, trains, shops, even in private homes one is surrounded by music, usually of an inane nature, and if one asks for it to be stopped, or at least for the volume to be reduced, one is often regarded as a miserable spoilsport. There are even instances where background music is compulsory—thus there seems to be an international agreement that music, preferably of a soothing nature, must be provided in aircraft during takeoff and landing. The psychology of this regulation is hard to understand. Do the airlines try to tell the passengers that they are going to be involved in some very dangerous maneuvers and that their nerves need to be steadied?!

The better understanding of musical instruments. The application of better measuring techniques and of the science of acoustics to the study of musical instruments has led to a better understanding of what determines the various qualities of the instruments and may open the way to improve them. Such a remark may come as a shock to those who believe that to imitate, let alone to improve on, the quality of the violins of the great masters like Stradivarius or Guarneri is a blasphemous pipe dream, and that such efforts should be discouraged and frowned upon. But let me quote from an article in the Scientific American[1] by Mrs. Carleen Hutchins who, with other like-minded scientists and musicians, founded in the United States the "Catgut Acoustical Society." She writes: "The Italian masters had brought the art of violin-making to such an extraordinarily high level that it is only now that one dreams of equalling or surpassing it. These early masters must have had an open mind towards the little that was known in their time about the physics of sound. Their successors deserve the credit for having lovingly preserved an art, but certainly not for advancing a science. In effect they have formed a cult that has been plagued with more peculiar notions and pseudoscience than even medicine."

But acoustics is an exact science and the generation and propagation of the stimuli that affect the sense of hearing can be described and analyzed with great precision. Similarly, physical measurements on the great violins of the past may help to discover what particular characteristics are responsible for the quality of the tone of those instruments. If the violin makers of the future, instead of being guided solely by tradition and intuition, will be helped by science in the pursuance of their art and craftsmanship and if, as a result, more people will be able to possess superb instruments, science will have performed a praiseworthy service to the arts.

Synthesizers. The concept of synthesizing music is very old since, strictly speaking, a musical instrument *and* its player can be called a synthesizer. But even if we discuss only synthesizers which use electricity, the history stretches back more than half a century. The amplification of music may be regarded as an early example of music synthesis since sound, emerging from a source, be it the human voice, a musical instrument or a symphony orchestra, is manipulated electrically on its way to the listener. The manipulation can be of two types. The aims may be to retain as faithfully as possible the original composition of the sound but bring it to a hugely increased audience. One of the shining examples of this beneficial use of technology is the sound amplification in the Hollywood Bowl in California. I once listened there to a flute concerto accompanied by a symphony orchestra and the amplification was so good that I had the impression of sitting in a moderate-sized concert hall rather than in a vast, open-air amphitheatre seating some 25,000 people. Unfortunately, sound amplification not infrequently goes hand in hand with a distortion of the original composition, not necessarily of its tunes and harmonies but of its *timbre*. Thus, for me, the enjoyment of Papp's otherwise delightful production of *The Pirates of Penzance* (I am not a dyed-in-the wool D'Oyly Carte follower) was marred by the drastic variation in the proportion of treble to bass.

Synthesizers, as we know them now, really started in the mid-1960's with Robert Moog's patented invention.[2] The inventor's surname has become almost a generic term for all similar instruments, although, unlike "Hoover" and "Xerox," it does not appear in dictionaries. What then is a "Moog"? Briefly, it is a machine equipped with a keyboard which, using solid-state (i.e., not thermionic) electronic circuitry, can produce a wide variety of sounds, musical or non-musical, and, in particular, can reproduce the sounds of the standard orchestral instruments. Thus, if it is presented with the ringing sound obtained by striking a glass, the Moog can reproduce it faithfully or can transform it to make it sound like a church bell. The following are some of the applications of these synthesizers.

Since they can imitate the orchestral instruments and superimpose them, synthesizers can be a great aid both in teaching and in practicing orchestration. In fact, a musician sitting at the keyboard of a Moog is, in many ways, more powerful than if he were conducting a whole orchestra. He can control *directly*, and with immediate effect, the relative strengths of the various sections of the orchestra and, if he so wishes, can add further sections. Synthesizers are therefore particularly well-adapted to transcribing compositions. The object may be didactic, e.g., to bring out more clearly parts which, in a conventional recording, may be submerged in the general sound. It could be artistic and lead to a reinterpretation of well-known examples of the musical repertoire. It can also be a powerful tool for musical satire, especially since a slight change in the character of some of the instruments can fundamentally alter the character of the piece. Examples of these uses can be found in the synthesizer recordings of Walter Carlos, whose transcription of Bach's Brandenburg Concertos, Chorales and Inventions fall into the didactic/artistic category while, on the other hand, his "Variations and Fantasy on a theme of Elgar" entitled "Pompous Circumstances" is an amusing parody of "Land of Hope and Glory," and his "Wedding March" pokes fun at the Bridal Chorus from Lohengrin.

Since synthesizers can produce pretty well any sound, an infinite reservoir of musical instruments is now at the composer's disposal. In the past composers have been known to invent new instruments to suit some particular composition. If synthesizers had existed at the beginning of the 20th century Richard Strauss would not have had to go to the trouble of constructing the thunder machine for his Alpine Symphony, or the wind machine for Don Quixote. The Moog could have provided them. The following brief list of sounds, not usually available in an orchestra, shows the great potentialities of the synthesizer: taxi horns, train whistles, the "circuit for beach" to recreate the sound of the surf on the shore and the all too familiar "space sound" which, together with the all-pervasive barbecue smell of hot dog and hamburger stalls, is high up on the list of curses bestowed by technology upon our civilization.

The discussion of synthesizers quite naturally leads to the question of *Composing by Computers*. This will undoubtedly be dealt with by Professor David Charles in his paper, "Music and Technology Today," and all I propose to do is to express some random thoughts on the subject and to stimulate the discussion by some provocative statements.

The connection between music and mathematics has often been discussed and, indeed, many of the classical rules of harmony have a mathematical base. Thus, for four-part compositions there was—perhaps still is—a sacrosanct rule about avoiding "consecutive octaves." The reason is simple: the frequencies of the two sounds of the octave are in the ratio 2 : 1, they appear very similar to the ear, consequently if two voices in a four-part composition move parallel to each other an octave apart they sing the same *tune* and the four-part composition is downgraded to a three-part one—a clear case of cheating. The same rule applies to "consecutive fifths" with a frequency ratio of 3 : 2. To discourage this practice it used to be said, "Murder is a heinous crime but writing consecutive fifths is worse."

In asking a few questions about the relation between the synthesizer/computer and the composer I am not merely parading my ignorance but hope to stimulate discussion and thus clear up misconceptions and dispel doubts and allay—or possibly create—fears. Does a "contemporary" composer—i.e., one who believes in and uses the modern idiom and the latest techniques, regard the electronic instruments as a source of inspiration or an aid to it?

As long as he regards the synthesizer as a reservoir of hitherto unavailable sounds, and the computer as an aid for applying certain rules, the contemporary composer will be not much different from his predecessor as regards his artistic outlook. It is true that his compositions will bear little resemblance to those of earlier times and will be disliked by many, but one must remember that tastes change. It is a sobering thought that Mozart's

Symphony No. 40 in G Minor (K. 550) was described by a contemporary critic as "abscheulich" (detestable).

But what if the composer severs his ties with the past and, instead of being the master of the electronic instruments, becomes their partner, drawing inspiration from them? We may then see the development of an entirely new art form and the question is whether or not it should be called "music." I am not being insulting or passing a value judgment when I ask this. "Musical" sound can be defined by considering the spectrum of acoustic vibrations associated with it and can be distinguished from "noise." So it is legitimate to suggest that if a certain type of contemporary composition is far removed from what still goes by the name of "music" a new word should be coined for it.

Having raised a number of controversial points I should like to give my personal views. I strongly hope that the composer will continue to be the master of the instruments and that computer experts, electronic enthusiasts, mathematicians, etc., will not become, or pretend to have become, the composers of the future. I believe that science and technology form an essential part of our culture, but we must strive to establish a harmonious symbiosis between science and technology on the one hand and the arts and humanities on the other. I am using the word "symbiosis" advisedly, rather than the illusory term "unity," because I feel strongly that to pretend that some fundamental unity binds together the large variety of intellectual and artistic human endeavors hinders rather than helps a true understanding of the connections between them.

Seeing

Unlike in the section about HEARING I shall not refer to the influence of science and technology on artistic creativity, but will restrict myself to the more mundane, but nonetheless important, subjects of: conservation, reproduction and the use of novel scientific methods for the study of art objects and for their restoration.

Conservation is an important and certainly a continuous preoccupation of gallery curators. Thanks to improved techniques, control of temperature, of humidity and, most importantly, of illumination, have reached very high standards.[3] It is interesting to note how the illumination of paintings, together with a well known theatrical event, led to a sudden surge of electric lighting in London, as recently described by J.E. Harris.[4]

The new Savoy Theatre, which opened in October 1881 with Gilbert and Sullivan's "Patience," was the first in Britain to be lit by electricity. On the other hand we have the famous duet in the second act of "Patience," which gently ridicules the artists and poets who formed the clientele of Sir Coutts Lindsay's Grosvenor Gallery in Bond Street:

> 'A pallid and thin young man—
> A haggard and lank young man—
> A greenery-yallery, Grosvenor Gallery,
> Foot-in-the-grave young man!'

As a result of this publicity the business of the Grosvenor Gallery expanded and, in 1883, Sir Coutts Lindsay installed an electrical generator to provide lighting for his pictures. This was so successful that other galleries and residents asked to be put on the supply. By 1887 the London Electricity Supply Corporation, which still generates electricity in the heart of London, was founded. Within a few years this whole industrial activity was transferred to the newly constructed Deptford Power Station which, with its ninety megawatt output was, at that time, the largest in the world. A clear case of painting, music and poetry aiding technology.

Science also leads to a better understanding of the causes of the deterioration of paintings and other art objects, and to the development of methods of repairing the damage and preventing further deterioration. The latter is particularly important in the case of medieval stained glass. As a result of increasing atmospheric pollution some priceless examples of medieval art have suffered more damage in the last five decades than in the previous five centuries.

Reproductions of works of art, especially of paintings, have made great strides in recent times, mainly thanks to the improvements in color photography. Many more people can enjoy the great paintings than has been possible heretofore, and I think this will, if anything, stimulate more frequent visits to museums and picture galleries so as to see the real thing. But color reproductions are not the same thing as true *replicas*, and the time has perhaps come to consider seriously the plea made by Dr. Helmut Ruhemann[5] in his book "The Cleaning of Paintings," namely that full size replicas should be made of all the greatest paintings so that, if they were destroyed in some sudden natural or man-made cataclysm, posterity could still be able to appreciate them. True replicas must include attributes of paintings that are missing from color photographs, such as the unevenness of the painted surface, the texture left by the brushwork, etc. Perhaps a two-pronged approach will be needed. First a painter, using his artistic perception and sensitivity, must make a copy which is *objectively* as true as possible. This must then be followed by a technical comparison of the original and the copy by scanning the physical properties, such as the spectrum of the reflected light, or the surface properties of the painting and by a correction of any differences. Would such a project be worthy of UNESCO's attention?

There exists a good example of painstaking scientific work resulting in the production of a replica, almost the re-creation of the original, namely the story of the Lycurgus Cup in the British Museum, as described by R.H. Brill.[6] The most striking feature of this cup made of glass is its unusual color: it has a jade-like appearance when viewed by reflected light but shows a deep magenta color under transmitted light. This dichroism is known to be due to the diffraction of light from minute quantities of colloidal metal particles dissolved in the glass. The availability of a tiny chip (less than 10 mg) which had been found when part of the cup's mounting was removed, allowed a chemical analysis of the material to be made giving the composition of the glass and the concentration of the silver and gold particles as well as of other elements. As a result of a long series of experiments involving several glasses undergoing a variety of heat treatments, samples were finally obtained which reproduced faithfully the "Lycurgus" effect.

The non-destructive physical analysis of paintings with the help of X-rays and also of other electromagnetic radiations such as infra-red, is a relatively new development. The method relies on the fact that the absorption of X-rays by paint depends on the nature of the pigment. Thus metallic pigments absorb X-rays much more strongly than non-metallic ones. Furthermore the rate of absorption depends on the atomic mass of the element in the paint. For instance, the two most widely used white pigments can be distinguished easily since the absorption of X-rays by lead oxide is considerably stronger than by titanium oxide.

Although X-rays were discovered in 1895 it was more than 50

years later that the use of X-rays to detect features hidden by overpainting became an established technique. The explanation of this delay was given by Charles F. Bridgman[7] in his article "The Amazing Patent of the Radiography of Paintings." The first X-ray examinations of paintings were carried out within a year of Röntgen's discovery by W. König, and interest in this method grew both in Europe and in the United States. Then, in 1913, Alexander Faber, a radiologist in Weimar, Germany, radiographed several paintings and, apparently unaware of the previous work, stated: "It seems almost strange that they (X-rays) have not yet been applied in the field of artistic paintings." He patented the method in Germany on March 15, 1914 with an all-embracing patent claim, "Procedure for the determination of overpaintings in oil paintings and similar objects by the preparation of roentgenograms." The effect of the patent was "definitely detrimental to the development of radiography of paintings in Germany and as late as 1932 no German museum had been able to raise the funds for a patent license." The patent expired in 1937 and, certainly since the end of the 1939–1945 war, X-ray examination of paintings has become almost routine, especially before embarking on major cleaning or restoring work.

I shall give three examples of the use of the X-ray technique.

Pittoni's "Nativity," painted around 1740 and acquired by the National Gallery, London in 1958,[8] showed in the lower part of the picture, the Holy Family with the globe in the background (Fig 1). X-ray examination disclosed various hidden details in the upper part of the picture, e.g., a hand resting on the globe (Fig 2). Cleaning, and the removal of the over-painting, revealed the Eternal Father, His left hand resting on the globe and His right raised in Blessing, and also a white dove representing the Holy Ghost (Fig 3). Unfortunately, little is known about the history of the painting except that it was probably bought by Lord Willoughby for Grimsthorpe Castle (it is mentioned in an inventory of 1865), and that "both the Eternal Father and the Dove (actually visible in Fig. 1) were painted over, probably in the 19th century. Since they have been found in good condition under the overpaint this was presumably because they were thought unsuitable."

While in the case of Pittoni's painting we can only conjecture whether the "unsuitability" was on aesthetic or iconoclastic grounds, we can be pretty sure of what prompted the over-painting of Verelst's portrait of Nell Gwynn.

Mr. Alfred Wiggins of London bought the picture at a sale in the late 1940's. He gave it to a picture restorer to have it cleaned, but it was not until the cleaning was well advanced that it was discovered that Nell Gwynn's dress was an overpainting. When the dress was "cleaned" away Nell Gwynn appeared naked to the waist, and it is safe to assume that some private owner in Victorian times found it improper to display the portrait of a half-naked lady in a family drawing room. Alas, the painting was not photographed before cleaning and consequently we do not have documentary evidence of what the Victorians regarded as a suitable dress for the mistress of Charles II.

X-rays played an important part in solving the mystery of the famous painting by Hans Holbein the Younger, "King Henry VIII and the Barber-Surgeons,"[10] presumed to have been commissioned to mark the union of the Barber's Company with the Guild of Surgeons in 1540. It is a vast panel measuring 180 by 312 cms., and when the Surgeons seceded from the Barbers in 1745 to form their own Company, the painting remained with the Barbers. Then, in 1786 the Company of Surgeons, according to their Minute Book, bought a picture "being a cartoon painted by Hans Holbein representing King Henry VIIIth delivering the charter to the Barber Surgeons . . ." and that painting, which is almost identical with the one in the Barbers' Hall, now adorns the Hall of the Royal College of Surgeons. Is this second painting a later copy, as was held by many authorities until recently, or, as the librarian of the College stated in 1960, should they be considered "contemporary variants of the same composition, neither being a copy of the other"? This second view has been vindicated; the painting in the Surgeons' Hall was indeed executed largely by Holbein himself using the cartoon as a "canvas." X-ray examination proved this in the following way. The cartoon in those days was a full scale working plan drawn on paper. This plan was then transferred onto the wooden panel by first affixing the cartoon to the panel, marking the outline of the working by drawing thousands of small perforations and dusting the paper with powdered charcoal. The picture was thus transferred onto the panel with the perforated cartoon acting as a stencil.

Later the actual cartoon was strengthened by a canvas backing and used by Holbein as the basis of the second painting. X-ray examination proved this to be the case when the thousands of pinpricks under the painted surface showed up clearly in the X-ray picture.

My third example refers to a more recent discovery. *The London Times*, on January 5, 1972, announced that a hitherto unknown work by Van Gogh was shown by X-rays to be hidden under the paint surface of another of his pictures variously called "Vicarage Garden under Snow," "Garden under Snow" and "L'Hiver" (Fig 4). The X-ray picture (Fig 5) showing a woman at a spinning wheel seems to be in excellent condition and, according to Mr. Peter Hopkirk who reported the discovery, the London picture restorer Mr. Stavros Michelarias thought that it might be possible to separate the two works. The painting, belonging to Mr. Norton Simon, was at the time of the discovery on loan to the Los Angeles County Museum of Art but is now housed in the Norton Simon Museum in Pasadena. When I last saw it in the spring of 1982 it was still in its original condition, with no attempts at separating the two pictures having apparently been made.

This is perhaps just as well. According to Mr. Hopkirk: "It is probable that Van Gogh never intended the world to see his 'woman at the spinning wheel'. Unlike many artists he was never short of canvasses, having a brother who worked for an art dealer. He was extremely self-critical and may not have liked the work and therefore covered it."

Scientific examinations are also of great aid in the detection or proof of forgeries.[12] One of the most famous cases is that of Van Meegeren who, after the 1939–1945 war, was accused in The Netherlands of collaboration with the Nazis, the charge being that he had been involved in the sale of Vermeer's "The Woman Taken in Adultery" to Goering. The crime of collaboration was punishable by death but Van Meegeren claimed that, far from helping, he was fooling the Nazis because the painting was his own work and, similarly, several "Vermeers" in Dutch museums were his forgeries. As the result of a series of scientific investigations the charge of collaboration was dropped in 1947, but he received a one year prison sentence for forging artists' signatures which, however, was never carried out since he died shortly afterwards. Although the legal authorities accepted Van Meegeren's claim, many people in the art world did not, and maintained for the next 20 years that "Christ and His Disciples at Emmaus," claimed to be a forgery, was in fact genuine. The doubts were dispelled in 1968, thanks to a technique based on the minute radioactive content of any white lead pigment. Lead oxide is obtained from the ore galena which contains, in addition to the stable form of lead, traces of radioactive lead-210 and radium-226. During the smelting process the lead-210 accompanies the metal, while most of the radium remains in the waste slag. Thus, by careful radioactive measurements which can be carried out on samples as small as one microgram, it is possible to determine, albeit roughly, the date at which the white lead was made. The tests proved conclusively that the pigment could not have been used by Vermeer since it was certainly much less than 300 years old.

A discussion of the impact of technology on the visual arts should, I believe, also touch on the question to what extent technology, with its novel demands on buildings and engineering structures, can offend our aesthetic sense and lessen the enjoyment of our surroundings, be it town or countryside. Few modern developments have aroused so much passion. In fairness to present-day critics I must admit that this is not an entirely new attitude. I have heard that when windmills were first introduced from the Low Countries into East Anglia in the 14th century there were protests about these engineering contraptions disfiguring the countryside. Let us also remember that William Blake wrote his much quoted lines about the "dark Satanic mills" and "England's green and pleasant land" nearly two hundred years ago. Some of those same mills have now become cherished examples of industrial archaeology, and most of the remaining windmills are listed as historic monuments and cannot be pulled down.

But, fortunately, our forefathers were rather tolerant towards the visible signs of their civilization. If they had had the same attitude towards purely utilitarian buildings and engineering works as is prevalent today, we probably would have no fortified castles, no windmills, no bridges, no viaducts, etc.

It is often said that vast industrial buildings such as power stations, while acceptable aesthetically, are too large for their surroundings. But some of our most cherished architectural monuments are similarly out of scale with their environment. Take Ely Cathedral, towering above the open, flat countryside. I have been trying to get a photograph of it and asked a friend in Cambridge whose views on this subject are not much different from mine. He replied (with his tongue in his cheek):[13] "We have made an extensive search of the photographic literature relevant to Ely Cathedral and I regret to inform you that we can find no picture of its hideous bulk ruining the charming and characteristic Fenland landscape. Maybe the manufacturers of picture postcards agree with the conservationists in deploring the intrusion of man into the wonders of nature."

We have come to regard architectural monuments of the distant, or not too distant, past as part of the countryside and admire them irrespective of their nature or purpose. Future generations may feel likewise about some of our contributions to the landscape. The pylons of overhead power transmission lines are usually attacked whenever or wherever they are planned. We should remember what Sir Peter Medawar said in his presidential address to the British Association for the Advancement of Science in 1969:[14] "When the pylons are dismantled and the cables finally go underground people will think again of those majestic catenary curves and remind each other how giants once marched across the countryside in dead silence and in single file."

I hope that I have given a reasonably balanced view of the effects and influence of science and technology on the arts. There is no doubt that there exists some antagonism between those who believe that science and technology should stay clear of the arts except as modest providers of assistance, and those who are convinced that the arts will become fossilized unless they are revitalized by science and technology. This schism is not restricted to the practice of the arts. It can be found in other human activities in which science and technology play an increasingly important part. As I said earlier in this paper, I do not believe in the magic of "unity," but I do feel strongly about the need for tolerance in expressing and defending one's views. Scientists are often impatient with adversaries who refuse to accept "facts" without questioning; they are also often insensitive to human emotions such as anxiety, fear, pleasure or joy. But, conversely, the non-scientist should try to understand the scientist's way of seeking the truth and his trust in experimentally demonstrable facts. I believe that only by cultivating tolerance and mutual understanding can we avoid a polarization which would bode ill for the future.

Notes

1. Carleen M. Hutchins, *Scientific American*, Nov. 1962, pp. 78–93.
2. See e.g., John Jenkins and Jon Smith, *Electric Music* (David and Charles, Vancouver B.C., 1973).
3. See e.g., Ian Rawlins *Rep. Progr. in Physics*, Vol. IX 1942–43, p. 334, *Endeavour* 7, 1948, p. 107.

4. J. E. Harris. *Physics Bulletin.* Dec. 1982, p. 435.
5. H. Ruhemann. *The Cleaning of Paintings.* (1968) p. 117.
6. R. H. Brill. *Proc. 7th Int. Conf. on Glass,* Brussels 1964, paper 233.
7. C. F. Bridgman. *Studies in Conservation,* Vol. IX 1964, p. 135.
8. Michael Levey, *Natl. Gallery Catalogue, 17th and 18th Century Italian Schools.* (London 1971).
9. Personal communication from Mr. Wiggins, 1972.
10. Roy C. Strong, Burlington Magazine, *105*, 1963. Bertram Cohen, *Annuals of the Royal College of Surgeons of England, 40*, 1967, p. 179.
11. Peter Hopkirk, *The London Times,* 5 Jan. 1972, p. 1.
12. See e.g., Stuart J. Fleming, *Authenticity in Art.* (Inst. of Physics, London 1975).
13. Personal communication from Professor Sir Brian Pippard, 1972.
14. Sir Peter Medawar, "Presidential Address to the British Association for the Advancement of Science," Exeter, Sept. 1969. *Nature, 223*, 1969, p. 986.

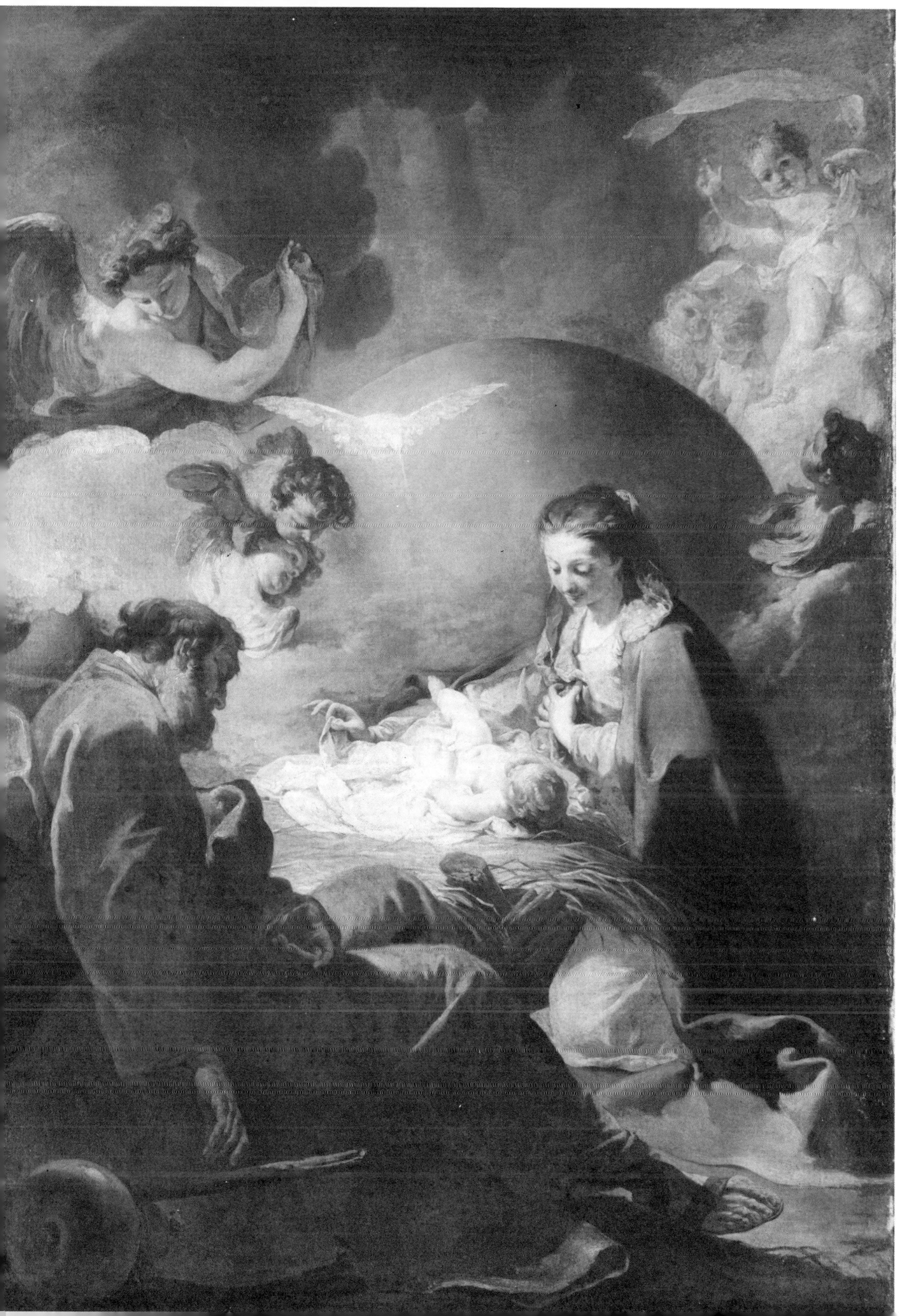

Fig. 1-1: Pittoni's "Nativity," before cleaning (National Gallery, London).

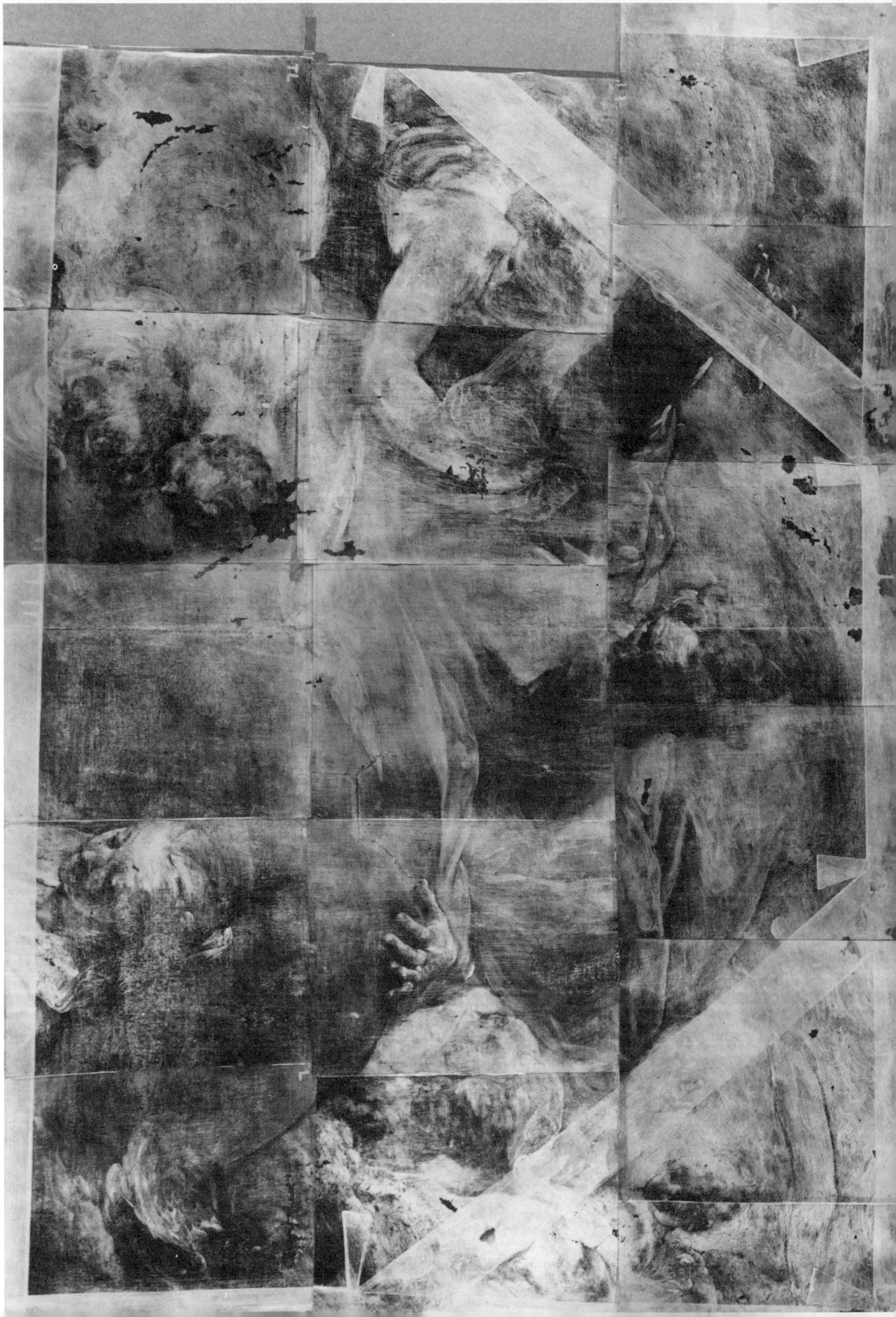

Fig. 1-2: Pittoni's "Nativity," during cleaning (National Gallery, London).

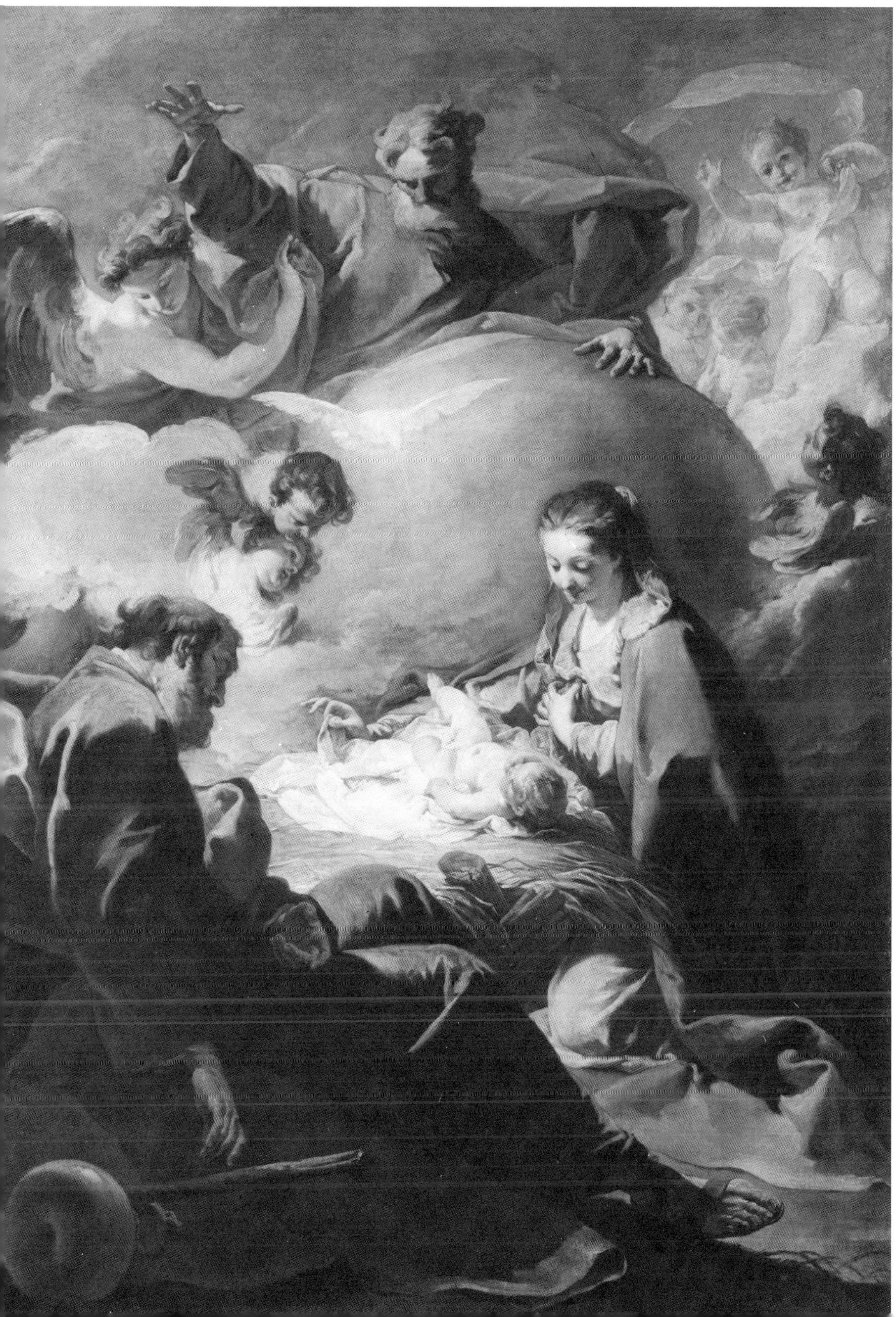

Fig. 1-3: Pittoni's "Nativity," after cleaning (National Gallery, London).

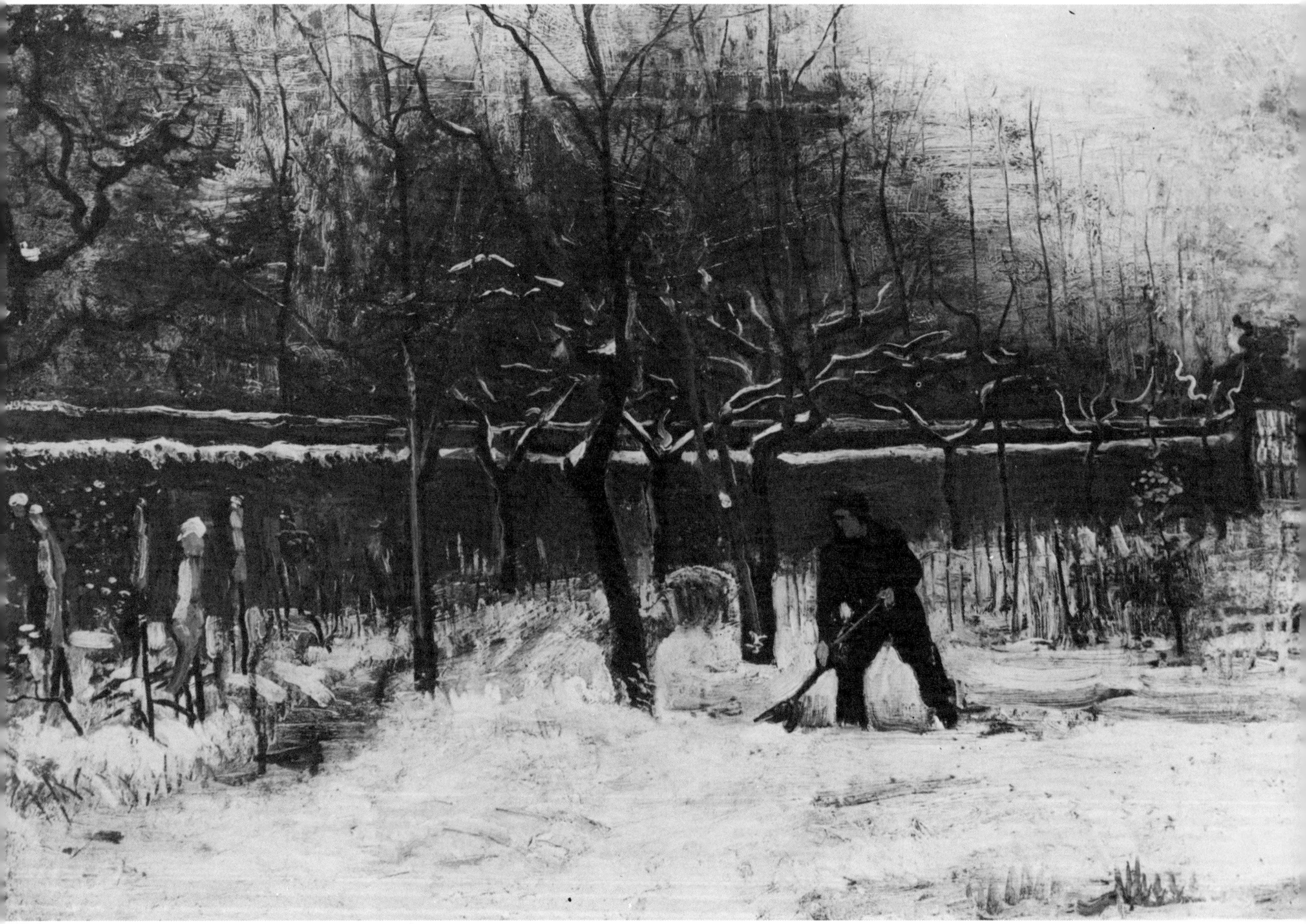

Fig. 1-4: Vincent Van Gogh, "Winter (January 1885)" (The Norton Simon Foundation, Los Angeles, F.69.39.2.P).

Science and Technology: Blessing or Curse to the Arts?

Nicholas J. Phillips

In this discussion I shall look at the links between the arts and the physical sciences at the present state of knowledge. The real truths are complex and show that in many respects the power of science and technology is not yet subtle enough to influence dramatically the creative core of the arts. Undoubtedly, the progress of electronics, acoustics and visual display is so rapid that the scene may well change with considerable rapidity, but for the moment we shall look at our real world situation and take stock of that.

The human being could be seen as essentially a machine with sensors and an information processing center, the brain. Such is the complexity of the brain that many people experience sensations based upon external stimuli that are so subtle as to defy clear description. If we consider the methods by which we sense external stimuli, we get some strong clues as to the answer to our primary question.

The arts as we know them are essentially media by which emotional experiences may be transferred from a creator to us, the observers. The "creator" will be, for example, a writer, an artist or a musician. Those connections between creator and observer are made by audible or visual means. In modern society, the media of radio and television probably dominate the communication scene, and by examining these media we find some glimmer of truth with respect to our fundamental inquiry.

The majority of us will at some time have been transfixed by the beauty of the experience of listening to a play or a stunning orchestral concert on the radio. But very few of us derive more than a superficial degree of pleasure from watching television programs, and this is not always due to the banal interjection of advertising. Fundamentally, the reason why radio can be relatively so compelling is the realism with which the acoustic presence can be recreated. Technological advances in studio techniques and methods of broadcasting have made possible the recreation in ordinary domestic environments of a sound stage of such reality as to be almost beyond the imagination of our civilization only some twenty years ago. It is, of course, relatively easy to fabricate the loudspeakers that reconstruct the sound stage using advanced, yet simple, technology. Stereophonic or multi-dimensional sound is now so good as to almost convey a better impression of orchestral music or a play than if the listener actually attended the performance. Most of us who have listened to the radio or observed the performance on TV in a 'simulcast' situation will prefer to have the TV switched off, as its puny visual display—even though it is in color—is no match for the dramatic quality of the stereophonic sound of the radio.

Cinema may be different from radio and TV; in cinema the roles of sound and vision may be reversed so that the technical quality of the cinema sound is often well below the visual quality of the film. But we can certainly experience strong emotional stimuli from visually bad films so long as the story is of sufficient caliber. Perhaps one could actually argue that some films benefit from technical inadequacies such that the mind switches off from trying to glimpse reality, and simply concentrates on the emotional fabric.

As a scientist, my view of the connection between science and technology and the arts is to look to science and technology to assist in conveying with the greatest clarity the artistic message. We are privileged, for example, to be able to listen to Lipatti perform the works of Liszt and Chopin many years after Lipatti's death. Enormous genius shines through the technical inadequacies of the sound recording available during this great pianist's lifetime. Yet there is no doubt that were Lipatti alive today, we could appreciate his great gifts with more real and emotional clarity just because of the advances in technology.

However, the progress of science and technology is such that earlier simplistic views of presentation of art to the viewer or listener have to be supplanted. Already, the quality of the artificial acoustic world is so profoundly interesting as to be able to transport us to the theater or concert hall.

If there is a shortcoming in the ability of science and technology with respect to the arts, it concerns the technology of visual representation. Fundamentally, the need to feed different information to our two eyes is very difficult to satisfy in an artificial way. Short of going to the theater or the cinema or the art gallery, there is no really competitive method of creating an equivalent stimulus. Progress in visual communication is rapid in a two-dimensional sense, but still erratic when we consider methods which convey three-dimensional information.

As a scientist who has been concerned with developments of the techniques of holography and three-dimensional imagery, I have increasingly been presented with the following question:

> If one is able to draw the real object or objects, is it worthwhile to create a picture of the object, albeit in three dimensions?

Thus we ask:

> is the object or its image more important?

My answer is that the image often assumes greater importance. This is probably due to our ability to view the image rather than the object in conditions free from distractions. Would Van Gogh's immortalized "pont d'Arles" make the same impact on us surrounded by traffic noise and modern litter?

Technology cannot yet create the necessary three-dimensional visual stimulus to compare with the remarkable advances in acoustics. I am convinced that when it does have this ability, we shall step forward into a world of staggering communicative ability.

The arts are often too unsophisticated to be assisted by technology. Take the case of Debussy's exquisite creations and compare these to works of Beethoven. The comparison may not at first appear to be sensible, but some simple facts emerge. Debussy's music is one of the few classical forms capable of rewarding synthesis into electronic form. Why? These complex sketches are, in fact, impeccably detailed studies with their form trimmed and modified to give uncanny impressions of life and the spiritual world. On the other hand, Beethoven's work is too rounded, too explicit to permit impressionistic representation. This is probably a weakness rather than a strength. The scientist accepts that the statement, "A has a greater chance of happening than B" is a more precise statement than "A will definitely happen and B will definitely not." Thus, what may ostensibly appear to be inexact is actually more exact.

Therefore, aside from providing just technical assistance, science can only enhance art when that art is of a suitably sophisticated nature. Otherwise, science and technology can only assist the conveyance of an impression of art from one place to another, from one person to another.

Though I am essentially critical of the role played by science and technology in the conveyance of visual forms of art, serious questions are posed by the technical breakthroughs which we may expect to actually take place.

Suppose we do reach such a state of advancement that we can experience artificial audible and visual stimuli of such realism that we no longer need to search out the source, no longer need to visit the great galleries and museums, no longer need to attend a play or concert. When such a state is reached, the lifestyle of the human being is likely to change dramatically. Whether this change will be beneficial remains to be seen. At the moment, the depressing truth about humanity seems to be that its psychological advancement lags alarmingly behind its technological abilities. Those of us in society who enjoy the arts too often forget the presence of reality in the form of the real world external to our own psyche.

Will science and technology irrevocably cause the detachment of the human reality from the "earthly" reality? It seems most likely that this will in fact happen. It will be all too easy for us to live cocooned in a false blanket of sonic and visual communication and forget that just down the road is a field with some cows and sheep, a form of reality which has not changed since biblical times.

Thus, if the arts are to really benefit from science and technology, it seems most important to ensure that the quality of human life is seen as part of the overall balance. We are at risk of living in a sophisticated cultural world which is at the same time a physical desert.

Let us summarize some of the ideas touched on here by simply stating the case for the belief that the arts may be enhanced by science and technology. But let us also look to the scientific future with very great caution and remember that the arts are a fundamental manifestation of true human life. Let us hope that humanity and its appreciative abilities are not destroyed by the very technology which could provide such potential advancement.

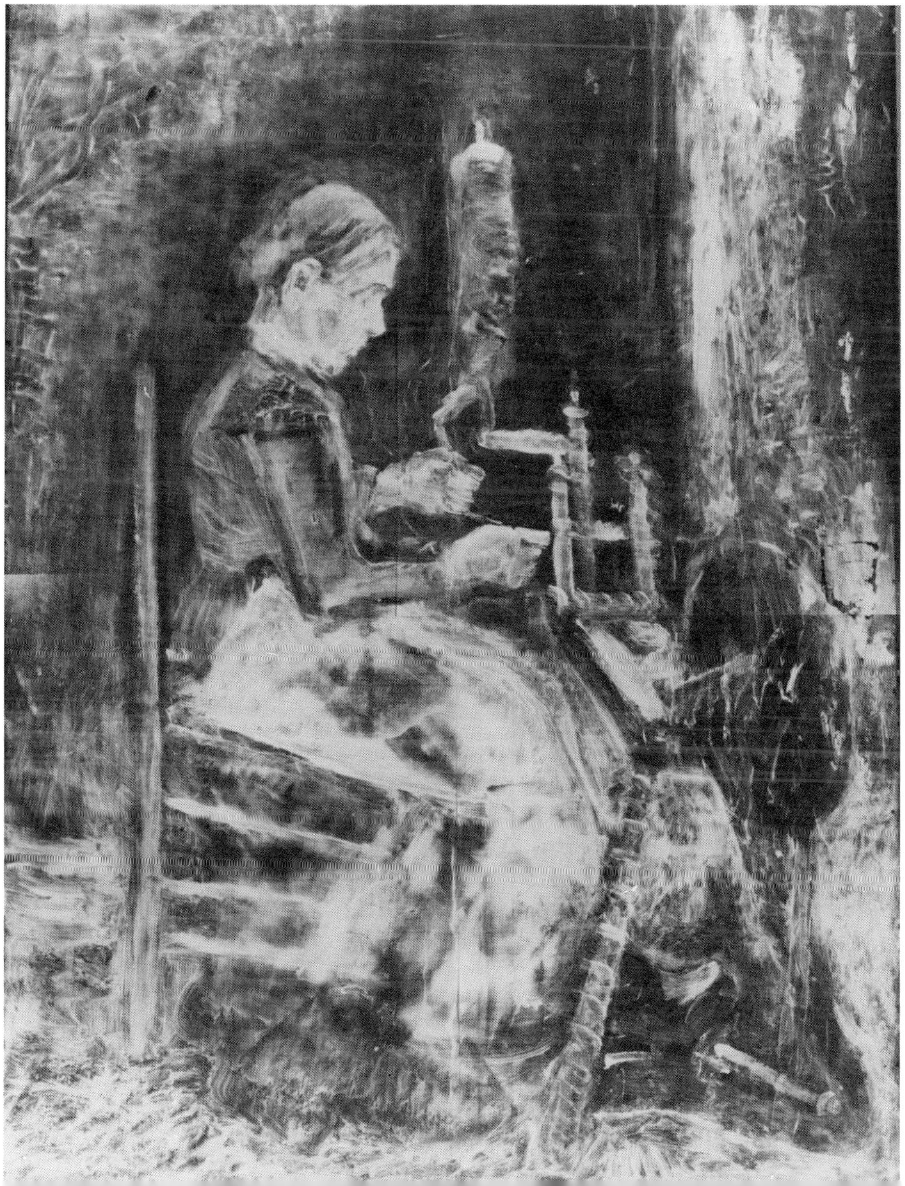

Fig. 1-5: Vincent Van Gogh, "Winter (January 1885)," X-ray view of picture (The Norton Simon Foundation, Los Angeles).

Chapter 2

The City: Organism or Artifact?

The City: Organism or Artifact?

Ervin Y. Galantay

The built form of human settlements is often described as an artifact, a mechanism or a large-scale work of art—the result of conscious design decisions.

By contrast the genesis and growth of the urban structure is generally explained by analogy with living systems: organisms capable of homeostasis or self-organization.

Neither explanation is fully satisfactory: the artifact or the work of art is a static object, immutable. A machine is an assemblage of parts for the transmission of motion or energy but evidently incapable of modifying or renewing itself. On the other hand, the organic analogy cannot account for the evidence of conscious design interventions.

Can the natural and artificial aspects of urban form be reconciled in a single, credible model?

These are some of the pertinent questions with respect to the city that can be asked within the framework of an interdisciplinary discussion group focusing on the relation of "Art and Technology." Since all technology is based on science, an inquiry into the nature and meaning of urban form ultimately confronts theories of the Arts and the Sciences both on the level of the description of the phenomenon, and on the level of the possibilities of its modification and control.

René Berger asked me in particular to examine to what extent the creative processes of urban design have been modified by the availability of new techniques, and to speculate about the changing perception and cognition of urban form resulting from technological change.

It is well known that to its inhabitants and users the physical pattern of cities constitutes a semantic system—an interrelated set of perceptual cues. However, it appears that the "legibility" of the built environment is size-dependent: while the image of the small-scale historic towns is clearly understood, the pattern of modern agglomerations is confusing to the inhabitants who are familiar only with its fragments without grasping how they relate to the whole. Can modern technologies be of help in explaining the pattern of the metropolis to re-establish a coherent "image," familiar to all?

In order to disentangle this set of interlaced questions, I divided this paper into four separate chapters:

1. a critical review of the debate about the nature of urban form leading to the metaphor of the "city as an ecological system";
2. a brief survey of the evolution of the perception and cognition of urban form and of the representation of the "image" of the city;
3. an appraisal of recent trends toward the use of quantitative methods and models as aids in planning and designing urban form;
4. a critical comparison of the "performance" of the urban structure of two recently built new towns—Brasilia and Ciudad Guayana—to explore whether computer-assisted designs have yielded a more satisfactory environment than designs based on intuition and conjecture.

The Nature of Urban Form

The debate on what makes a city has been as divisive within the design professions as the "Nature versus Nurture" controversy in the life sciences.

All interpretations seem to be value-loaded, rooted in deeply held convictions of the authors ranging from cosmology to social ethics and including sub-conscious biases. It is interesting to note that few descriptions are intrinsic to the phenomenon, and most authors return to metaphors and concepts borrowed from other fields of science.

Personally, I am convinced that it is false to attempt to discuss urban form as an "autonomous" design system, i.e., without taking into account the relation of the built environment to the powerful socio-economic and political systems as determinants. After all, a population and its habitat form an indissoluble entity. Nevertheless, within the limits of this paper I am compelled to focus my inquiry on the physical and perceptual aspects of the city.

The historian Arnold Toynbee observed that "the first ideas that the word city evokes in our mind are visual images." Although he goes on to define the city by characteristics of its society, he returns to built form by stating: "the existence of a genuine community is invariably proclaimed by the presence of public buildings . . ."

J. Rykwert concurs; to him "the city is an artifact, an artifact of a curious kind, compounded of willed and random elements, imperfectly controlled . . ." Lewis Mumford disagrees, he sees the city as "a fact of nature . . . like a nest or an anthill," yet he concedes that it is "also a conscious work of art which includes in a collective structure numerous simpler and distinct forms of art . . ."

Most architects cherish the concept of the city as a "work of art" since it implies that the individual designer has an important role to play in shaping the environment. Thus Edmund Bacon —former director of the Philadelphia City Commission—crusades to "dispel the idea . . . that cities are a kind of grand accident and they respond to some immutable law." He sees the city as an "Act of Will," an urban form determined by sequential design choices of individual decision makers. The architect Aldo Rossi goes a step further. To him the "city is an artifact which gives concrete form to society," and he claims that the built form enjoys autonomous existence: "as the city ages it acquires a conscience and a memory." Through the presence of its monuments and the permanency of great design ideas "the city guides and modifies itself." Both Rossi and Bacon seem to assume a built-in tendency toward architectural perfection—a teleological mechanism to account for a postulated finalism like Teilhard de Chardin's "Omega principle." It is a bit like reading the polemics of 19th century theologians arguing the merits of a "Divine Grand Design"

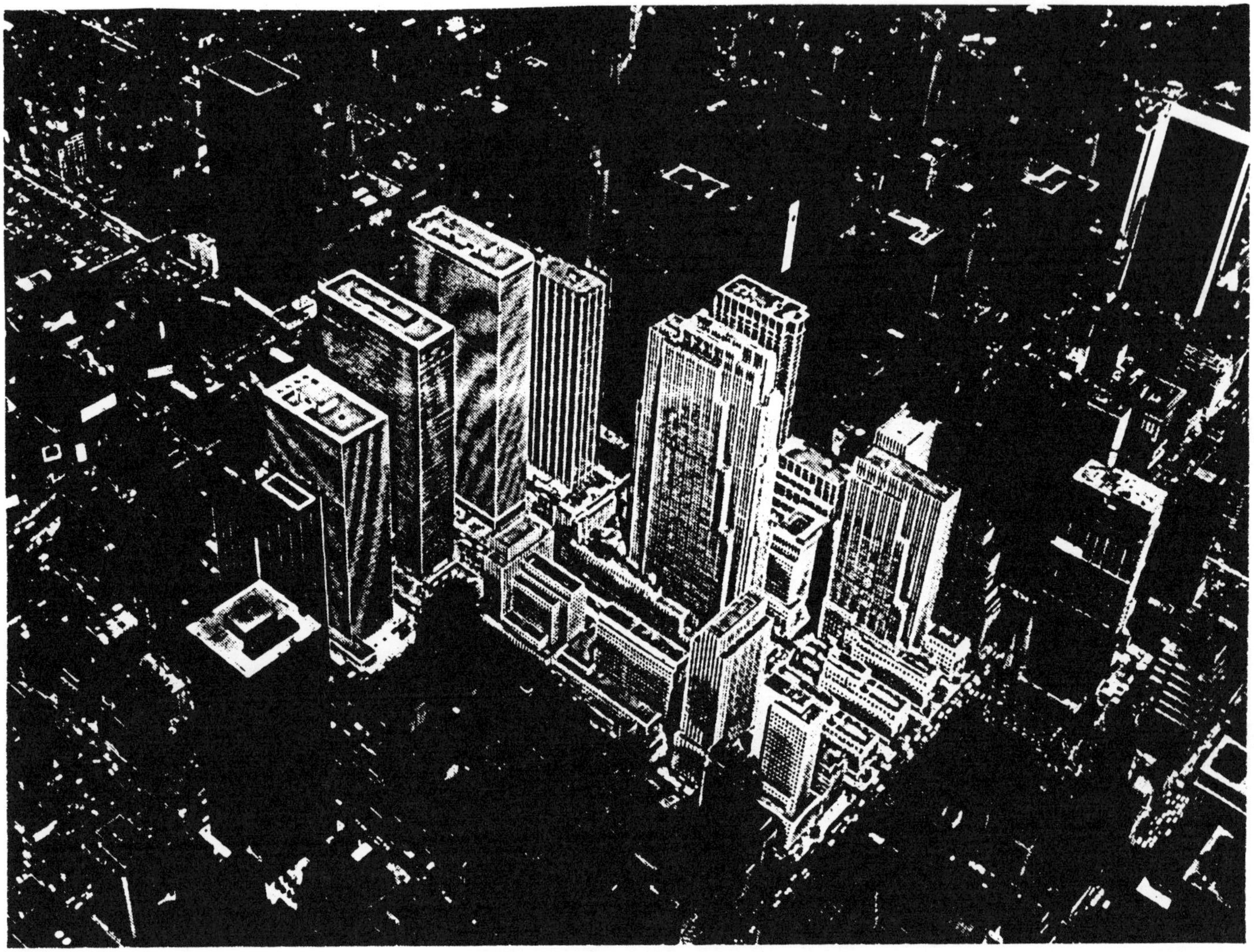

against the notion of living systems emerging by the "recklessly opportunistic" mechanism of natural selection.

Not every architect agrees with these visions of the "Grand Design" and of heroic leadership. To Sir Leslie Martin, the city is but "a logical pattern of entities and relationships built around activities." He even claims that this would as well do as a definition of architecture!

Jonathan Barnett also protests that "cities are not designed by making pictures . . . but created by a decision-making process that goes on continuously . . ." Rapoport tries to mediate by

Fig. 2-1: Aerial view and map of Rockefeller Center, New York.

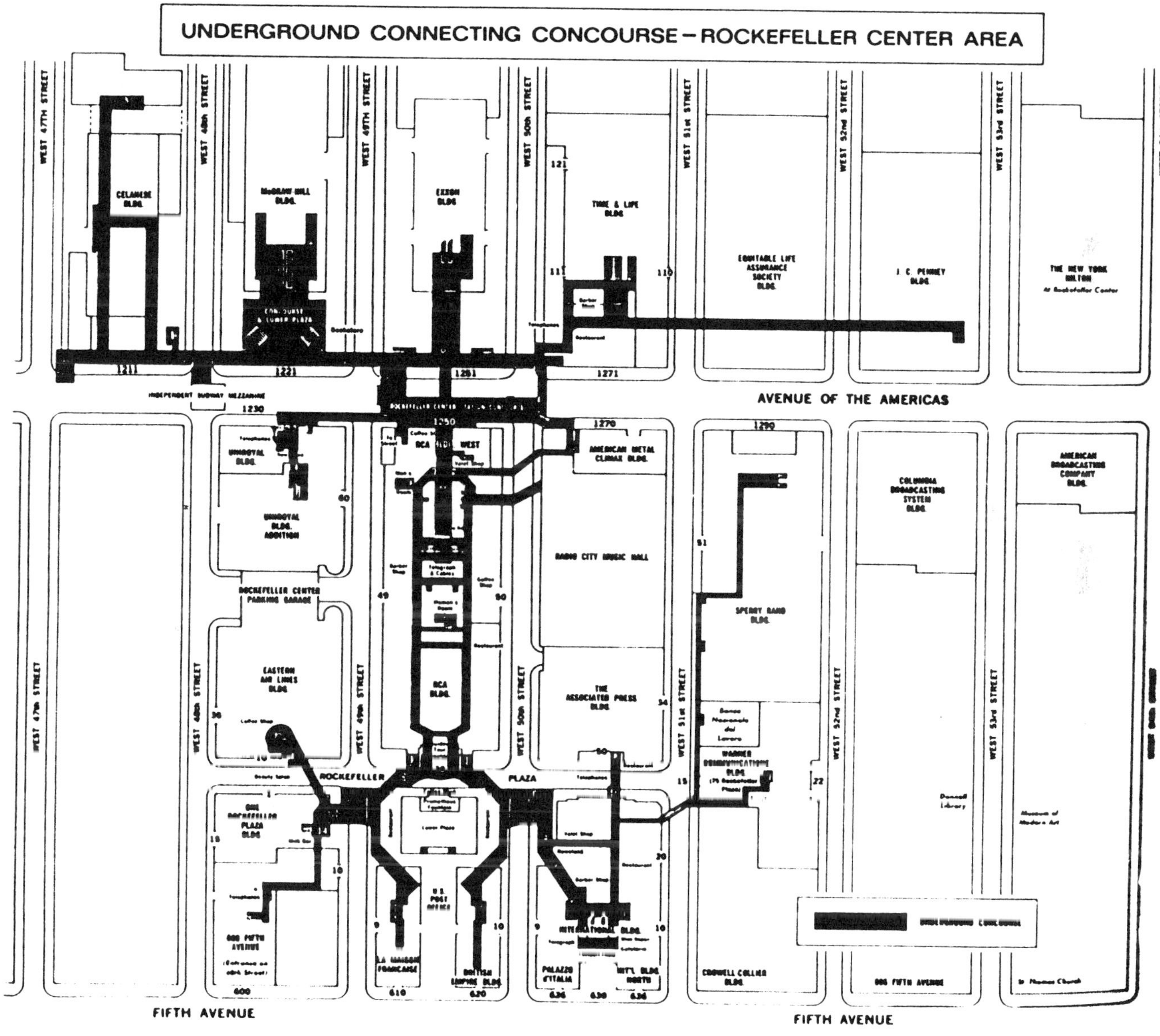

offering a broader definition of design as a "choice process and conflict resolution implicit in all decision-making" and can thus claim that "any artifact, whether a pot or the environment" is the outcome of a series of choices among alternatives and thus: "design."

Kevin Lynch disagrees. To him cities are "simply huge facts of nature to which we might adapt"—"vast natural phenomena beyond our ability to change." Lynch also rejects the notion of the city as a "practical machine" and points out that the city has some of the characteristics that distinguish living creatures from machines by possessing "a homeostatic dynamism which will establish a new balanced state whenever the organism has been disturbed by an outside force."

Finally, let us cite the most eloquent spokesman for the organic analogy, Constantinos Doxiades—founder of "Ekistics: the Science of Human Settlements"—to him "cities are biological individuals of a higher order than cells and organisms . . ." and he refers to the classification system of Sir Julian Huxley, who grouped human settlements with beehives and termite hills as "biological individuals of the third order." However, Lynch observes that a similar objection can be made to the hypothesis of the city as an organism as to that of the city as a mechanism: biological systems are made up of "unthinking" organisms unable to modify the system in any fundamental way. Indeed, if the organic analogy gives an adequate description of a spontaneous human agglomeration like certain squatter settlements—it cannot apply to the pattern of planned cities, whether colonial towns or modern New Communities.

Having briefly surveyed these contradictory postulates and opinions, let me now set forth my own concept of the "city as an ecological system."

I am impressed by Levy-Strauss' insight that the "city is at the confluence of nature and artifact" and Dobzhanski's observation that "cultural evolution is a unique process by which man *to some extent* shapes his environment."

Clearly, what is needed to formulate a good descriptive model of the city is an evolutionary system which can also account for the impact of the voluntary interventions of man. This brings me to the metaphor of the "city as an ecosystem."

An ecosystem is a set of biological individuals in relation with the inorganic environment where each organism is in competitive or cooperative relation with other individuals of its own kind as well as those of other species.

Within an ecosystem, morphological evolution is neither a chance phenomenon nor a deterministic phenomenon, but a two-step tandem process combining both. If an ecosystem includes individuals of the human species, the voluntary actions of man can be introduced as external constraints or disturbances affecting the growth and shape of other organisms and these constraints can be either stimulating or inhibiting.

Let us compare the pattern of human settlements with a natural forest or with an orchard. The natural forest, the wilderness, the meadow are ecological systems relatively undisturbed by man. By contrast, the orchard is an ecological setting, but it is also a man-made environment. In the natural forest, the localization of individual trees is somewhat due to chance—the seed may have dropped from a bird's beak. While the growth force of each tree is determined by its genotype, the shape of the phenotype will be influenced by its position and locational assets and liabilities in the competition for light and water, exposure to microorganisms, parasites, etc.

In our model orchard of pear trees, the position of each tree is carefully determined by the gardener providing optimal separation between trees planted in parallel rows. The gardener also improves the quality of the soil, irrigates, fights weeds and parasites. Genetic forces would drive the tree to assume a symmetrical shape with respect to the axis of its trunk. However our gardener trains the trees in geometrical patterns by bending, pruning and tying its branches against a guiding two-dimensional framework. The resulting shape is a compromise between the intentions of the gardener and the genetic forces programming natural growth. The gardener may also graft the tree, in which case he is tampering with the genotype and creating a hybrid. The actions of the gardener represent external constraints influencing the morphology of the plant. However, these constraints are mainly stimulating and contribute to the reproductive success of the tree and of the orchard as a larger entity (although simultaneously depleting the richness of the ecosystem having drastically reduced the variety of species within the set: weeds, parasites, other trees, etc.). More abundant and healthier fruit will be produced and, as far as the gardener is concerned, the flattened out geometric shape of the trees provides easier access for the picking of fruit. From an aesthetic standpoint some may prefer the regular form of the trained trees, others may prefer the natural shape, and the forest to the orchard. Let us compare the ecosystem of the forest and orchard with a settlement system within which the localization of individual towns may be "spontaneous" (the forest) or planned (the orchard) or mixed—as in the case of a grove which includes wild as well as planted trees. In the case of a single human settlement the "genetype" is represented by the gene-pool of the inhabitants and provides the driving force for their actions affecting the spatial pattern of the environment. The "phenotype" in this case is largely determined by the site. In a spontaneous agglomeration, the actions of the residents are generally limited to small-scale modifications in response to elementary economic and social forces, such as the attraction of the resource combination, the pressure of increasing density and compatibility/incompatibility relations. The result is a "natural" and somewhat predictible growth pattern. In more advanced societies, the majority of inhabitants also engage in the same sort of random small-scale activities modifying the environment as above, but there also emerges a small group of designers and decision-makers whose actions have a major impact on the physical environment. These changes may reinforce, halt, or, to some extent, reverse the direction and speed of the natural growth tendencies. In saying "to some extent," we take into account the conflict between natural growth forces and imposed constraints, i.e., a twig tied down to point toward the ground will strive to reverse its growth direction upward, and it may atrophy if excessively restrained.

The actions of the designers and decision-makers may be representative of the collective will or consensus of the population or just of the whims of a power elite, depending on the type of the social systems. In either case formal order is imposed to channel natural growth very much like the frame to which the branches of the pear tree are tied. The extent to which the imposed geometric forms will dominate natural growth tendencies depends on the amount of control. If the gardener stops pruning, bending and tying the branches, the tree will soon assume a more "natural" shape, and yet its major branches will always retain "the memory" of the imposed geometry. Similarly, in a planned city any relaxation of control will result in a fast decay of formal order. There is plenty of historical evidence for this, i.e., the orthogonal street pattern of Roman colonial towns decayed rapidly in the turbulent times of the late Empire, and in many cases became so overgrown by random medieval tissue as to be barely traceable (Hofer).

Without wanting to stretch the analogy too far, let me mention that the grafting of a tree which alters its genotype can be compared to the results of an abrupt change in the composition of the population of a city. One may cite as examples the Spanish conquest of Andalusian cities of the Moors—like Cordova and Sevilla—or the conquest of Byzantine Constantinople by the Anatolian Turks.

Here we can make a concession to Aldo Rossi's idea of the city as "collective memory." As M. Halbwachs puts it, "whenever a population appropriates a space, it tries to transform it to its own image, but at the same time it also becomes captive of the physical frame of the city which acts as the 'locus' of collective memory." The newly grafted population with its different value system and behavior pattern adds a new direction to the genetic forces shaping the city, but the already existing structure of street patterns and configurations of monuments will remain a determinant of further growth, just as in the case of a trained tree, which has been grafted and left to grow unassisted, i.e., without further constraints.

In developing this line of thought, I have not forgotten Christopher Alexander's seminal article, "The City is *not* a Tree." However, his argument consists essentially of a criticism of the hierarchical method of planning, and a plea in favor of considering topological relations in the spatial configuration of urban activities. This is in no way in contradiction with my analogy of the city and an ecosystem.

I hope even my metaphor of the orchard of pear trees has some explanatory value-shedding light on the intricate interrelation of urban morphology and design. The understanding of this relation is essential to the following discussion of the city as a communications system and to the inquiry into the means of controlling urban growth and form.

Perception and Cognition of the City

From the traces of the earliest civilization to the modern metropolis, we find confirmation for the coexistence of two archetypical patterns:

spontaneous agglomeration	in which the relations between components are determined by autocatalytic processes or self-organization;
and	
planned towns	exhibiting evidence of formal composition.

In spontaneous agglomerations, the pattern of the settlement is determined by the shape and configuration of the cells, or "territories" claimed by individuals or family groups. Public space is interstitial "left over space" between the cells.

By contrast, in planned settlements, public use has priority. Space is reserved for the street grid, for buildings and open areas for the use of the community. "Left over" space between the streets—and not reserved for public use—is subdivided for private development.

Hybrid forms may result from the growth of both the spontaneous and of the planned settlements. Disorderly agglomerations may gradually acquire some formally ordered internal structure by renewal action and their expansion may also take place by the addition of planned suburbs. On the other hand, the internal structure of planned settlements may decay in times of relaxed control and their expansion may occur in the form of chaotic fringe development, such as squatter settlements, illegal manufacturing establishments, etc., forming an agglomerate pattern.

Most modern metropolises show this hybrid pattern in which islands of formal composition seem to float in vast areas of amorphous tissue. Take the example of *Caracas*, where the checkerboard patterns of the city center and villages of colonial origin (Catia, Baruta, Petare, etc.) and some formal 19th and 20th century extensions are submerged in vast spontaneously built-up areas by "ranchos." Superimposed on all this is the macroform of modern urban expressways—to which some large-scale tract developments are attached: the "Helicoide" shopping center, the University City, the Airport, the Racecourse, etc. Traffic engineering seems to have superceded town planning—the expressways being the components of the urban structure which have the greatest notional and conceptual validity. By contrast *Moscow* has as its core the historical city showing a classical agglomerate pattern resembling concentric tree rings, but this core area is surrounded by totally regimented large housing estates and planned satellite towns on its periphery.

Parallel with the change of scale and overall pattern of the city is the transformation of the image of the city. As Henry Wöllfflin observed in 1915, "beholding is not a mirror which remains the same, but a living power of apprehension which has its own inward history and has passed through many stages . . ."

It seems to me that the perception and cognition of urban form are size-related—i.e., influenced by the relation of the dimensions of the city to the size of the human being, the observer, user and shaper of urban form.

It is important to take note that the tremendous increase of the dimensions of cities has occurred quite recently. Few pre-industrial cities exceeded the size of one km^2 or at most one square mile (2.56 km^2). Even 19th century Paris (1834) with more than a million inhabitants covered a built-up area of only 33 km^2 and could be crossed in either direction in a good hour's walk.

By contrast modern metropolises spread over thousands of square kilometers. It takes more than an hour to cross Los Angeles by car on the freeways at a cruising speed which is at least 18 times faster than the speed of the "promeneur" in 19th century Paris (90 km/hr instead of 5 km/hr).

This implies that perception and cognition are size-related because they are also speed-related. Since no one would cross Los Angeles on foot, the overall image of the city emerges as seen from the freeways.

Let me now make the premise that the changing perception and cognition of major urban form in the historical context can be best examined on the basis of the evidence of the evolution of the techniques used for its description and representation. For this purpose it is convenient to distinguish between five basic paradigms of the image of human settlements:

1. agglomerate
2. symbolic
3. perspectivic
4. statistically ordered
5. topological

1. *Agglomerate;* We have already discussed the spontaneous agglomeration in which the cells are the dominant components of the pattern, with the movement system and other public use confined to the interstitial space between the cells. This is, without a doubt, the most primitive and ancient form of towns and villages, yet also familiar as the dominant pattern of spontaneous squatter settlements around the world.

   The graphic representation which corresponds to this type of simple spatial organization is the type of map drawn by children or by primitive men. Such maps are selective, i.e., limited to features that play a significant role for the observer, and are also "egocentrically" scale-distorted: the size of the represented phenomena indicates relative connotative significance—i.e., imageable places are exaggerated while trivial phenomena are omitted.

2. *The symbolic city;* The physical expression of speculation about cosmological order such as the vision of the "heavenly Jerusalem" or the Mandala. The desire to be in harmony with the universe provided the pattern for the most ancient and venerable examples of planned urban structure—true "cosmograms"—in which the relation of major buildings or precincts is determined more by theological speculation than by functional or aesthetic considerations.

   Early representations of the symbolic city are mere *ideograms* without any pretension of providing information about scale or building detail. At later stages of the differentiation of human society by division of labor, boundaries appear between private, semi-private and public use of land, and an *axonometric technique* of representation is developed which permits us to identify individual buildings. A further refinement is the invention of the *isometric* drawing in which the true scale relations of built form can be conveyed. In Europe such drawings first appear in the 15th century and become popular in the 16th (first known plan of Venice of 1479, of Paris 1524, etc.).

3. *The perspectivic city;* Concerned with *visual*, rather than symbolic, ordering of urban space. A set configuration of the principal buildings is established as seen from fixed viewpoints on a level with the eye of a horseman or pedestrian. Landmarks are interconnected and intervisible. This new pattern is expressive of social stratification and power relations. Its representation—the perspective—is the fragmented view of the town structure as seen by the individual. To complement this loss of vision of the city as an entity, interest arises in *aerial perspectives*, drawn from a bird's eye (see Leonardo's sketchbooks). The plan of the city is often represented on a wheel as a circular panorama as seen from the top of a high tower (Cortes Plan of Mexico 1524, Plan of Rottweil 1564).

   In the baroque city—and in analogy to Vesalius's anatomy—streets are increasingly seen as traffic arteries rather than mere visual corridors opening perspectives among monuments—but the spatial organization and the urban pattern remains "perspectivic" up to the beginning of the industrial revolution.

4. *The statistically ordered city;* The rapid expansion of the size of industrial cities brought about an interest in "efficient" city form and a more scientific analysis of the urban structure as a form of "machine." Nineteenth century urban theoreticians like Ildefonso Cerdã (1867) —like the contemporary physicist Lord Kelvin—were primarily interested in arithmetic relations: building height, coverage, density. Not visual principles but regulations, prescriptions, zoning should bring about a new statistically ordered urban pattern. At the same time, the compartmentalization and separation of urban functions became "the holy writ of town planning departments" (Krieg).

   However, notions that the city can be best described by a sort of Newtonian mechanics became gradually replaced by functionalist theory based on an analogy with natural organisms. Patrick Geddes (1905) was the first to argue that the structure of the city is "organic" and urban form is a response-to-function on the Darwinian model. This

Fig. 2-2: Computer simulation of southwest sector of Boston.

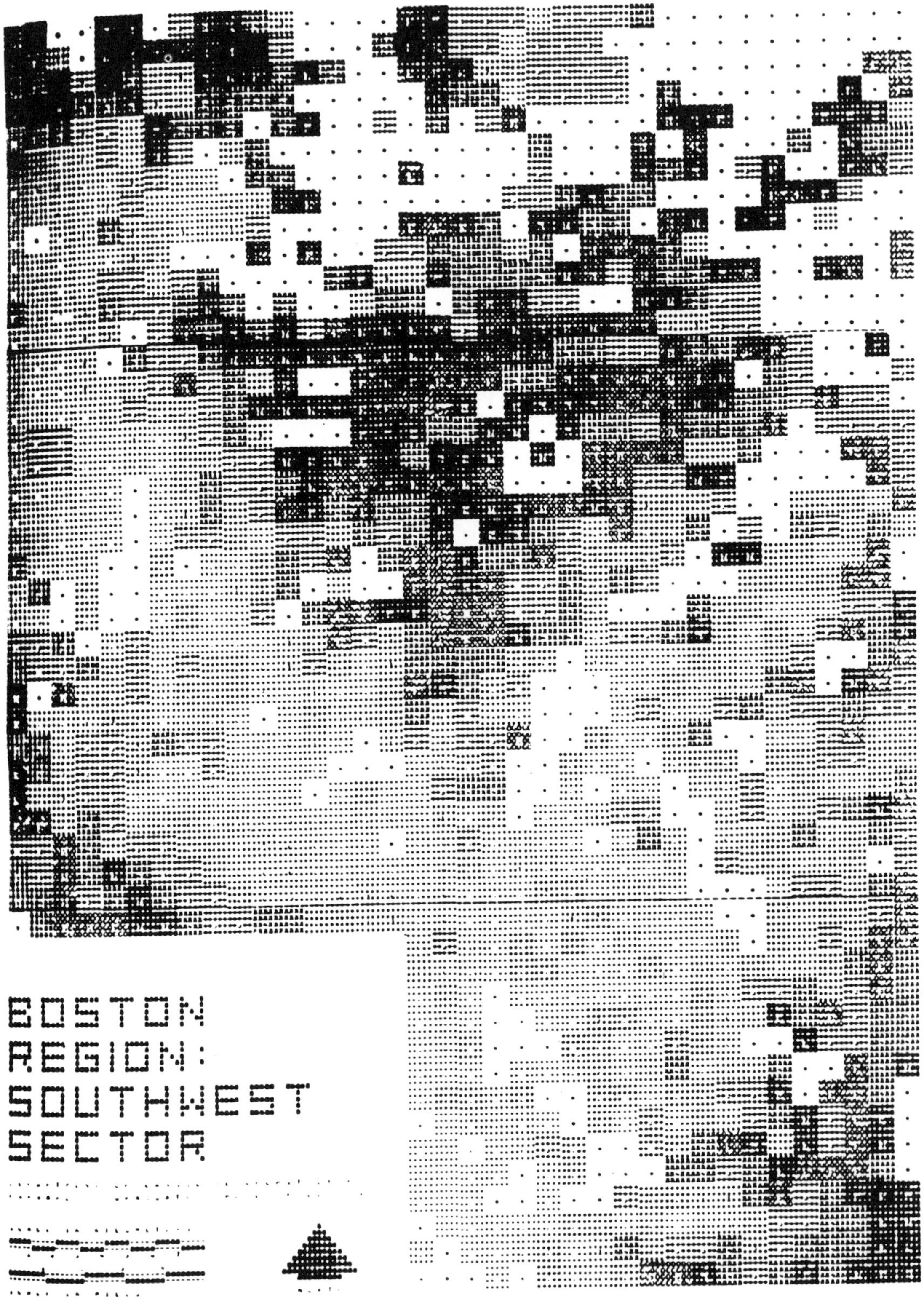

view was encouraged by the evidence of the unplanned, and apparently chaotic, growth of urban "tissue" on the edges of the modern metropolis. The new preoccupation with random phenomena gave rise to the increasing use of statistical methods of description and analysis of the urban pattern.

The graphic representations of the "statistically ordered city" include increasingly accurate cartography providing information of the measurable, arithmetic kind —as well as photographic inventorization of the built form—giving precise information on detail. It is worth noting that the photographic image cannot eliminate trivial detail in contrast to the artist's drawings—where even apparently meticulously accurate "vedute" like Canaletto's Venetian pictures of the 18th century show selectivity and concentration on major features (Corboz).

5. *The topological paradigm;* Meanwhile, technological change introduced a new geographical scale to the metropolis. Railroads and their terminals—and more recently urban expressways with the associated parking areas and vast shopping centers at their intersections —are structuring elements far surpassing the form-giving impact of individual buildings. Unlike the pre-industrial city with its pattern of public buildings interconnected by arterial roads, the pattern of the modern metropolis is based on transportation with the major activities and buildings "plugged in," a description inspired by electrical network theory which has been borrowed for the modeling of the urban system.

   The use of electricity has also contributed to a change in the perception of urban form. The unprecedented extent of streetlighting, illuminated advertising signs, floodlighting of buildings, etc., brought about a *nighttime image* equal in importance, but quite different from, the daytime appearance of the cities.

   Aviation has added a further dimension to the bird's eye view; *vertical aerial photographs* taken from increasingly higher altitudes permit once more an all encompassing view of the spreading metropolis. The urban structure as seen from the air—including its nighttime image—is becoming ever more familiar with the increase of air traffic and the popularity of helicopter sightseeing.[1] Note, however, that the use of high-level vertical photography is not only important for what it shows, but also what it doesn't show—namely detail. In contrast to the use of inclined aerial views which require artistic sensibility in the choice of the most favourable and characteristic angle, the preference for the "objective" vertical airview contributes to the depersonalization of the information on urban form (Boesch).

Another characteristic of the modern metropolis is that important networks of infrastructure (water and sewerage lines, gas and steam lines, telephone cables), as well as of movement systems (subway trains, pedestrian passageways, truck tunnels, etc.) are buried underground and their layout does not necessarily correspond to the pattern of the surface streets and building blocks. The layout of these networks cannot be comprehended by direct visual observation, yet their abstract image must be associated with the day- and night-time image of visible urban form.

The inadequacy of the fragmented and additive characteristics of the model of "the statistically ordered city" led to the recent preference for a topological description of urban spatial structure. The city is seen in its "Gestalt": a total configuration or pattern of phenomena so integrated as to constitute a functional unit with properties not derivable from its parts in summation. There is now less interest in numerical information on elements of urban form, and more in the relation between facts.

The topological understanding of urban form is strengthened by the increasing use and familiarity with images obtained by remote sensing. In addition to aerial photography this includes thermal infrared imagery acquired by the use of optical–mechanical scanners and by radar-images obtained from manned or unmanned satellites—a technique in which electromagnetic energy is used to "illuminate" the terrain and the back-scattered radiation is detected by radar which synthetizes an image from it (Sabins).

Such active imaging devices are independent of the light from the sun, and since clouds, fog and precipitation have little effect on microwaves, one can obtain images at night and in any kind of weather. By watching television weather reports, science fiction films, etc., images of highly abstract nature are becoming quite familiar and legible to the average person. Even conventional black and white photography transcribes colors in gradations of grey, but since infrared and ultraviolet radiation lie outside the visible spectrum, the colors that appear on the images are "symbolic" and bear no relation to natural colors. Radar-imagery is color-processed and the colors are assigned solely on the basis of the intensity of the "backscattered" microwave radiation: thus red corresponds to the built-up areas while lakes and rivers appear as black surfaces. The increasing use of imagery acquired by remote-sensing tends to further emphasize the macro-elements of urban form and downgrade the importance of the relatively small historic core areas ordered by symbolic or perspectivic principles. For example, on Landsat images, the metropolitan areas appear as grey mosslike blots advancing on the red surfaces of vegetation, emitting yellow streaks of pollution into the black rivers and bay areas and the urban expressways are the major "design" elements immediately discernible.

A further example of the increasingly abstract information influencing the cognition of urban form is the use of photogrammetric surveying transcribed by computerized drafting of the contour-maps. Another example is the use of scale-distorted maps such as those used to explain time-distance from a given number of points, for example, the isochronics denoting driving time from a constellation of residential suburbs to major employment or shopping centers.

Needless to say, the usefulness of the abundant availability of such abstract imagery is limited to people who grew up in the mental climate of technologically advanced societies. Much misunderstanding results from the attempts of American or European "experts" trying to elucidate the decision-makers in certain developing countries by the use of highly abstract information which they are unable to interpret correctly, having been brought up in a cultural tradition which has conditioned entirely different perceptive and cognitive images of urban space than the "mental maps" used as a reference system by the advisors.[2]

1. Recently, even works of art are being composed to be seen from the air, works which are not at all legible from a

ground level viewpoint, i.e., Christo's recent experiment with the plastic-wrapping of Caribbean islands.

2. I may cite the experience with the presentation of the development plan for the Nigerian City of Owerri, capital of Imo State. On a land use map using symbolic colors to denote different activities, the blue color indicating the area reserved for "institutional land use" was misread by the Nigerian partners as a large lake. In the following discussion of vertical aerial photographs of the city, it took a long time for the consultants to realize that these were interpreted by some of the Nigerians as pictures of a "big building with many windows." In fact, the "windows" were spots of sunlight reflected from the numerous corrugated iron roofs in the market area.

Quantitative Methods: Modeling

The great size and increasingly informal pattern of the modern metropolis led to the insight that "the act of design is no longer central in the form-making process" (Martin) and to the suspicion that formal design is a sort of luxury that can be afforded only in the city center and some other prestigious areas.

Rejecting Bacon's heroic view of the architect as the creator of urban form as "an act of will," Martin and March assert that "planning is no longer concerned with visual images," and call for new methods of rational analysis to be able to guide the forces that shape urban form.

In fact, during the last thirty years, a vast amount of theoretical work has been accomplished to help our understanding of spatial organization of the city by the use of quantitative methods and mathematical models. This research was much helped by the borrowing of concepts from other fields of science; gravity and hydraulic models were developed to explain traffic-flow; electrical circuits or networks used in modeling, and the information theory used in thermodynamics were found to present numerous analogies with certain aspects of the urban system.

Quantitative methods permit us to measure some facts with precision. These measurements permit valid comparisons in time and space and can serve for previsions. They constitute the framework for the conceptualization of a problem and its theoretical formulation. Mathematical formalization permits us to demonstrate and to analyze in depth the interaction of different factors (Merlin).

The topological description of the city turned out to be particularly useful as a framework for the theoretical formulation of urban dynamics as a sequence of disequilibrium/equilibrium transitions.

Urban structure can be viewed as the result of a homeostatic drive toward equilibrium between two opposing forces: the demand for space (amenity or territory) and for accessibility, which is a measure of the interchange between related activities. Such considerations form the basis for the modeling of the urban structure. Following Echenique and March we may distinguish descriptive, predictive, explorative and operation models:

1. *Descriptive models* are explanatory and provide an insight into the working of the city structure. In general, it is "difficult to obtain a good fit between model and reality"; if the model aims at accuracy in a particular situation, it will have a low degree of generality and vice-versa.

2. *Predictive models*, such as *extrapolative models* based on the continuation of present trends; and *conditional models* which require hypotheses of the mechanisms of cause and effect.

3. *Explorative models* permit us "to discover by speculation the realities that may be logically possible by systematically varying the basic parameters" (Echenique, 1972). For each parameter a specific range can be defined (i.e., density, speed, etc).

4. *Operational or planning models* in which "a measure of optimization is introduced in terms of chosen criteria to determine means to achieve stated planning goals" (I. Lowry, 1965).

We may briefly recall some of the statistical methods used in these models which can be grouped as: equation systems; simulation; and computer algorithms (Wilson, 1967).

If we think of the "city as an ecological system," the use of *stochastic techniques* is required, to take into account certain random choices. This can be done by simulating an experiment to determine some probabilistic property (Galantay & Ebenegger, 1981).

What is the predictive power of the models? Few would defend "the extreme point of view claiming that the objective structural analysis of the functional requirements of a social organization would, ipso facto, generate the design of the building or the environment to accommodate it" (Martin & March).

On the other side of the ideological curtain, we find the phalanx of those who consider practical reasoning and intuition as opposites.

These critics include such "non-interventionist" planners as Jane Jacobs, who believes that "the livable city emerges as naturally as a bed of oysters"—best left undisturbed—and clamors that design based on rational analysis will lead to "sterility."

More profound seems the observation of Albert Mayer that predictive models can become normative in the sense of "self-fulfilling prophecies" since decisions are frequently made in line with the trend projections and thus "trend becomes destiny." Or, as Edmund Bacon puts it, "with the enormous improvement in the techniques of mathematical manipulations of electronic computers applied to the problem of projecting past trends we are in danger of surrendering to mathematically extrapolated futures which at best can be nothing more than an extension of what existed before . . ."

In fact, the real danger of an excessive reliance on solutions provided by modeling lies in the fact that the building of models necessarily implies simplification in the selection of variables and their range, and in the process of ponderation. Determinants which can be easily quantified (transportation cost, land values, etc.) are overvalued in comparison to "imponderables" of psychological and emotional nature (such as the incompatibilities among certain social groups, preferences for a certain type of view, orientation, etc.), with the result that these factors are insufficiently considered or entirely neglected. Since few people make entirely rational choices and many act in an apparently arbitrary fashion, models assuming too much rationality may be misleading and even completely false.

Brasilia and Ciudad Guayana Compared

To gain a better understanding of the relative merits of creative intuition and rational planning methods, let us compare the "performance" of two recently built New Towns: Brasilia and Ciudad Guayana.

Brasilia—now just 25 years old—has over half a million inhabitants within the "Plano Piloto" (excluding the population of the five satellite towns). The Venezuelan industrial town of Ciudad Guayana is of comparable age and size. Its planning began in 1961, and in 22 years the city has acquired an impressive complex of heavy industries and a population of over 400,000 inhabitants.

The reason for selecting these two Latin American New Towns for our comparison is that their plans reflect radically different approaches to planning and urban design.

It is well known that the plan of Brasilia was the brainchild of a single designer, Lucio Costa. Rapidly sketched for a competition, the original plan was little more than an "ideogram" arrived at by a combination of reasoning and intuition, and Costa's desire to create a memorable image based on formal order, a "grand gesture" (Bacon).

By contrast, Ciudad Guayana was designed by an interdisciplinary team keen to try out a more scientific design process. To avoid arbitrary decisions, objective techniques were used in the form of mathematical models to simulate urban growth and intraurban travel. In particular, the spatial distribution of the residential population and the location of the City Center were chosen by the computer evaluation of numerous alternatives.

The plans of both Brasilia and of Ciudad Guayana were based on forecasts of rapid population growth. This has occurred faster than anticipated in Brasilia and somewhat slower than hoped for in Ciudad Guayana.

In both cases we might examine the extent of the "congruence," or match, of the actual spatial distribution of the population and its activities and the original plan and development strategy.

Another comparison can be made about the "efficiency" of the urban form in terms of the cost of transportation to its inhabitants.

A further comparison could concern the "legibility" of the urban structure in terms of its perception by the inhabitants.

Let us first examine the intentions of the planners and the methods used to achieve their design objectives.

The plan of Brasilia has a simple and pious ordering idea, it is

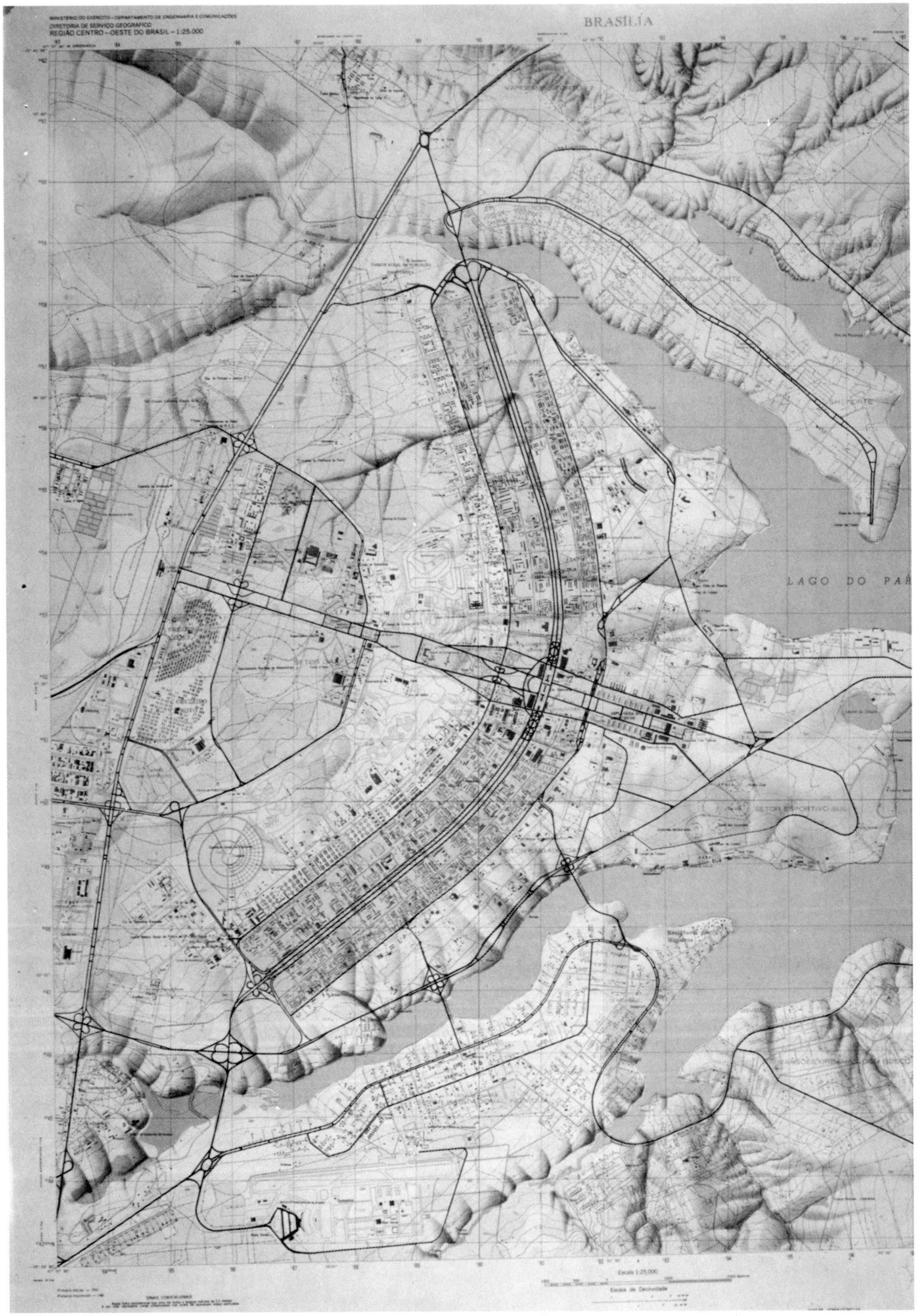

Fig. 2-4: Site-plan for the city of Brasilia.

Fig. 2-5: Site-plan for Ciudad Guayana.

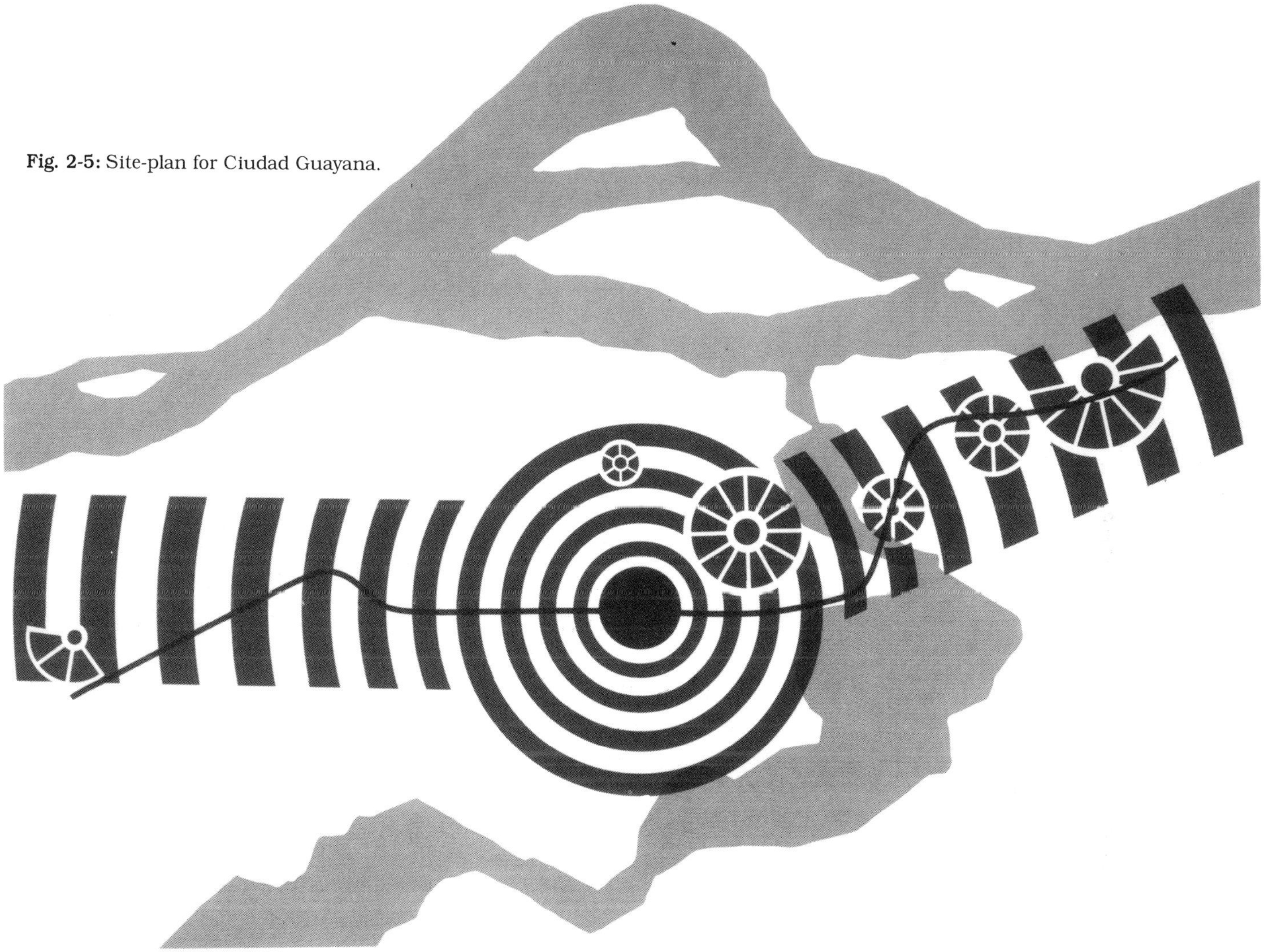

Plan proposed for Ciudad Guayana (Guayana City), a single town dominated by a downtown area, an appropriate level of services, and opportunities for all.

Ciudad Guayana is capable of offering substantial economic and social advantages. It has in fact already experienced an improvement as regards income and living standards since as recently as 1960. If we take the total national social cost for the new town into account, the net economic yield compares favorably with that of the overall Venezuelan economy, which shows that the new town will benefit the country as a whole.

Ciudad Guayana will facilitate development decentralization within the country and make it possible to deflect some of the rural urban migration current presently converging upon Caracas and other cities in the central region, which has caused great social and economic problems.

In order to furnish the necessary impetus to industrialize the nation by developing Guayana, an indispensable factor will be a great city combining the efforts of the public and private sectors.

"the sign of the Cross," with an east-west "institutional" axis bisected by the slightly curved north-south band of residential neighborhoods. In the intersection of the two axes is a transportation terminal surrounded by buildings of the commercial and entertainment center. An artificial lake has been created curving around the eastern leg of the institutional axis. It is the only major environmental feature in a flat, treeless and uninspiring site. Industries are located along a third axis, which is like a string to the bow of the curved "eje rodoviario." Residential areas are composed of neighborhoods of 12,000 inhabitants housed in nearly identical "superquadras" and rowhouses; single-standing residences are dispersed along the eastern shore of the lake.

In contrast to Brasilia, the site of Ciudad Guayana has unique topographic assets. It is located at the confluence of the fast-flowing Caroni River and the mighty Orinoco, which provides a natural boundary on its northern side. The designated area included a vast steel mill being built on its western extremity, some small population nodes on both sides of the Caroni, and a lively little town of colonial origin, San Felix, on the extreme northeast corner of the site at a distance of 23 km from the steel plant.

This indicated the first and obvious design idea of trying to tie together the various fragments by a major linear element, the "Avenida Guayana."

In defining its objectives the planning team was critical of the overwhelming monumentality of the plan of Brasilia, which was also considered too mechanistic and automobile oriented. For Ciudad Guayana, a more humane and open-ended urban structure was sought, fixing only certain key elements of the plan to leave room for a degree of personalization and ambiguity. The planners were also concerned about the social segregation built into the physical structure of Brasilia and sought to achieve a more "democratic" population mix of a wide range of income groups assigned to each development sector. Finally, an efficient road network was required, capable of dynamic expansion at minimum cost in development and maintenance, and in transportation cost to the population.

The planners didn't want to select the site of the future city center and of the residential areas on the basis of topographic and design considerations alone, and hoped to be aided in their

Fig. 2-6: Plan for Ciudad Guayana, showing the location of areas for specific utilization.

Dominating elements of the city-Ciudad Guayana will have four principal centers: the heavy-industry center in Matanzas; the downtown area in Alta Vista; the recreational center in Punta Vista; and the old Saint Felix area. These centers will be interconnected by Avenida Guayana (Guayana Avenue), the city's main artery.

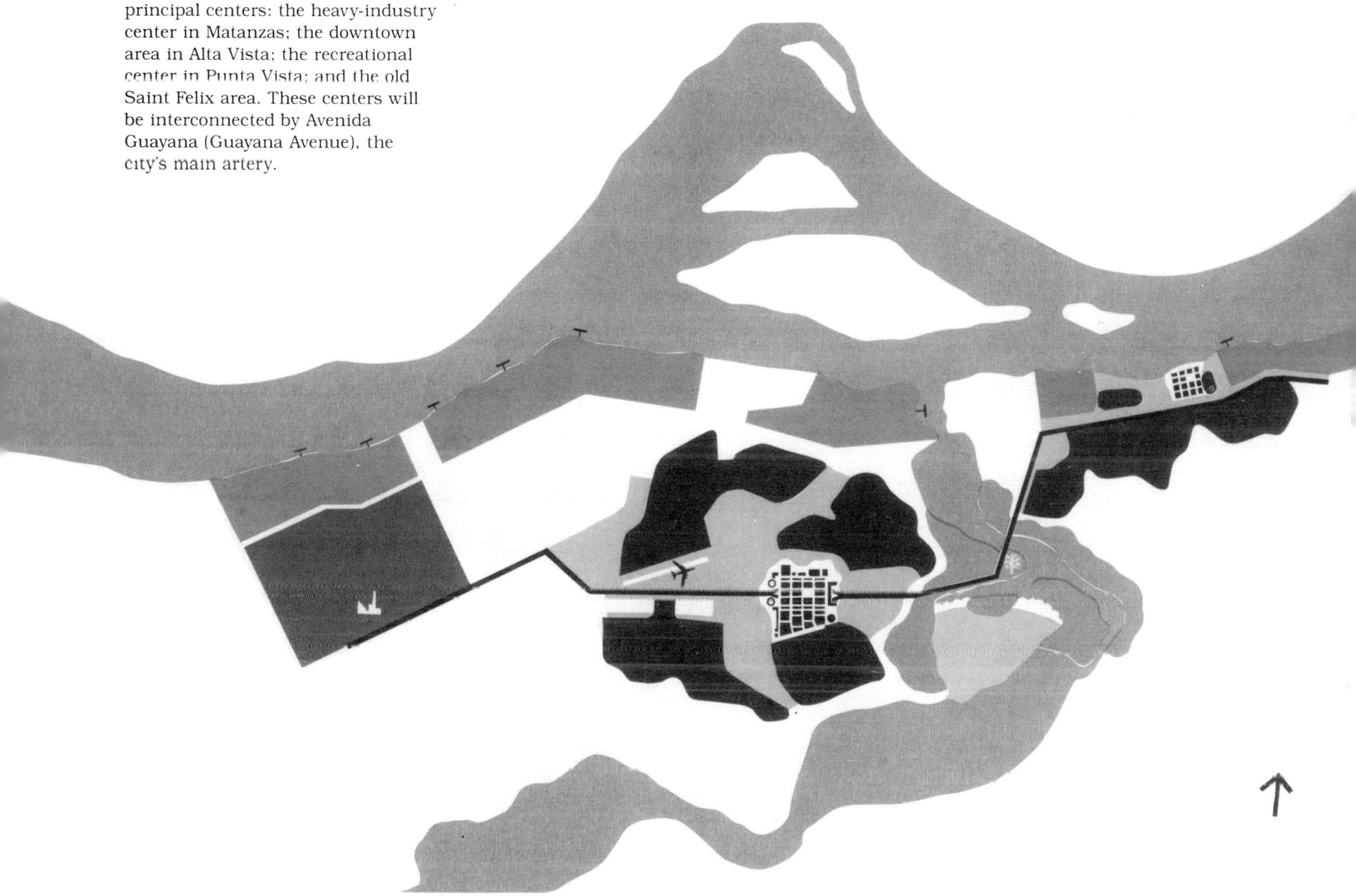

decisions by the objective methods of mathematical simulation.

Since the location of the heavy industry was known, it was possible to simulate the spatial distribution of the residential population by an accessibility model at discrete levels of employment in basic employment. This was achieved by dividing all disposable land into distinct development areas, each with its "terminal descriptor" representing the capacity and attractiveness of each cell. A distance function was then used to measure the accessibility of the cells to centers of employment.

Once the population was spatially distributed, the theoretical location of the business center was found by determining the point of highest accessibility to all population. The advantages of this location, "Alta Vista," seemed confirmed by a second model used for the comparative evaluation of four potential sites for the city center, again in terms of accessibility (Penfold).

All these models indicated a clear tendency for the residential population to concentrate around the new center on the western side of the Caroni, with only one–fourth of the population grouped around San Felix on the eastern side. It seemed clearly demonstrated that considerations of accessibility would shape the form of the city and, in consequence, a strategy of development was adopted to accelerate growth in accordance with the projected trend.

Now, twenty-two years later, it is clear that something was wrong with these projections. The actual distribution of the population is 3/4th on the eastern side (San Felix), and 1/4th in proximity of the new center (Alta Vista), exactly the reverse of what had been predicted. The social goals of the planners have not been achieved either: the Caroni splits the town in two societies—middle and upper income groups prefer the western side, while the "informal sector" is confined to the eastern half of the city. Contrary to the predictions of the "growth strategy," real growth was disjointed with subcenters "booming" at the expense of the major City Center.

The urban tissue is certainly less rigid and more "open-ended" than in Brasilia, but it does not favor pedestrian or public transportation. Due to the fragmented, dispersed, and low density development, the road network is too stretched out; bus services cannot be run efficiently, and the cost of transportation to the residents in terms of the percentage of disposable income (16%) is higher than in Brasilia (9%).[1]

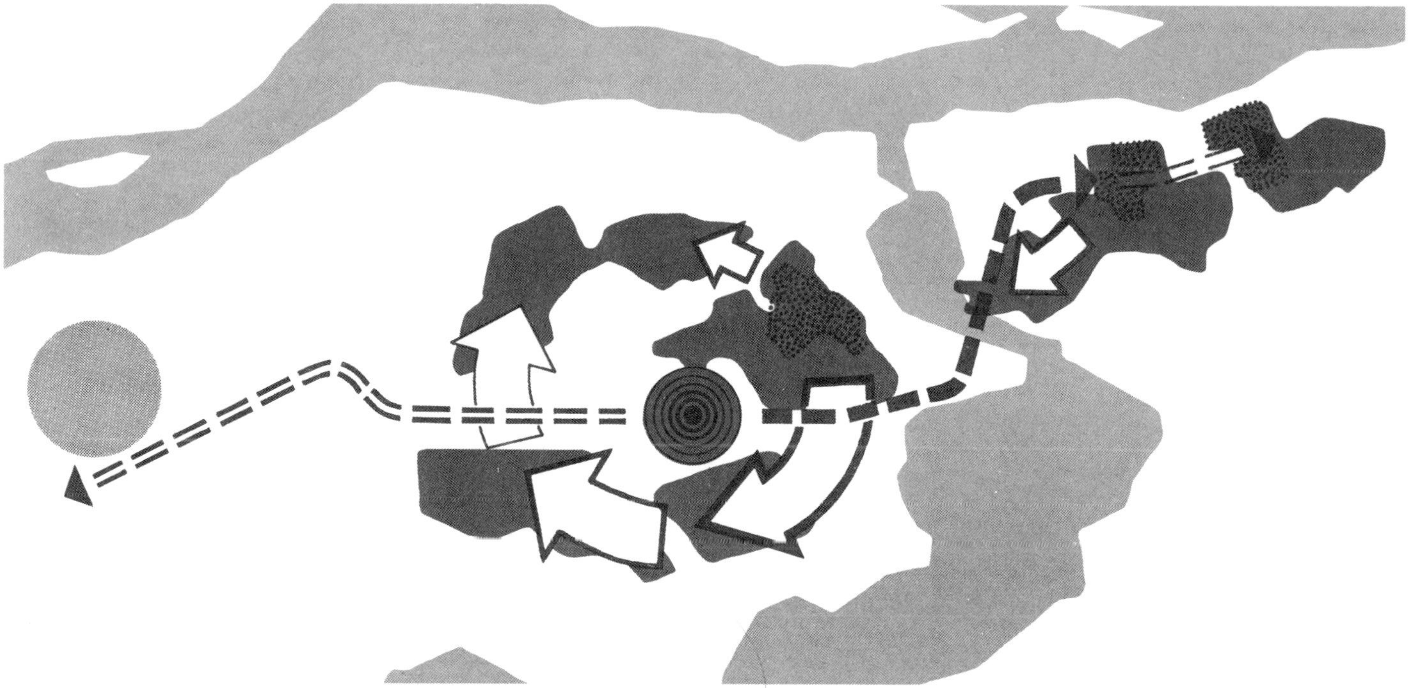

Fig. 2-7: Plan for future expansion and construction of Ciudad Guayana.

(two arrows)
Order and direction of development for the new residential communities

(dots)
First developments; progressive remodeling

(concentrated dots)
Intensive development in the city's central area

(arrow points)
Construction stages on Guayana Avenue

(lightly concentrated dots)
Concentration of new industrial developments.

Blueprint for development, future continued westward expansion, and concentration in Alta Vista.

The concentration of residential areas along relatively high density corridors permits the operation of an efficient bus system.

We may now compare the perceptual qualities of the two towns. "Sense of space," according to Lynch, is "the clarity with which the settlement can be clearly perceived and mentally differentiated by its residents and the ease with which its elements can be linked in a coherent representation of time and space."

In this respect Brasilia is clearly the winner; in spite of the uninspiring site a memorable city has risen due to the unique qualities of its built-form. The residents seem to master the intricacies of the many one-way loops of the road pattern, and the urban structure is of such clarity that it is impossible to get lost.

By contrast, the plan of Ciudad Guayana has certainly not exploited fully the potentials of the unique site. In this city surrounded by the waters of the Caroni and the Orinoco, the City Center of Alta Vista is located at a distance of 4 km from the nearest waterfront, a walk of about an hour for the pedestrian! This center, which was slow in developing, has now caught momentum and gradually its tall buildings began to form a "skyline" visible from a considerable distance. On the other hand, the spinal "Avenida Guayana" proved to be a far too weak design element to tie the fragmented elements of the city into a coherent whole. In fact, the residents are hardly aware of Ciudad Guayana as an entity, and keep referring to its parts as San Felix, Porto Ordaz, etc., names of the small settlements on the site predating the foundation of the new town.

This all too brief comparison of the present "performance" of these two New Towns should not be interpreted as a vindication of "Grand Design," but rather as a warning against overreliance on computer-assisted simulation models in decision-making with respect to urban form.

After all, a model is only the mathematical formulation of concepts or relations which first have to be defined by reasoning. "It is the qualitative intuition of the researcher which permits him to define the relations which he wants to express" (Merlin).

All elements for decision-making cannot be provided by computers. Furthermore, the information provided by the models may be misleading, since no model can take into account all the factors that influence people's aspirations and preferences. The simulation models of Ciudad Guayana were based on accessibility considerations but limited to accessibility to employment. One

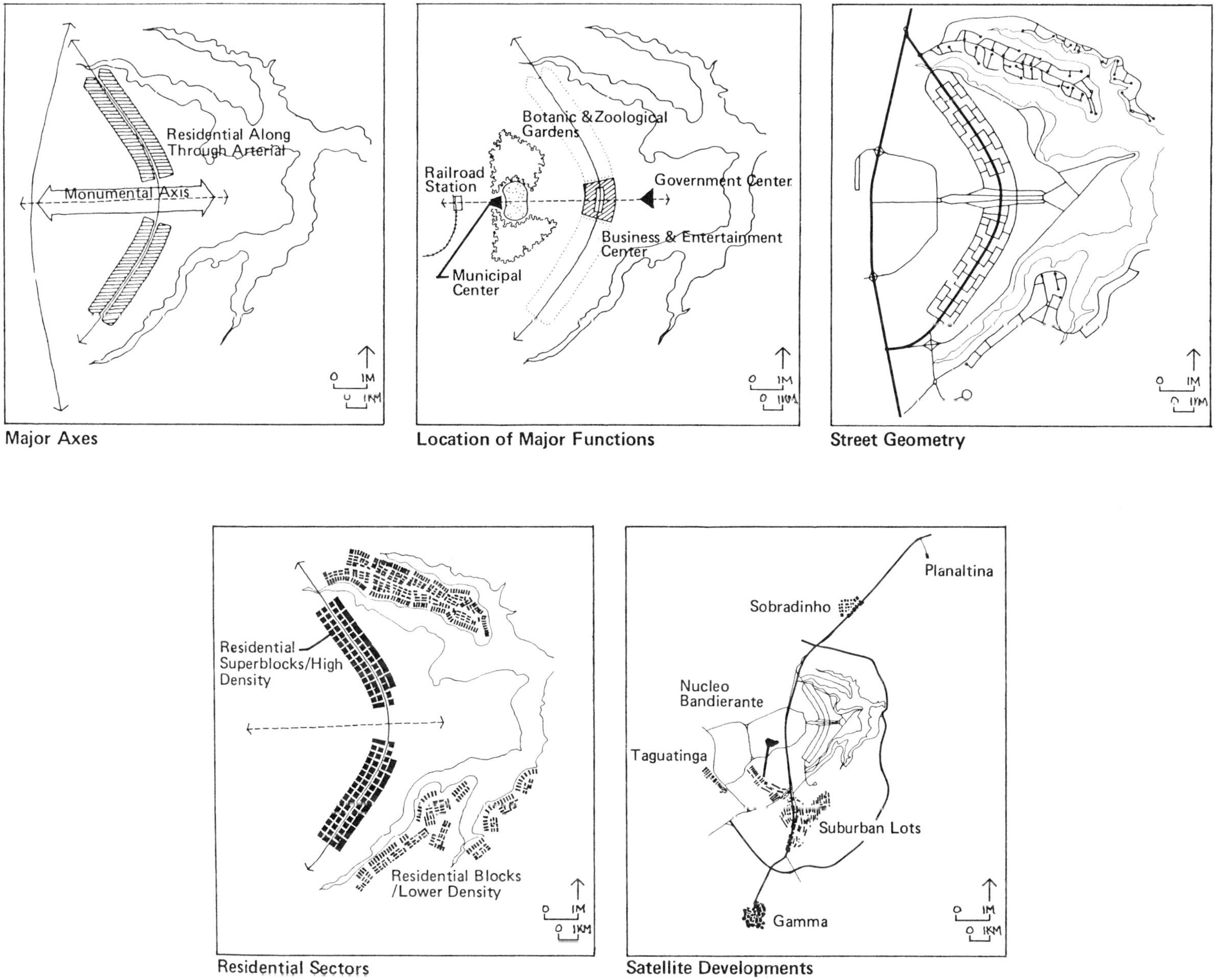

Fig. 2-8: Plan-maps for Brasilia, showing locations and geometry of various features of the city.

should also have included a "social interaction indicator" to measure the opportunity to encounter the maximum number of other residents (for example in lively plazas and shopping streets), and a "proximity to open space" indicator to measure the opportunity of access to the edges of the two rivers, to parks and playgrounds, and so forth.

My conclusion is that the predicted performance of a design can certainly be evaluated by simulation, but only if the model takes into account the interrelation of a sufficiently large number of factors to provide a good "fit" between model and reality.

Also, one should not underrate "common sense." If the results of a simulation by computer don't "look right," quite possibly they are false, and it is time to reexamine the basic parameters.

There will always be a place for intuition and conjecture in a rational design process. As Sir Leslie Martin puts it, "a strategy involving many successive steps in design formulation and evaluation permits the intervention of creative judgement and invention at a series of intermediate stages, but with the advantage that these are exercised in relation to a clear statement of the structure of the problem . . ."

In the cities of advanced Western societies which are in a transitory phase from an industrial to a post-industrial type of society, qualitative considerations begin to override the importance of purely quantitative reasoning. As in most cases, inner-city populations are no longer growing, problems of the further expansion of the city recede behind the problem of a meaningful reordering of the existing urban structure. In this task the emotional and psychological needs of the population for identity, self-fulfillment, and harmony become predominant, and city planning becomes an ongoing process of monitoring and sequential adjustments, opening up a great field of action for artistic sensitivity and imagination on the microscale of design, leaving room for the use of quantitative methods as aids in decision-making on the level of the design of urban macroform.

[1] By contrast, the growth of Brasilia shows remarkable "congruence" with the plan of Costa. Although the development of the southern wing of the residential axis was given priority, by now the right wing has also been completed, making the plan symmetrical as intended.

Bibliography

1. Alexander, C. *Notes on the Synthesis of Form*. Harvard University Press, 1964.
2. Alexander, C. 1965, "A City is Not a Tree," *Architectural Forum* 122, no. 1.
3. Akurgal, E. *Ancient Civilisations and Ruins of Turkey*, Istanbul, 1970.
4. Appleyard, D. *Planning a Pluralist City*, MIT Press, 1968.
5. Architectural Review: "Activities and Form," *AR*, August 1972.
6. Bacon, E. *Design of Cities*, Viking Press, New York, 1967.
7. Barnett, J. *An Introduction to Urban Design*, Harper & Row, New York, 1982.
8. Benevolo, L. *The Origins of Modern Town Planning*, MIT Press, 1971 (1967).
9. Blumenfeld, H. *The Modern Metropolis*, MIT Press, Cambridge, 1967.
10. Boesch, H. & Hofer, P. *Flugbild der Schweizer Stadt*, Kummerly & Frey, Bern, 1963.
11. Bronowski, J. *The Common Sense of Science*, Penguin Books, Baltimore, Md., 1966.
12. Bronowski, J. "The Discovery of Form" in *La Structure dans les Arts et les Sciences*, G. Kepes (ed.), La Convenance, Bruxelles, 1967.
13. Branch, M. *Comparative Urban Design*, University of California Press, Los Angeles, 1978.
14. Canter, D. & Lee, T. *Psychology and the Built Environment*, Architectural Press, London (1974).
15. Centre de recherche d'Urbanism, "Photographie aerienne et CRU, Paris, Urbanisme" (1970).
16. Chapin, S.F., *Human Activity Patterns in the City, . . .* John Wiley & Sons, New York, 1974.
17. Cerda, I. *Teoria general de la urbanizacion*, Ed. Lopez de Aberasturi, A., Madrid 1867, *La theorie generale de l'urbanisation*, Editions du Seuil, Paris, 1979.
18. Crane, D. "The City Symbolic," *Journal of the A.I.P.*, Vol. 26, Nov., 1960.
19. Dobzhansky, T., *Mankind Evolving*, Yale University Press, New Haven, 1962.
20. Downs, R. & Stea, D. *Image and Environment*, Aldine, Chicago, 1973.
21. Doxiades, C. *Ekistics*, Hutchinson & Co., London, 1968.
22. Echenique, M. "Models: A Discussion" in Martin, L. & March, L. *Urban Spaces and Structures*, Cambridge University Press, 1972.
23. Evenson, N. "Two Brasilian Capitals," Yale University Press, New Haven, 1973.
24. Fleury, M. & Derens, J. *Le Plan de Paris par Truschet et Hoyau 1550*, E. Seefeld, Zurich, 1980.
25. Fourastie, Jean *La Grande Metamorphose du XXe siecle*, Presses Universitaires, Paris, 1961.
26. Galantay, E. *New Towns from Antiquity to Present*, Braziller, New York, 1975.
27. Galantay, E. *Nuevas Ciudades*, Ed. G. Gili, Barcelona, 1977.
28. Galantay, E. & Ebenegger, C. "Self-organisation in spontaneous human agglomeration," *Proceedings of the 10th ICUS*, ICF Press, New York, 1981.

29. Galantay, E. & Haig, M. & Leaker, D. Eds. *New Towns in National Development*, The Open University Press, Milton Keynes, United Kingdom, 1980.
30. Galantay, E. "New Towns, Definitions, Goals & Policies," *Ekistics Magazine*, 277, July/Aug. 1979.
31. Galantay, E. "The Planning of Owerri, New Capital of Imo State in Nigeria," in *The Town Planning Review*,Vol. 49, Nr. 3, July 1978, pp.370–386.
32. Galantay, E. "The Urban Design Process," *Proceedings of the Symposium on Urban Design Education*, The University of Washington, Seattle, 1981.
33. Geddes, P. *Cities in Evolution* (1915), New York, Howard Fertig, 1968.
34. Gottman, J. *Megalopolis*, New York, The 20th Century Fund, 1961.
35. Halbwachs, M. *La Memoire Collective*, Presses Univ. de France, Paris, 1950.
36. Hofer, P. *The Zähringer New Towns*, ETH-Zurich, 1966.
37. Huxley, Sir Julian, *The Uniqueness of Man*, Chatto & Windus, London, 1960.
38. Jacobs, J. *The Death and Life of Great American Cities*, Random House, New York, 1961.
39. Krier, R. *Stadtraum in Theorie und Praxis* 1973, Karl Kramer, Stuttgart, 1975.
40. Lang, S. "The Ideal City from Plato to Howard," *Architectural Review*, Vol.112, No. 668, Aug. 1952.
41. Levy-Strauss, C. *Tristes Tropiques*, Plon, Paris, 1955.
42. Lowry, I. "A Model of Metropolis," Rand Corp., Santa Monica, Calif. 1964.
43. Lynch, K. "The Pattern of the Metropolis" in *Daedalus*, Winter, 1961.
44. Lynch, K. *A Theory of Good City Form*, MIT Press, 1981.
45. Maki, F. & Ohtaka, M., "Reflections on Group Form" in G. Kepes, *Structure in the Arts and in the Sciences*, G. Braziller, New York, 1965.
46. March, L. & Steadman, P. *The Geometry of the Environment*, Methuen & Co., Ltd., RIBA, London, 1971.
47. Martin, Sir Leslie, & March, L. *Urban Space and Structures*, Cambridge University Press, 1972.
48. McHarg, I. *Design with Nature*, Doubleday, New York, 1969.
49. Meier, R. *A Communication Theory of Urban Growth*, MIT Press, Cambridge, 1962.
50. Merlin, P. *Methodes quantitatives et espace urbain*, Masson & Cie, Paris, 1973.
51. Moles. A. *Information Theory and Esthetic Perception*, University of Illinois Press, Urbana, Ill., 1966.
52. Montroll, E. & Badger W. *Introduction to Quantitative Aspects of Social Phenomena*, Gordon & Breach Science Publishers, New York, 1974.
53. Mumford, L. *The Culture of Cities*, Harcourt, Brace, New York, 1938.
54. Pahl, J. *Die Stadt im Aufbruch der Perspektivischen Welt*, Ullstein, Berlin, 1963.
55. Parr, A.E. "City and Psyche" in the *Yale Review*, Vol. LV, no. 1, 1965.
56. Penfold, A. 1966 "Ciudad Guayana," in *Town Planning Review* 35, no. 4b.

57. Penfold, A. "Urban Transportation" in Rodwin, L. et al. *Planning Urban Growth . . . ,* MIT Press, 1969.
58. Piaget, J. *The Child's Conception of Space,* New York, Norton, 1962.
59. Preziosi, D. *The Semiotics of the Build Environment,* Indiana University Press, 1978.
60. Rapoport, A. *Human Aspects of Urban Form,* Pergamon Press, Oxford, United Kingdom, 1977.
61. Ratzel, F. *Anthropogeographie* 1882 (Englehorn, Stuttgart).
62. Romanelli, G. & Biadene, S. *Venezia, Piante e Vedute,* Museo Correr, Venezia, 1982.
63. Rossi, A. *L'architecture de la Ville* (1966) l'Equerre, Paris, 1981.
64. Rykwert, J. *The Idea of a Town,* Princeton University Press, 1976.
65. Sabins, F.Jr., *Remote Sensing,* W.H. Freeman, San Francisco, 1978.
66. Summerson J. "Urban Forms," pp. 165-176 in Handlin, O. & Burchard, J. *The Historian and the City,* MIT Press, 1963.
67. Steadman, P. & March, L. *The Geometry of the Environment* (1961) Methuen & Co., London, 1974.
68. The Royal College of Art: *Automatic Cartography and Planning,* Architectural Press, London, 1971.
69. Toynbee, A. "Cities in History" in *Cities of Destiny,* McGraw-Hill, New York, 1967.
70. Vesalius, A. *De Humani Corporis Fabrica,* 1543.
71. Webber, M. "Order in Diversity" in L. Wingo (ed.) *Cities and Space,* John Hopkins Press, Baltimore, 1963.
72. Wilhelm, J. *Brasilia 1960—Uma Interpretacao,* Acropole, Brasilia, 1960.
73. Wilson, A.G., "Mathematical Models for Planning," *Arena,* London, 1967.
74. Wölfflin, H. *Principles of Art History* (1915) republished by Dover Inc., New York, 1965.
75. Wurman, R.S. *Cities: Comparisons of Form and Scale,* MIT Press, Cambridge, Mass., 1974.

Concerning Galantay's "The City: Organism or Artifact?"

Dietrich E. Kunckel

This paper is intended to comment on some of the points raised by Prof. Ervin Galantay about the nature of urban form and the influence of technological change on the design and perception of the urban environment. Following the structure of E. Galantay's paper, I will discuss the following aspects:

1. The concept of the city as an ecological system.
2. The role of technology in the perception of the city.
3. The use of new technology in the design process.
4. The use of computer-aided modeling in Ciudad Guayana.

The Concept of the City as an Ecological System

The theory of the city as an ecological system has been put forward by other authors in different contexts. To my knowledge, the first such definition was used by Park, Burgess and McKenzie of the University of Chicago's School of Sociology.[1] As described by Don Martindale in his prefatory remarks to the 1958 edition of Max Weber's *The City*, "Park provided the general framework for the ecological theory of the city." Martindale then goes on to criticize the concept because of its orientation towards geophysical aspects of the city rather than to social life: "social life is a structure of interaction, not a structure of stone, steel, cement, asphalt, etc." But then he admits that "it (the ecological theory of the city) facilitated the examination both of the external physical and internal social aspects of urban phenomena."

Still, this view of the city as an ecological system pays very little or no attention to the actions of the individuals or institutions that willfully modify the urban environment. The theory identifies a number of processes "whose operation creates 'natural areas' that form the physical structure of the city." This model concentrates on social interactions and does not consider the interventions of a designer or a planner.

In the theory set forth by E. Galantay and illustrated by the example of the orchard, the role of planned action has been incorporated into the system, forming part of what he calls the "two-step tandem process" which combines chance and deterministic phenomena.

While the initial theory was probably modeled too closely on the examples of plant-ecology, in my view it is the illustration—rather than the basic concept of Prof. Galantay's theory—which oversimplifies the complex processes involved in the shaping of cities. The city comprises different species and communities, and these react in different ways to the constraints and stimuli of the natural and artificial environment. Ripley and Buechner have pointed out that "populations of human beings are uniquely characterized by their highly complex social behavior, which emerges as human society. The societies of man, together with their total environments, form the highest level of biological integration" (ecological system). And, "Ecosystems at the highest level of biological integration are not only more complex than we think they are, but more complex than we *can* think."[2]

If we agree with this statement, then any act of planning and designing of the urban environment or, more generally, the human environment, will have to be an act of simplification, as the workings of the system are beyond human comprehension. It is important to make the correct kind of abstractions of reality, as will be seen in the context of the use of computer-aided models. Nevertheless, there will always have to be acts of deliberate intervention to create the structures needed for the development of complex social organizations.

If we look at the physical structure of cities from the earliest examples to our day we can distinguish in all of them the existence of a man-made, deliberate order, easily recognizable in some, less evident in others, but certainly present. It could be stated that the existence of such an ordering principle is what sets cities aside from mere villages and small towns, which might have developed randomly over time.

This physical structure reflects to a certain degree the social organization of the city and its possibilities of social interaction, offering to its inhabitants an environment which is no longer "natural," with "a somewhat predictable growth pattern"[3] as found in spontaneous agglomerations, but probably a more versatile spatial structure, which permits and supports different, and more complex, social interactions.

Edgar Morin has compared the "grande ville" with the brain, where complex organizational structures communicate at random in a polycentric environment of little hierarchy: "It is through this deployment of apparent disorder of individual movements that the hypercomplexities emerge; these constitute, at the same time, spheres of personal liberty—physically, economically, sexually, intellectually and eventually politically."[4] This "sociocultural ecosystem," as a characteristic of cities, is the environment which offers the spheres of personal freedom and lack of constraint which stimulate creativity, innovation and new ideas in arts and sciences.

Maybe we will see a new development in our big cities, based on the ever-increasing organizational abilities, which will lead towards a reduced necessity of controls and to the general relaxation of physical constraints of city-form, giving rise to urban environments in constant transformation, adaptation and reorganization.

Perception of the City

If we discuss the perception of the city by examining the ways of its representation, as Prof. Galantay suggests, we recognize the role of technology in widening both the range of perception and the ways of creating images of our urban environment. The way we represent our environment has, of course, to do with our experience, and technology has contributed decisively in increasing the total scope of our experience of the environment.

This amplification of the range of our experiences takes place not only in the direct way by using new means of transportation like the car or the railroad, or by elevating ourselves above the ground by airplane and rocketship, but also in an indirect way by exposing us continuously to the images of movies and the television screen. Many of these images show different places and cultures, while others create imaginary places and environments such as the space age planets and cities of science fiction. These images form part of our daily experience and even more of the experience of our children. I can only speculate about how much the perception of our immediate urban environment might be influenced by this. David Canter cites experiments in which kindergarten level children are able to recognize correctly various elements of an aerial photograph; this is an amazing example of the capacity of abstraction and interpretation even at this level. We can count much more on the trained adult's capacity to "read" correctly abstract images of urban space.

There remains, however, the question whether there exists any significant relationship between this power of abstraction and the way that spatial organization is experienced by the individual. The individual person experiences space and spatial organization through the senses and through movement through space. Thanks to what E.T. Hall has termed "extensions" of man's organism,[5] we have amplified the range of our experiences, becoming able to move farther and faster, to record and replay, to see in the dark, and so forth. It seems reasonable to assume that the use of these extensions has altered our understanding, and therefore our perception, of the environment.

On the other hand, the "man in the street" still very much relies on his "unextended" senses to perceive and judge his urban environment. This fact is all too often forgotten by today's urban planner. City-form looked at from the satellite or the airplane is quite different from city-form that is walked on, driven through, or lived in. While general ordering principles might be better dealt with at the scale of aerial photography, it is an inadequate tool to deal with the perspectival, the symbolic and the cultural aspects of city-form, which are the basic factors influencing our understanding and use of space.

The Use of Technology in the Design Process

While technological advances have helped our understanding of many aspects of the urban environment, they have also helped create a trend towards generalization. At a certain level of abstraction, many features of cities of different cultures become similar, or at least comparable.

When we build a model of reality we have to abstract from reality, simplifying and reducing the number of variables in order to be able to identify the effect of a change of one variable on the others. As Prof. Galantay has pointed out, the use of computer-aided modeling has become more and more prevalent in city-planning, lending itself especially to handling quantitative aspects on the macro-scale.

Success in using these models depends very much on correct interpretation or on a level of abstraction which, in many cases, is directly proportional to the availability and reliability of the data describing the reality to be modeled. In developing countries the data base is often extremely shaky, inducing the planner to use assumptions or "educated guesses" based on his personal experience. This personal experience might include cultural factors different from those present in the environment to be planned, and therefore might lead to an erroneous "weighting" of the variables, as we will see in the example of Ciudad Guayana.

In my opinion, the area where new technologies of computer modeling and computer graphics will become a tremendously useful tool is in the simulation of the three-dimensional environment. We now have programs at our disposal which allow us to simulate three-dimensional space, including environmental aspects like the degree of isolation, shadows and so forth. We can let the computer move us around these spaces and volumes, and we can simulate the movement through space at different speeds. It seems to me that these techniques will become especially useful in aiding participation by the users in the design process by simulating different environments in a three-dimensional, "understandable" way. With the new technologies available the designer acquires the opportunity to control and check aspects which are normally beyond the abilities of imagination, even of the designer.

It is my experience with laymen that, while they may be able to read a plan or a map, correctly interpreting the different elements represented, they very rarely have the ability to transform this information into a spatial image; it seems to require special training to perform this act of "imagination." But understanding things on a computer screen is different. Since we have been trained to perceive and judge the environment by looking at its images on the screen of the moviehouse or television set, we are less likely, when looking at a computer, to commit errors as to how these spaces will look in reality.

The Example of Ciudad Guayana

The comparison of Brasilia and Ciudad Guayana as examples of different approaches to designing a new city illuminates very well the possible dilemma of the designer/planner when faced with the results of quantitative models. Shall he disregard any such tool and concentrate on the visual environment, confident that the activities will later somehow adapt to the given physical structure, or shall he follow the numbers and design the structure of the city so as best to fit the results of the model?

In the planning of Ciudad Guayana a lot of thought was given to qualitative and social aspects of the urban form, location of different income groups and ways of creating adequate housing for the poor. In addition, transportation was a major issue to be dealt with because of the initial poles of development—the steel plant to the west and the village of San Felix to the east.

When designing a model for the future growth of the city, the planners wanted to keep the model relatively simple in order to be able to judge the effect of cost, distance from work and other easily measurable factors on the location of the population. The model assumed certain specific policies to be implemented by the authorities, such as the promotion of low income housing in the western part and restrictions to unchecked development in the eastern part of the city.

The majority of these policies were never implemented, leading eventually to what M.P. Garcia and R.L. Blumberg have called, "The Unplanned Ecology of a Planned Industrial City,"[6] claiming that the underlying reason for the growth of the city in opposite direction to the plan was the failure of the planners to "adequately consider salient sociological, legal and ecological factors" in their models and in their policies concerning the spatial distribution of the poor.

The fact is that, while the planners were quite aware of many of the social variables, their model was a tool to measure efficiency. A failure to incorporate the social constraints as an integral part of the model meant that the results would only become reality if certain policies were adapted. As these policies were not implemented by the authorities, reality increasingly moved away from the predictions of the model.

M.P. Garcia and R.E. Blumberg cite five principal factors which were not sufficiently taken into account producing this disequilibrium. These factors are:

1. Structural factors and social class—referring to the type of employment (few stable jobs) and high unemployment rate.
2. Lifestyle and social relations—indicating the need of the poor for housing without fixed payment, stores that sell small quantities and give credit and proximity to structures of support provided by family and friends.
3. Legal factors referring to laws which alter the behavior of the population—like the one that guarantees the worker compensation for his travel time to work.
4. Ecology—which provides very favorable climatic conditions that reduce the need for a well-built home.
5. Planners' housing policies—which showed a strong prejudice against the "rancho," the shacks of the poor, prohibiting their construction in the western half of the city.

It is apparent that it will be extremely difficult to incorporate some of these variables into a mathematical model, because "No matter how much information can be collected by techniques of the hard sciences and mathematics, we are a long way from being able to computerize the interplay of forces that operate in social conditions . . ."[7]

In addition, it can be argued that the authorities in charge of executing and monitoring the initial plan of the city did not react in time to reduce the effect of the factors that were working against the plan, or to adapt the plan to the obvious trend.

But then, would it have been conceivable and politically feasible to radically alter the plan of the city so as to develop mainly on the eastern side of the river Caroni, not taking into account the enormous investment in industry, land and infrastructure in the western part?

In all, Ciudad Guayana, as the result of both planned and unplanned actions, is an interesting case study demonstrating in many ways the inherent difficulties of the planning process in aspects like:

–the danger in taking the results of planning models for granted. Models not only simplify reality but in most cases are at least partly based on assumptions (which can be wrong).

–the dynamics of urban growth against preconceived plans. If the plans do not fit the needs of the population and are not enforced, the individual "cells" of the ecosystem will try to modify the system to better adjust it to their preferences.

–cities are an environment in constant change. Urban form is varying over time with a tendency to an acceleration of this process. Thus the city at best is "situation art."

–cities, in order to project a strong image, have to be designed in three dimensions; two-dimensional concepts are only recognizable on paper.

–the need to follow and monitor more closely the "two-step tandem mechanism" between the planned and the unplanned in order to avoid plans that do not suit the inhabitants and to provide the citizens with more than one option to locate and participate in city life.

–the danger of projecting inadequate images of different cultures into a reality neither able nor suited to adapt these.

It also demonstrates that a continuous planning effort is needed, that our techniques have to be improved and adapted to the needs of the social as well as the physical environment, and that cities, in order to mature, need both planned and spontaneous action over time.

1. The ecological idea was developed by these authors in various publications between 1921 and 1927. Among others: Robert E. Park, Ernest W. Burgess and Rodwick D. McKenzie, *The City* (Chicago: The University of Chicago Press, 1925); and McKenzie, "The Ecological Approach to the Study of the Human Community," *The American Journal of Sociology*, XXX.
2. Dillon Ripley and Helmut K. Buechner in their article, "Ecosystem Science as a Point of Synthesis," in *Daedalus*, Fall 1967, propose the "study of the whole ecosystem of human-society-plus-environment and the interrelationships between these systems on a global basis." While their concern is mainly directed towards the natural environment and man's effect on it, the man-made environment can be readily incorporated into this concept.
3. E. Galantay
4. Edgar Morin, *Le Paradigme perdu: La nature humaine*, Editions du Seuil, 1973.
5. E.T. Hall, *The Hidden Dimension*.
6. Published in "Urbanization in the Americas: From Its Beginnings to the Present," Ed. by Richard Schaedel, *et. al.*, Aldine Publishing Co., 1977.
7. Leonard J. Duhl in his article, "Planning and Predicting," *Daedalus*, Summer 1967, suggests the incorporation of *processes* in the theory of prediction.

Concerning Prof. Galantay's "The City: Organism or Artifact?"

Tao Ho

This discussion of Professor Ervin Y. Galantay's paper, "The City: Organism or Artifact," is based on some of the findings of my recent research, entitled "Search for An Orderly Chaos," where I attempt to approach urban planning and design from non-conventional multidisciplinary points of view.

To Galantay, the genesis of a city is either a "static work of art produced by conscious design decisions" or "an organism capable of self-organization." We can only accept this, however, as a very broad description of the origin of human settlements, a definition that represents contradictory characteristics of humanity: the rational versus the irrational, order versus chaos, the tangible versus the intangible. There is a logic behind each of these definitions, as each is an expression of the human predicament. But there is no need, in my opinion, to search for a single credible model.

I find the statements by renowned scholars quoted by Galantay in the section "The Nature of Urban Form" interesting but short-sighted. To me, a city is an ever-changing dynamic interaction of many tangible and intangible forces. No single statement is capable of defining the true nature of a city. The trouble with many academic scholars is that in giving a definition of what they think a city should be, they often push their arguments to a short-sighted dead end for the sake of being consistent with their own philosophy which bears little relationship to reality. The saying by the Taoist philosopher Lao Tzu that "The Name that can be named is not the Eternal Name" serves to illustrate my point.

Galantay advocates the concept of "the city as an ecological system." I am very sympathetic to such an approach because the idea of "ecosystem" brings in human as well as natural factors as important forces in shaping urban form. Galantay's example of a natural forest (undisturbed by man, spontaneous) versus an orchard (manmade environment, planned) is an interesting analogy, but it deals only with the physical formal appearance of a city and has little to do with the invisible form. The invisible form, in my opinion, is the social economic structure of the city and the life force pattern of the inhabitants who are like particles of a subatomic world: dynamic, restless and unpredictable.

The discovery of quantum mechanics by Max Planck at the turn of the century has dramatically changed our outlook that reality is static. The unpredictable and dynamic behavioral pattern of subatomic particles is quite similar to most of our urban phenomena today. The main reason why we have failed so badly in solving urban problems is because we are still applying static planning theories—like the laws of Newtonian physics—to solving dynamic urban problems in isolation. Physical constraints proposed by planners are limited by time and can only temporarily solve limited tangible problems. Life forces, on the other hand, have always demonstrated that they can overcome physical constraints. The physical form of a city has to change with time, to be in time with its changing needs. A city without life is a dead city, and life is what we should strive to create in a city.

In the section "Perception and Cognition of the City" Galantay points out that there are two archetypical patterns: (a) spontaneous agglomeration, and (b) the planned town. From there on there are hybrid forms of human settlement. In this section Galantay attempts to define different city forms by visual perception. Much as I consider this analytical approach valid, I question the ultimate validity of taking the visual quality of a city as the most important aspect of human settlement. Many great works of art of human civilization, such as painting, architecture, sculpture, movie, dance, video art, etc., have been created for the appreciation of people who can see, and much effort of artistic creation has been channeled towards creating something that is visually exciting at the cost of neglecting the inner meaning which is invisible to human eyes. I do not mean that a painter or a dancer should create for the blind because these creative media unfortunately cannot communicate without visual perception. But a city is different; the fabric of a city which is generated by the life force of the people is for the blind as well as the sighted. The invisible form of a city, which has been neglected by many architects and planners alike, should be given equal importance. Too many designers of cities have over-indulged themselves in creating their personal visual images of what they think a city should be without allowing future flexibility for the invisible force of life patterns to grow and develop. There is only form without content, and lack of content is what makes a city dead.

Dealing with different aspects of a city form in isolation such as those forms outlined by Galantay—agglomerate form, symbolic form, perspective form, statistically ordered form, topological paradigm—is good as an analytical approach, but trying to understand the true nature of a city in this isolated way is like looking at the state of painting before the development of cubism by Picasso in 1907. Each painting represents only a part of reality from a fixed viewpoint. In order to have a more comprehensive understanding of a city, one has to combine different aspects of the urban form into one multiple image. The message we can discover from cubism is that interpretation of reality from multiple viewpoints can be the source of inspiration of new city design and planning strategy.

"Quantitative Methods: Modelling," as discussed in Galantay's paper, is a useful tool to understand rationally the isolated aspects of a city. But, in my opinion, rational model analysis is a rather impersonal approach to the understanding of a city. Quantitative methodology does not give enough indication as to the real nature of a city. It is like looking at a set of quantitative measurements of a woman; no matter how perfect it is, it does not give any indication that the person is beautiful and lively. Quantitative data can be useful for academic purposes, and as a means to classify available information, but planning and designing a city on the basis of quantitative data alone will be inadequate. It is inevitable that our city will become more and more complicated in the future due to the increasingly complex nature of our society. If life itself is irrational why should the physical environment of human settlement be so rational? To me, the forces that shape a city are qualitative rather than quantitative.

There is no question that the advancement of the computer will be very useful in aiding quantitative analysis of various urban problems. There is, however, a serious hidden danger which is threatening human civilization. If we accept the Big Bang of 15 billion years ago as the beginning of this universe, then it has taken several billion years for the radiation era to evolve into the matter era, and another several billion years for the matter era to evolve into the life era. *Homo Erectus* did not come into existence until only a couple of million years ago. Ape-like human beings walked on this planet for two million years before they started to record history less than ten thousand years ago. For the last ten thousand years, one of the greatest things that happened on earth was the development of the human brain. The nature of this fantastic carbon-based human brain is that it is capable of being rational and emotional at the same time. Mechanically, it is far from being perfect. Qualitatively, however, it has created much joy and beauty for human life. It is only within the last fifty years of cosmic evolution that a new "creature," far superior in a technical sense to the carbon-based human brain, has been created by man. It is the silicon-based computer. Many believe that the silicon-based computer can never replace the carbon-based human brain, despite the fact that some scientists today are working on the ultimate creation of a robot with artificial intelligence. But the threat is still there.

The invention of tools to aid the advancement of human civilization is a part of history. Thus the development of a super computer to aid human affairs is a natural development in our technologically oriented society. The threat that comes along as a by-product of improving the capability of the computer is a gradual diminishing of the value of human judgement. In fact, much of today's social patterns and policies is directly or indirectly being controlled in one form or another by computerized decision-making processes. This gradual "siliconization" of humanity, made inevitable by its increasing overdependence on impersonal analysis by computer, is, to me, most alarming. It has changed our lives as it effects, for better or worse, the way we think. The constant reassertion of human qualities is most important if the human-dominating life era is not to be replaced by a robot-dominating silicon era in the next phase of cosmic evolution.

Without a doubt, a computer-aided planning program is a useful tool to facilitate the understanding of some of the quantitative aspects of urban problems, and computer graphics will help in visualization of complex spatial conditions. But ultimate decision making must be determined by qualitative human factors. After all, cities should be inhabited by carbon-based humans instead of siliconized robots. I hold the view that if God wanted perfection in His creation of man, He would have created a silicon-based robot instead of an ambiguous carbon-based human. Thank God we are carbon-based so we have the joy of love and the enjoyment of the arts, and we have dreams.

The comparison of Brasilia and Ciudad Guayana in Galantay's paper is interesting in that it shows how twentieth century men have failed in making both the architect-designed city and the computer-analyzed city. In the architect-designed city of Brasilia the designer was concerned only with visual importance, whereas in the computer-analyzed city of Ciudad Guayana the impersonal quantitative data analysis created a lifeless city. What was ignored in both of these is the importance of the needs of people as the basic force that shapes the city, including the city of the future if we want to remain human.

Hong Kong is an example of a city that has grown prosperous without a plan. Over one hundred years ago, Hong Kong was a small fishing village, turned later into a colonial city under British control. In 1949, there was a great influx of people into Hong Kong after Mainland China changed governments. The need for survival pushed the people of Hong Kong not only to work hard, but to be flexible, and Hong Kong started to grow and grow. Due to the uncertainty of Hong Kong's future, a certain unique pattern of growth has developed which defies the logic of our traditional static planning methodology. The only law that governs this pattern of growth is flexible adaptation to changing situations. Action and new solutions for survival from the people often precede governmental policy. This is the reverse of the traditional planning approach, which still deals with static elements in isolation.

Today, Hong Kong is one of the great cities, and the third most important financial center in the world. Looking at the urban form of Hong Kong from a distance, no one can deny its fascination and visual excitement. Socio-economically speaking, Hong Kong is full of life and activity, yet the city of Hong Kong was not planned by government planners, nor was it designed by practicing architects. It was dictated by the law of supply and demand created by speculative developers. Hong Kong has demonstrated that the most important element in the course of urbanization is the invisible life force of the people, rather than the physical environment they live in. Adjustment to circumstances is an art of survival characteristic of the Chinese people in Hong Kong, which is perhaps the basic strength of the invisible urban fabric of that city.

There were, however, planned Chinese villages in the New Territories before Hong Kong was colonized by the British a century ago. Most of those villages are in the form of rectilinear "walled cities." Within a wall of the city there is a rectilinear grid on which passages and houses are built. Normally each walled city belongs to one family clan consisting of about 50–60 house units. It is worth observing that the spontaneous development style in the urban district of Hong Kong today sets a strong contrast to the rigid architectural environment of the past. Today, these walled cities are no longer active. Most of the descendants of these walled cities are either taking up employment in the urban areas or settling overseas. Some of the walled cities are now being preserved by the government as tourist attractions.

The Hong Kong government did, in the past, plan several "Industrial New Towns" in the New Territories. These Industrial New Towns usually consist of government housing schemes together with land leased for private development for industrial, commercial and residential usage. Since Hong Kong does not have heavy industry as such, there are, instead, many smaller scale manufacturing enterprises. Industrial development in Hong Kong really means the provision of a great number of high rise "flatted factory" blocks grouped in one area which is called an "Industrial Town." A "flatted factory" means a block of buildings with many open floors catering to the individual manufacturers of plastic flowers, electronic watches, printing and garments. The development of high rise flatted factories in Hong Kong can be regarded as a unique Hong Kong creation, reflecting the versatile mentality of the people working to create a flexible building type to meet the unexpected economic demands. Since the flatted factory blocks are either for sale or for rent within one building, it makes management tremendously difficult. As a result, the environmental quality within and around the factory blocks is usually very poor. In a city like Hong Kong, the development of flatted factories fulfills a social economic need. Architecturally speaking, it has, in principle, created a new building type in an urban environment.

Mixed use within a building is another characteristic of Hong Kong urban architecture. It is not uncommon to find a newly constructed twenty story building turned into a complex, vertical, social and economic structure within a matter of a couple of months after the tenants occupy their individual premises. For example, the ground floor may turn into a shopping arcade with different tenants selling the latest French fashions or Chinese sausages. The mezzanine floor may be shared by a girlie ballroom and a foreign language school. The floors above may be taken up by tailors, Chinese bone setters, a Bible school, guest houses and residences. This structure of mixed use is by no means static, but changes constantly in order to meet the unpredictable social and economic demands. I personally believe mixed use buildings are one of the important ingredients that makes Hong Kong city life active and exciting. Segregating the activities of city life may achieve apparent architectural and social order, but it is often an order without much life.

Galantay's paper does not mention the strong influence that the profit orientation of commercial developers has had in the shaping of urban environments. In our increasingly commercial and marketing oriented society, at least in the capitalistic world, decision making related to urban development is very much dictated by cost benefit analysis by the developers. Hong Kong is a classic example, where the short-term mentality of the speculative developers and the laissez-faire government policy have contributed much to the city's growth and excitement.

But there is also a high risk factor in such an uncontrolled approach to urban development. Currently Hong Kong is going through a very depressed period caused mainly by three factors: (a) the uncertainty of Hong Kong's political future after 1997, when the lease between Britain and Imperial China will expire, (b) an over-supply of high priced, but low quality, property developments as a result of the recent boom two years ago, and (c) the general world economic depression. This depressed situation has already caused the collapse of several major real estate empires. On the other hand, even if there have been careful marketing research and feasibility studies prior to any development, as normally conducted by corporations in the West, success is still uncertain. As one very successful developer in Hong Kong says, if he had studied at Harvard Business School, he would not have dared to venture into any of his highly speculative but successful projects. He further says that he believes in his intuition rather than in marketing experts or computer analysis.

Singapore is also facing a similar predicament at the moment. The Urban Renewal Authority (URA) from time to time releases land for private development through a tender process. Each tender must submit a tender price plus a design scheme. The Singapore government is very conscious of environmental quality and "good" architectural design. Award of tender depends a great deal on the design merit of the scheme. As a result, many developers invite international superstar architects to prepare design proposals. Because of the competitive nature, many winning schemes are so visionary and spectacular that they have little to do with local conditions. Critics call these ambitious futuristic schemes "Singapore's Space City." Some of these huge schemes are now under construction but face tremendous financial difficulties and doubtful market demand at completion. These fantastic structures will no doubt change the nostalgic Singapore, with its human scale, into a zoo of superstars' architectural fantasies.

Something, however, is missing in both uncontrolled development, such as in the city of Hong Kong, and in an architect– or computer-designed city. What is missing is what I am searching for: "An Orderly Chaos." What sort of city do we want anyway, an aesthetically pleasing form or a rational machine? I think that what we want is a beautiful, functional and livable human environment.

Chapter 3

Reel to Real: The Cinema as Technological Co-Reality

Reel to Real: The Cinema as Technological Co-Reality

Pierre Rouve

> *Mnogo trudov predstoit teoretikam.*
> Much work lies ahead of the theoreticians.
>
> —Einsenstein

Doubts Patterned as Prolegomena

It all seems child's play: a practicing filmmaker turned theoretician in his ripe old age proposes to outline the relation between the technological core of films and ambient "reality." But is it? The inverted commas flanking the colloquial concept of reality cast a first doubt that all may not be plain academic sailing. And this is only the start of a growing incertitude.

Is reality an inviolable and immutable absolute, shielded forever by the armor of its logical consistence? But does not precisely this logical consistency imply the notion of possibility? The possible dominates the real—thus speaks Heidegger, tantalizingly. Possibility must then entail action—and action must inevitably result in some sort of relation between the agent and the entity upon which he acts. Reality, then, must be an active compound of existential relations or, to put it differently, of logically consistent material or ideal *modalities of being* bridging the gulf between actuality and possibility. And the fate of all human endeavors is to extract the possible from the actual.[1]

But this inquiry is consciously confined to a rigidly circumscribed field of highly specialized cognitive concerns. Surely all such restricted cognitions meant to constitute self-contained sciences are only minute specks in the motley multitude of our mental quests. In this desparate mosaic, the time-honored notion of Science loses its monolithic majesty and shrinks to the modest rank of a somewhat vague taxonomic class label. Utilitarian expediency accepts only sciences: the plural seems imperative and inexorable.

Yet, the conference for which this groping inquiry is intended quietly states its determination to debate some (possible? probable? inevitable?) Unity of the Sciences. Unobtrusively, this conference aspires at bringing Science in the singular back to its former ontological dignity. Inevitably, this calm but crucial declaration of intent is bound to inhibit the researcher proposing to explain—be it in a theoretical key—the impact of excessively specific technological imperatives on the constitutive procedures of motion pictures. Tamed, but not teemed, by the customary expulsion of philosophical meditations from all technological pursuits, he stops in his tracks, intimidated by the sudden realization that his technological commentary will have to unfold under the crossfire of two worrying world views: the Pluralist belief in a congenitally splintered universe and the Monist certitude in the essential oneness of this universe. On one side—the dogma that truth and/or knowledge are attainable only through *diversity without unity*.[2] On the other—the conviction that the only path to truth and/or knowledge leads to *unity without diversity*.[3] At the heart of this ontological (hence methodological) conflagration nestles a modest technological product—the motion picture. The disproportion is indeed much too gigantic not to be intimidating.

Yet, it is precisely this incongruity that proves stimulating, exhilarating, even illuminating. The film, a much-maligned technological David, firmly grips its cinematic sling–minute and well nigh impotent in its overspecialization. But the analytical pebble which it is about to hurl at the haughty and intolerant philosophical Goliaths is not without pertinence: does not the film, both empirically and semiotically, transmute necessary discontinuity into inevitable continuity? And in so doing, does it not transcend the antinomy between pluralist and monist excess to present a pragmatically viable and logically unassailable paradigm of *unity in diversity*?

Such mediation between traditionally irreconcilable contradictions is not to be mistaken for a hasty recourse to some sort of

Hegelian palliative. Had this been the case, both antagonists would have been compelled to relinquish their identities in favor of a monolitic entity utterly alien to both of them. But, as things stand in filmic structures, both conjunction and disjunction, monist convergence and pluralist divergence preserve intact their respective ontological cores while joining forces in the structuring of a composite entity whose all–pervading characteristic is the relational equipoise between unity and diversity. Again, in tune with this inherent dual essentiality of its structure, the film does not imply any subservience of its technological devices to superior philosophical considerations that would inevitably result in a Hegelian type of philosophy *OF* technology. It is imperative to examine what a film is, and is not, not only as a particularized instance of such a philosophy of technology, but as a generalized test case revealing the inextricable presence of philosophy *in* technology: in its methodological strategy, logical construction, epistemological de-construction and ontological self-assertion.

A Touch of Taxonomy

Familiarity may not necessarily breed contempt, but it certainly fosters indifference, and the fate of films in our society is here to prove it. Stultified by an incessant gale of films—screened, televised, and now canned for domestic consumption—the ordinary spectator has long lost the will to think about film as film. What is worse, some self-appointed cinematic mentors manifest the unmistakable symptoms of such mental paralysis. A case in point is the voluminous "Oxford Companion to Film," which proudly proclaims that its "aim is to answer any query which may occur to the amateur of film." And yet, for all its avalanche of biographical irrelevancies, clumsily-potted plots and skin-deep ideological verdicts, this portentous publication does not devote a single line to the basic question: what is film? Obviously, it never occurred to the editors of this curious Companion that the quaintly named "amateur of film" may want to learn something less superficial about his paramour.[5]

Regretfully, in many quarters, such insubstantial film chatter is the rule rather than the exception. A touch of taxonomy is therefore in order, however elementary and approximative its methodological bearings may seem to the educated eye and the searching mind—a twin rarity among film technicians reluctant to see beyond their empirical noses and omniscient scribblers blissfully skating on the motley surface of the (once) silvery screen.

For self-evident reasons of expositive clarity, the global grasp of the film has to be effected from two distinct and consecutive viewpoints: initially, films must be apprehended as *objects*, and subsequently they can be evaluated as *events*. With film objects the analytical approach of the spectator is essentially *structural*; with film events it is predominantly *sociological*. Film objects display, first of all, their empirical substratum: a tangible material strip bearing the mimetic imprint of motion in the ambient achieved reality through appropriate technological (chemical, mechanical, electronic) manipulations which ensure the mimetic adequacy of the imprint and its chosen manner of reproduction. Emerging from such a technological matrix, these strips undergo a subsequent but no less *technologically-conditioned structuring* in view of punctual semiotic aims. The barrier between technology and semiology is thus abolished.

Once fully and properly structured, film objects may be envisaged as impact-generating *film events*—theoretical compounds of already constituted film objects and their environmental exposure, behavioral responses and societal reactions to the messages which they convey, be it overtly or covertly. In the light of cybernetic theories of information, these two profiles of cinematic occurrences are clearly outlined: film objects program the transmitting potential while film events mirror the receptional efficacy of a given cinematic line of communication. It follows that, whatever allowances may be made for fertile feedbacks in such communicational circuits, technological film objects precede and preordain the interpersonal reverberation of sociological film events. To put it with a dash of Romantic flourish: the film object is the semiological seed that will blossom in ideological impact.

Yet it is the film object that has been largely ignored—certainly by the mass media that enlighten public taste or, more likely, buttress public ignorance. Unashamedly, they put the cart before the horse and focus all their attention on the narrative/ideological aspect of film events. However regrettable, this worldwide aberration is quite understandable. It is inevitable in totalitarian regimes where a single state ideology is imposed by administrative compulsion. In democratic countries a similar lopsided situation is the outcome of epistemological limitations. Films are measured with non-filmic criteria lifted lock, stock and barrel from literary criticism: narrative concatenation, realistic life-likeness, social significance. Specific cinematic structures and matrices of visual meanings are ignored. And because film objects are charged with a much heavier load of cinematic specificity than film events, they, too, are ignored.

Such structural specificity, a prerogative of the film object, is technological: the *Real has been recorded on a Reel*. To dismiss this idea in a cavalier fashion means tacitly to restate the traditional superiority of psyschology and ideology, prime movers of the mind, over technology, the passive handmaiden of the mind. And therefore, any endeavor to stress the semiological stature of the film object is ultimately an effort to confirm the philosophical credentials of technology. This paper is but such a plea.

The Birth of Co-Reality

John Dewey is both right and wrong when he claims that the real esse of things is neither their *percipi*, nor only their *intelligi* but their *experiri*.[6] He is wrong if his statement leads us to believe, in a fit of psychological reductivism, that experiences are natural occurrences impervious to our considerate control. But he is right if experiences can be logical constructs. They do not mirror the reality, they constitute a reality. But this teleologically—and therefore logically—restructured reality comes a step closer to a deeper cognition of the Aristotelian *ti-esti*—the tantilizingly simple "what-it-is" of reality, its elusive esse-nce.

This analytical strategy has already presided over our deconstruction of the existentially monolithic film experience into empirically conjoined, yet epistemologically diversified, experiences of the film as object and as event. This exploratory model not only draws an illuminating demarcation line between, on the one hand, structuring modes and generative processes (spotlit in film objects), and, on the other hand, narrative ways and receptive by-products (central in film events), but it also hints at the methodological key of our analysis taken as a whole: the opposition Unity/Diversity.

In turn, this fertile syntheton mutates into a host of analogous *binomials required for the unraveling of the constitutive core of the film object*. Such are the ontological syntheta of motion/rest, time/space, identity/difference, conjunction/disjunction, presence/absence, visuality/aurality, etc. They operate as logical functors activating the subtended technological features: Photogram/Sequence, Fixed Frame/Variable Frame, Shallow Focus/Deep Focus, Tilt/Roll, Panoramic Shot/Tracking Shot, Close Up/Long Shot, High Angle/Low Angle, Single Exposure/Multiple Exposure, Slow Motion/Speeded Motion, Direct Light Impact/Laboratory Processing, Color/Black and White, Direct Sound Track/Post-

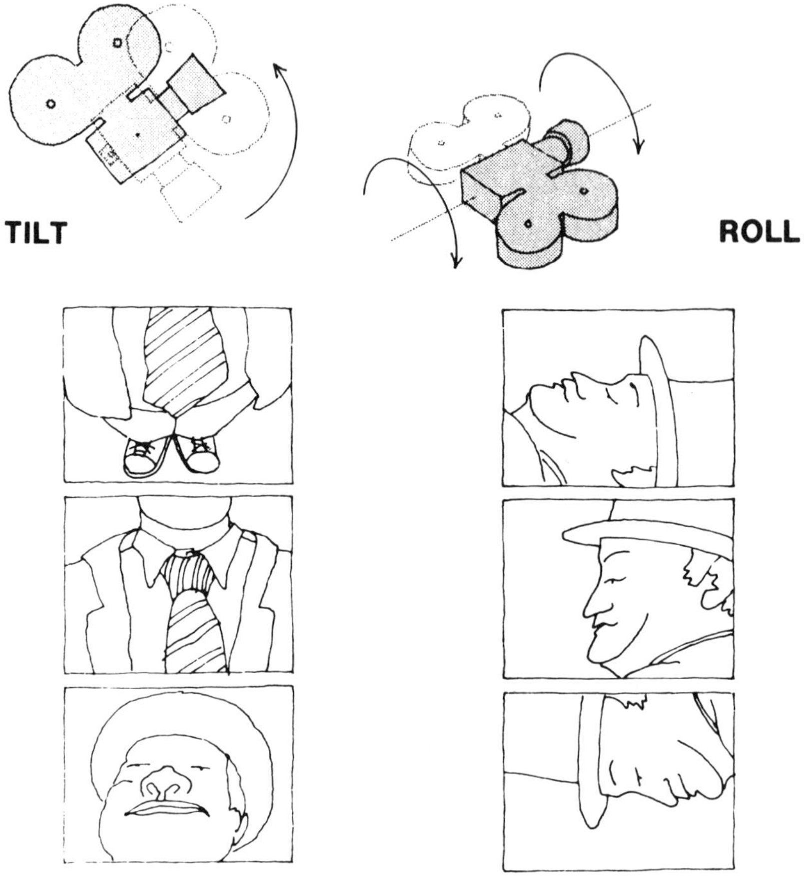

Fig. 3-1: Tilt and roll camera movements.

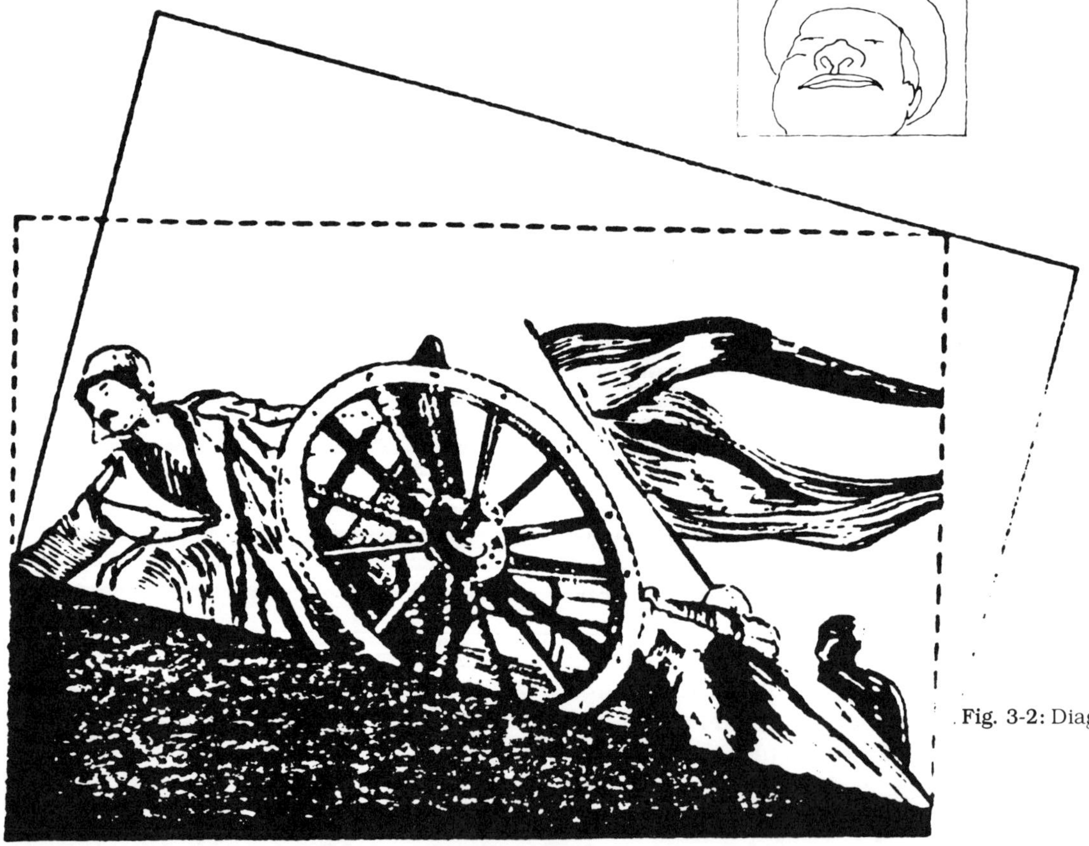

Fig. 3-2: Diagram for Eisenstein's "October" (1928).

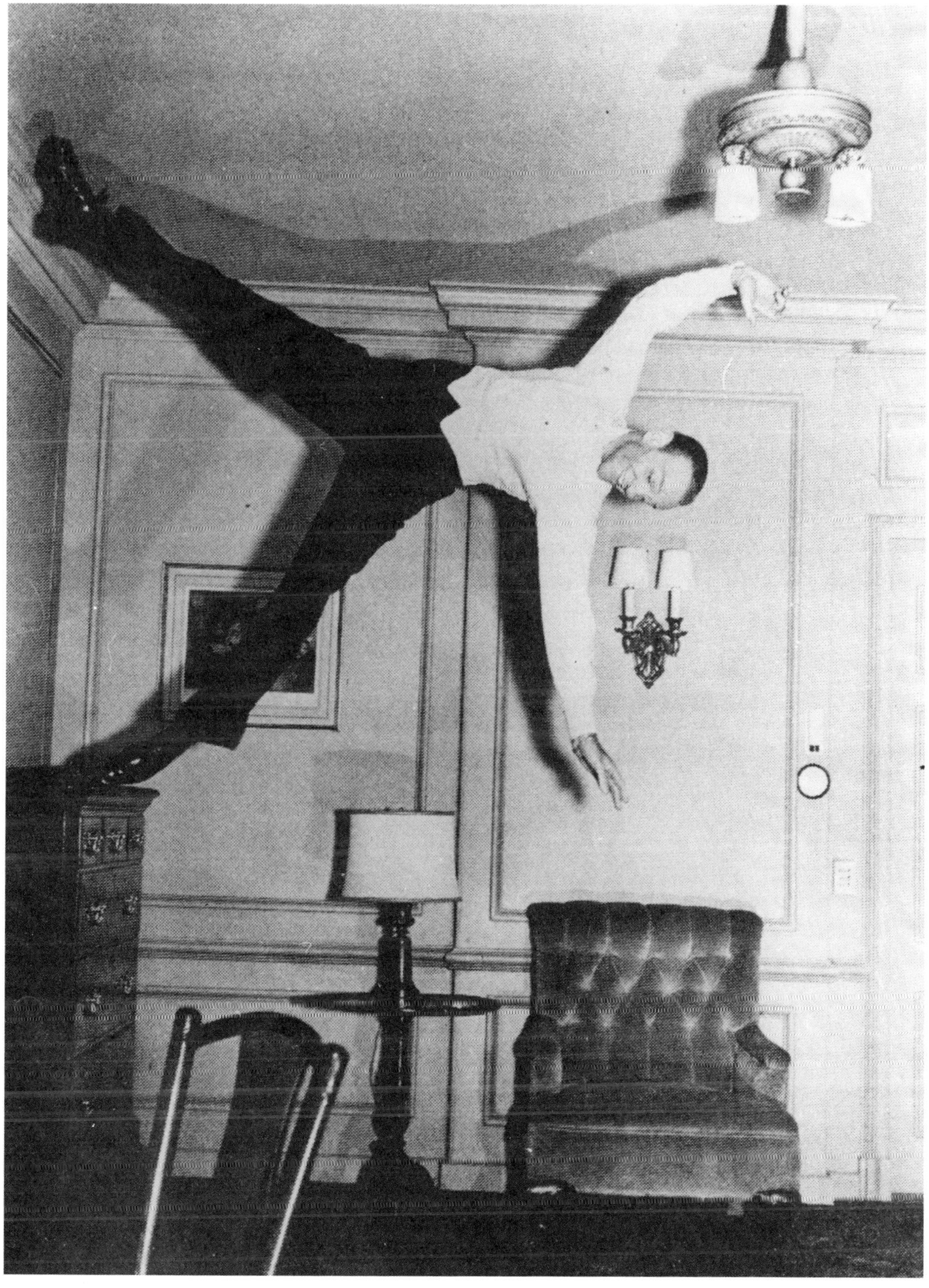

Fig. 3-3: Fred Astair in Stanley Donen's "Royal Wedding" (1951).

synching, Synchronous Sound/Asynchronous Sound, Cut/Dissolve, etc.

This arid enumeration is bound to appear quite forbidding. It could hardly be otherwise: the span and speed of this paper impose a heavy servitude on its writer. Still, he may be permitted, hoping against hope, to clothe this inventorial skeleton with the livelier flesh of a few illustrations. Their photographic quality is bound to be exceedingly poor—but then they are not meant to be paragons of typographic perfection. Their aim is to provide a first, fleeting glance at the eminently technological modes in which the allegedly neutral (and therefore slavishly mimetic) film camera processes the visible reality by a blatant or surreptitious infusion of purely visual spatial inflexions. Thus begins the charting of the adventurous journey of the Real to the Reel.

By way of introduction to the constitutive relation between the camera position and its screen results, the first images illustrating the binomial Tilt/Roll are of a somewhat diagrammatic nature. (Fig.1) Pointing downwards, the camera begins by recording a foreshortened image of a man's legs, then tilts upwards to a transitional level shot of his chest and comes to rest on his face, once again distorted, this time because of the upwards angulation. An Ingres-like portrayal of the chest is inserted between two Caravaggioesque foreshortenings. Or, to put it differently, a realistic mirror-like image is compressed between two *technologicallyobtained distortions*, dear to any expressionist worth his traditional label.

A cardinal question surges forward, to be dealt with in some detail in due course: what is the purpose of this *transmutation, but not disidentification*, of the visible reality? A foretaste of the answer to come is to be found in the diagram of a tilted, static camera shot in Eisenstein's film, "October," of 1928. (Fig.2) It shows that the tilt is conditioned not by an actual gradient of the road, but by the intent to signify the obstacles encountered by the combatants and their determination to overcome them. If the tilt had been predetermined by the configuration of the terrain, it would have been no more than a demotative reproduction of a given tract of road taken by some armed men. But the deliberate tilt confers to the image a new connotative dimension. If this tilted shot "portrays" anything it is the ineffable figure of speech, "uphill struggle." Semiotically enriched, *the Reel enriches the Real.*

The same occurs with the Roll, though in a quite different

manner. When a camera rotates 180° around its horizontal axis, all animate or inanimate components appear to shed their positional fixity: up and down and left and right cease to be gravitational bearings. The result is the complete disorientation of the spectator—a curious case of what may be called pure Technological surrealism. The most startling evidence of such purely visual scorn for the laws of gravity remains the famous wall and ceiling dance of Fred Astaire in Stanley Donen's 1951 film, "Royal Wedding." (Fig.3). There the camera roll was synchronized with the revolving of the set mounted within a drum, with a camera and furniture fixed to it. This made it possible for Astaire to begin his dance on the floor, then move up the left wall, continue dancing upside down across the ceiling and come down along the right wall—the whole routine being photographed in one, long, uninterrupted take.

What happens there is not mere "trick photography," and is not to be confused with the animated model shots complete with front or back projections, a staple diet of all science fiction extravaganzas from *King Kong* to *Tom Thumb* to *2001*, and after. They display a breathtaking array of "special effects" harnessed in the service of an imaginary irreality striving, for reasons of credibility, to mimic our daily reality. The Astaire dance, for all its visual modesty, achieves a "rhetorical effect" never envisaged by the strictly denotative science fiction spectaculars. They endeavor to suffuse blatantly unreal situations with an air of reality. The Astaire dance heightens trivial reality with a dash of unreality. It is a matter of conferring a halo of strangeness on familiar occurrences, aptly termed as "ostranenye" by Victor Shklovsky in the early twenties.

A similar rhetorical redemption of banality can result from yet another technological procedure—the so-called Multiple Exposure consisting in the printing of two or more negatives onto a single negative. In the remarkable screen version of Poe's "The Fall of the House of Usher," made in 1928 by the sadly underrated French director Jean Epstein (Fig.4), a close shot of slender, white candles has been superimposed onto the long shot of a wooded landscape so that not only do candles and trees appear equally tall, but both converge on the blurred silhouettes of three minute figures outlined against the distant horizon. Taken separately, these images would have been banal and inexpressive. Combined by dint of a double exposure, they appear strange, enigmatic, macabre. While remaining the same, they become different. Triv-

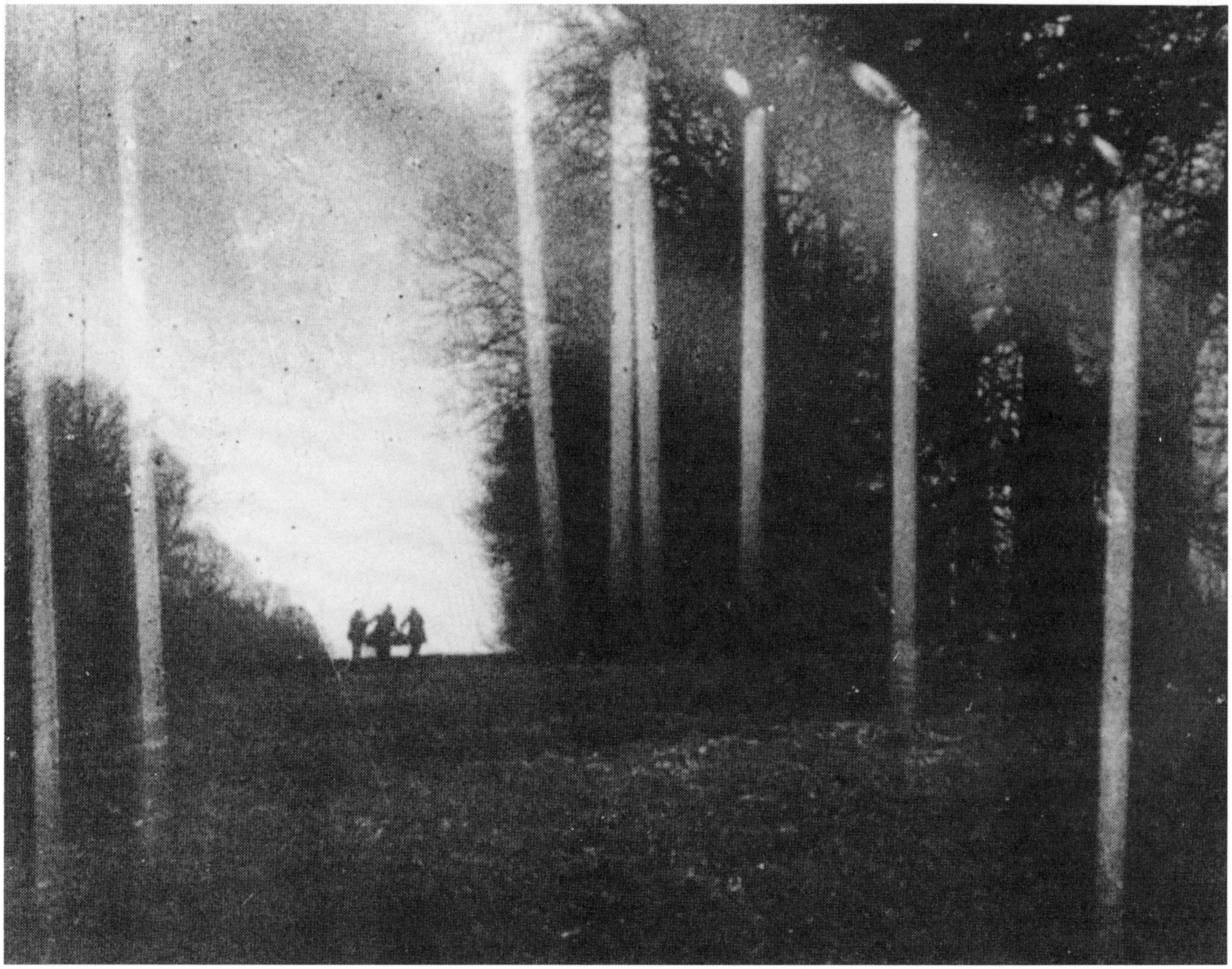

Fig. 3-4: "The Fall of the House of Usher" directed by Jean Epstein (1928).

Fig. 3-6: Motion photography: man sneezing.

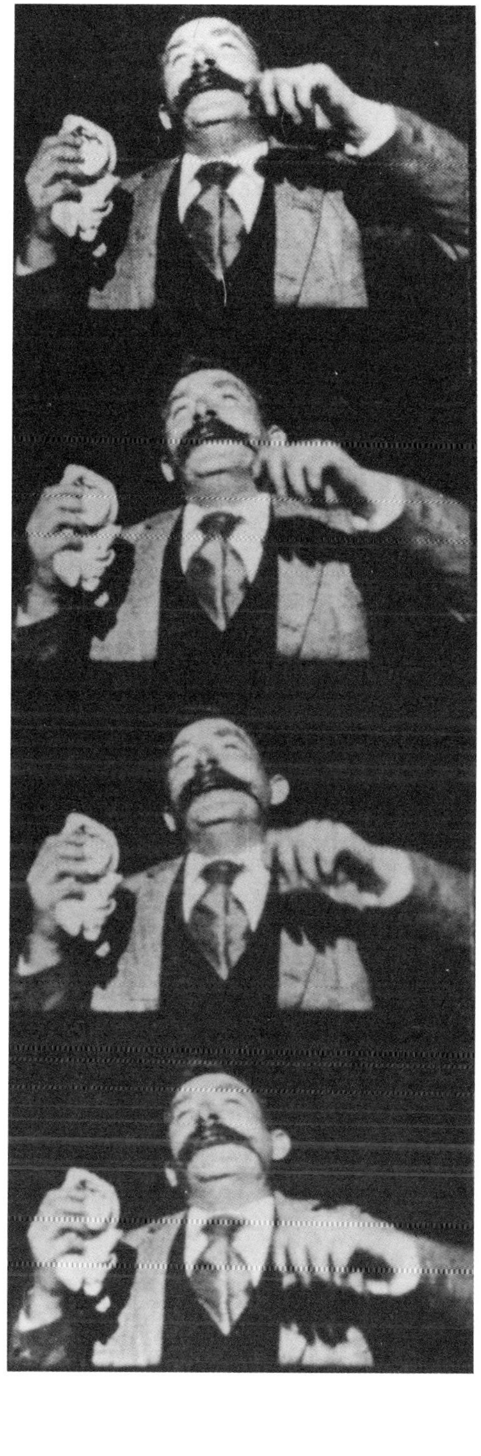

Fig. 3-5: Motion photography: running horse.

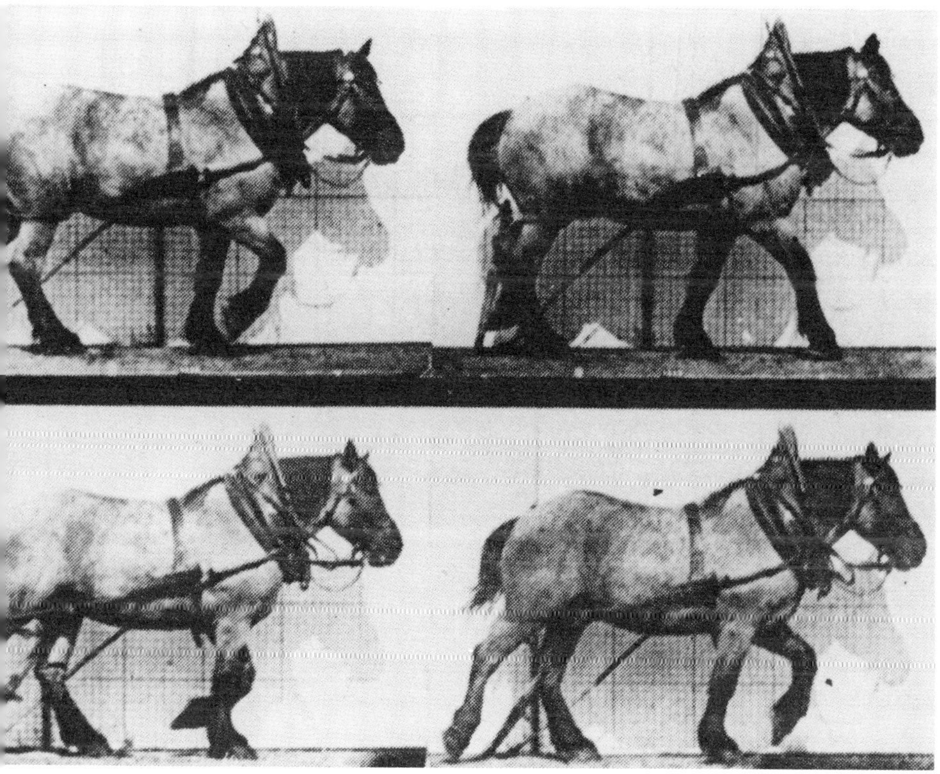

ial denotations are transmuted into unusual connotations, and this semiological alchemy is unleashed by technological devices.

It is now clear why any structural scrutiny of the film object must give pride of place to the *generative potential of cinematic technology* in preference to Determinist causality restrictively grounded in subliminal compulsions (Freud) or societal pressures (Taine, Marx, Adorno). If this is done, the methodological and telelogical cornerstone of this inquiry stands out in all its inexorable clarity: FILMS DO NOT MIRROR EMPIRICAL REALITY BUT STRUCTURE A SUI GENERIS TECHNOLOGICAL REALITY.

A first hint of such semiotically generative technology has already been given by the analytical de-construction of mechanically constructed elementary film constituents like Camera Tilts or Rolls and Multiple Exposure Shots. Examined at close quarters, they have all revealed the same modulation from things perceived to ideas conceived by virtue of punctual technological manipulation. It has thus transpired that such material starters have the distinctive features of *Syntactic Structures*, meaningless in themselves, but indispensable for the ulterior emergence of meanings. And indeed, submitted to further technological transformations, these Syntactic entities yield punctual *Semantic Structures*—clusters or systems or messages which can refer jointly or severally to: a) any given film fully constituted and interpersonally experienced as a coherent whole; b) its discrete constituents envisaged as self-contained items assembled according to definite "grammatical" rules. And, if the film is a mode of communication operating through Syntactic Structures, Semantic Messages and Grammatical Rules, then surely the film is a language.[7]

Film as Language

The validity of the linguistic model for a firmer grip on the structural substance of a vast array of epistemological pursuits has long been accepted. Arguably, it may even be the least tortuous route to a plausible unity of all sciences. And yet, its application to film studies still calls for the elaboration of a *systematic film grammar*. Only patient efforts to work out such a coherent all-embracing system can help clear the air and complement, decant and streamline the praiseworthy pioneering endeavors of Lotman and Metz,[8] often debased and distorted by semiliterate pillagers.

The main objection to the transfer of a suitably adjusted model of verbal intercourse to the structural modalities of visual intercourse may be seen as a marginal manifestation of the general inflexibility of philosophical Pluralism. Analytical methods evolved within a given epistemological field are prescriptively confined within its borders: anthropology, psychology or political science are not permitted to make use of the cognitive insights gained by linguistics. But the undeniable achievements of Claude Levi-Strauss, Jacques Lacan and Louis Althusser have proved the absurdity of this ban on interdisciplinary cross-fertilization.

It is time to redeem visual communication which has lagged behind such achievements, with the notable exception of Dora Vallier's systematic studies in visual syntax.[9] And because no attempt has been made to work out a systematic visual grammar, it has been glibly assumed that such a system is inconceivable. But this is palpably not the case. To the once decried but now universally accepted verbal chain: phoneme—morpheme —phrase—sentence corresponds an equally consistent visual chain: opteme—morpheme—sight phrase—sight sentence.[10] And, as in all grammatical systems, the stability of this newly erected edifice of visual grammar rests on the plausibility of its fundamental root unit—the opteme.

The same methodological principle should govern the constitution of a systematic film grammar, which is but a sub-class of the more generalized visual grammar. This means that due care should be taken to establish with a sufficient degree of proven veracity the nuclear minimal unit on which the prospective film grammar is to be built. On grounds of logical assonance with currently accepted terminology this syntactic root unit may be called *filmeme*. But to coin a term does not necessarily mean that one has discovered the specific properties of the entity to which the name refers. It is precisely this quest for the distinctive features of the minimal filmic unit that has proven the least satisfactory and most vulnerable aspect of current film theory. An attempt to rethink the matter afresh is more than necessary.

In Search of the Filmeme

The opening move in such a critical reevaluation in a structural vein would be the endeavor to retrace, on a strictly empirical level, the trajectory *from Real to Reel*—i.e., the specific technological ways which make it possible to record uninterrupted motion and present it as such. This can be achieved only if the continuous flow of Real motion is deconstructed into a succession of discontinuous static pictures, and the Reel on which they are imprinted is projected under appropriate technological conditions which optically reconstruct the original natural motion.[11] Unbeknownst to them, Edison and the brothers Lumiere, Friese Greene and Max Skladanovsky rephrase Zeno's paradox in technological terms: spatial mensurations and temporal segmentations become convertible. The Empiricist conviction that "there is nothing which moves and does not move at the same time," canonized by Quine's self-assured formula:

$$\sim (\exists x)(x \text{ moves} . \sim (x \text{ moves}))$$

appears to be seriously subverted. And this is no mean event, because this subversion undermines the principle of non-contradiction, main pillar of our reasoning habits. This problematic state of affairs should have been sufficient to alert any researcher anxious to pinpoint the minimal unit, cornerstone of any specifically filmic structure. But theoreticians have, by and large, failed to meet this intellectual challenge. Hence, their misconception of such cardinal minimal units due to the inability or reluctance to define them according to structural rather than empirical coordinates.

The confusion originates with the otherwise dazzling "Poetics of the Cinema" of 1927 by Boris Eichenbaum,[12] which fails to sense the far-reaching implications of the relationship between synchrony and diachrony for all modes of interpersonal communication. In terms of the cinema, this systematic scrutiny of simultaneity and successiveness, due to Tinyanov and Jakobson, puts time and space on equal structural footing and provides the logical foundation of Bela Balasz's innovative concept of cinematic "Zeitraum."[13] This agglutinative synalephe, blending on equal terms the notion of time and space, is perhaps best rendered in English as "Chronotype." But for all his analytical acumen, Eichenbaum somehow fails to sense the entire structuring domain of this bifocal cinematic chronotopy. As a result, he hardly gives to Time what is due to Time. Space becomes the exclusive determinant in his search for the structural nucleus of cinematic syntaxis. He centers his attention on the optical record of "real" spatial denotations. Consequently, for him, the minimal unit is the *single static photogram, isolated and identified by virtue of the spatial location of its components*. For Eichenbaum (and for those who consciously or unconsciously take their cue from him) these consecutive images of a running horse (Fig. 5) constitute *two* distinct filmemes. And so do the two consecutive portraits of a sneezing man (Fig.6). This decisive differentiation is justified merely on extra-cinematic grounds: the precise positioning of the horse's legs or the exact location of the man's facial muscles and hand.

It is seldom realized how much this assumption that single, immobile pictures are the root units of subsequent motion pictures has affected filmmaking at large, particularly in the last three decades or so. Its resonance is most blatantly evident in the single-shot products of a certain self-proclaimed avant-garde, such as Andy Warhol's eight hours long "Empire" of 1964 (Fig. 7 A & B)—though it may be argued that atmospheric and environmental changes make it a photographic reminiscence of Monet's celebrated pictures of the Rouen Cathedral from sunrise to sunset. What is, however, much more relevant than such occasional "experimental" whims is the fact that the one-shot filmeme has provided the theoretical foundation of the one-shot sequence (plan sequence) theory which has divided the film world into two implacably opposed camps—and of which more will be said in due course.

What matters now is to consider at closer quarters Eichenbaum's proposition that motionless photograms constitute the root units of moving pictures—a widespread belief which seems to be supported by a common sense feeding on empirical verifiability. Such an attentive examination is bound to reveal certain logical inconsistencies. If single photograms are taken for filmemes there is no reason why any agglomeration of such static shots would be any different. In that case, there should be no difference between an album of cherished family shots and a home movie taken on our holiday in Normandy—or, if you prefer, between the stills displayed on the front of a cinema and the movie projected on its screen. Cinematographic motion and photographic rest would be then identical. But this cannot be the case. It follows that if the set *a.n* ontologically differs from the set *b.n*, then its root constituent *a* must differ from *b*. Which means that, logically and semiologically speaking, no static *a* can be the root unit of the dynamic *b.n*. We are back at Aristotle's and Quine's joint admonition: there is nothing that moves and does not move at the same time. But, incurable "amateurs of the film"

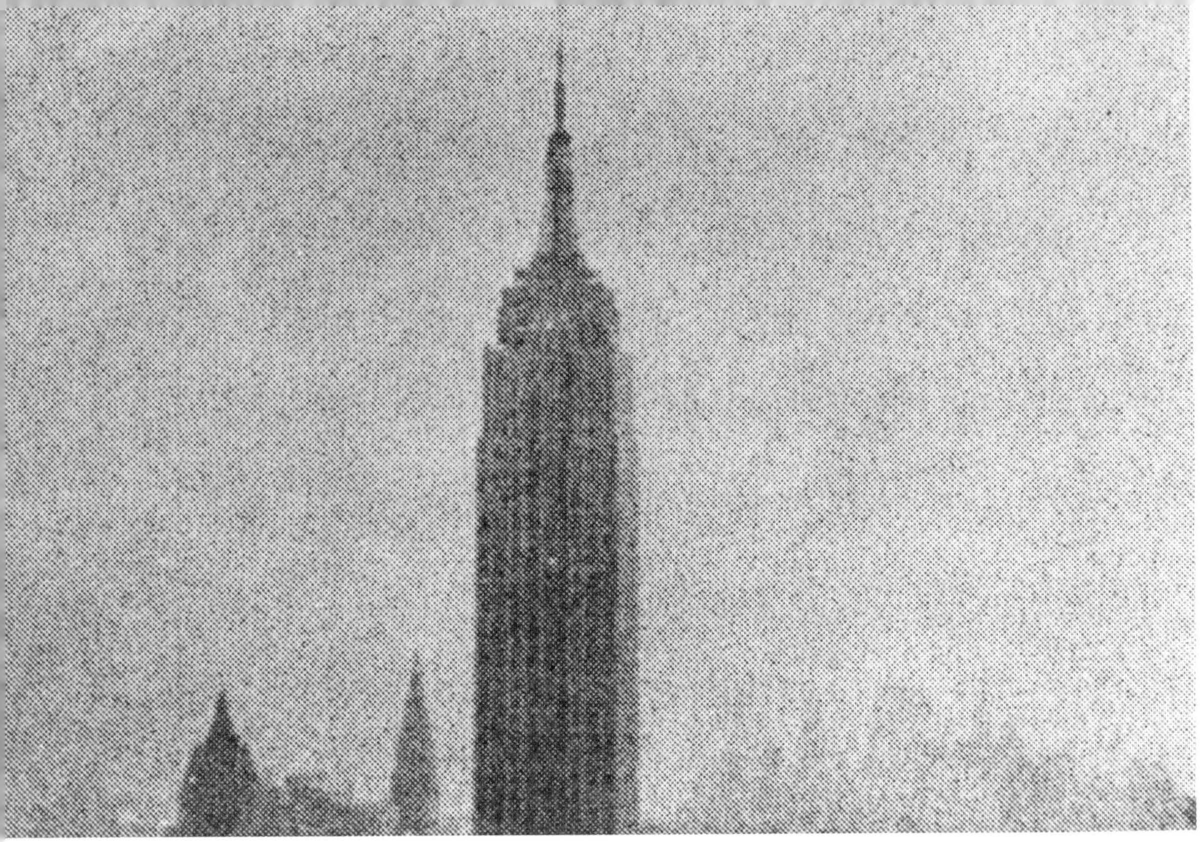

Fig. 3-7A: Andy Warhol's "Empire" (1964).

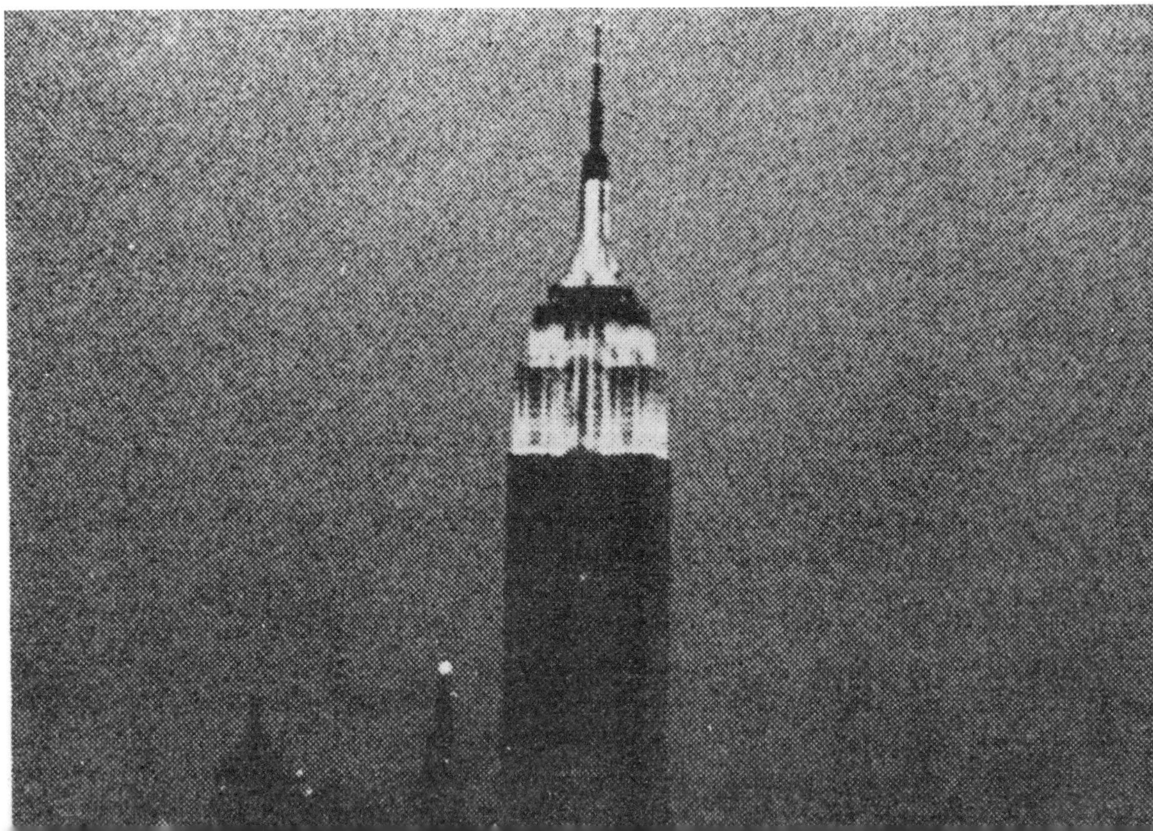

Fig. 3-7B: Andy Warhol's "Empire" (1964).

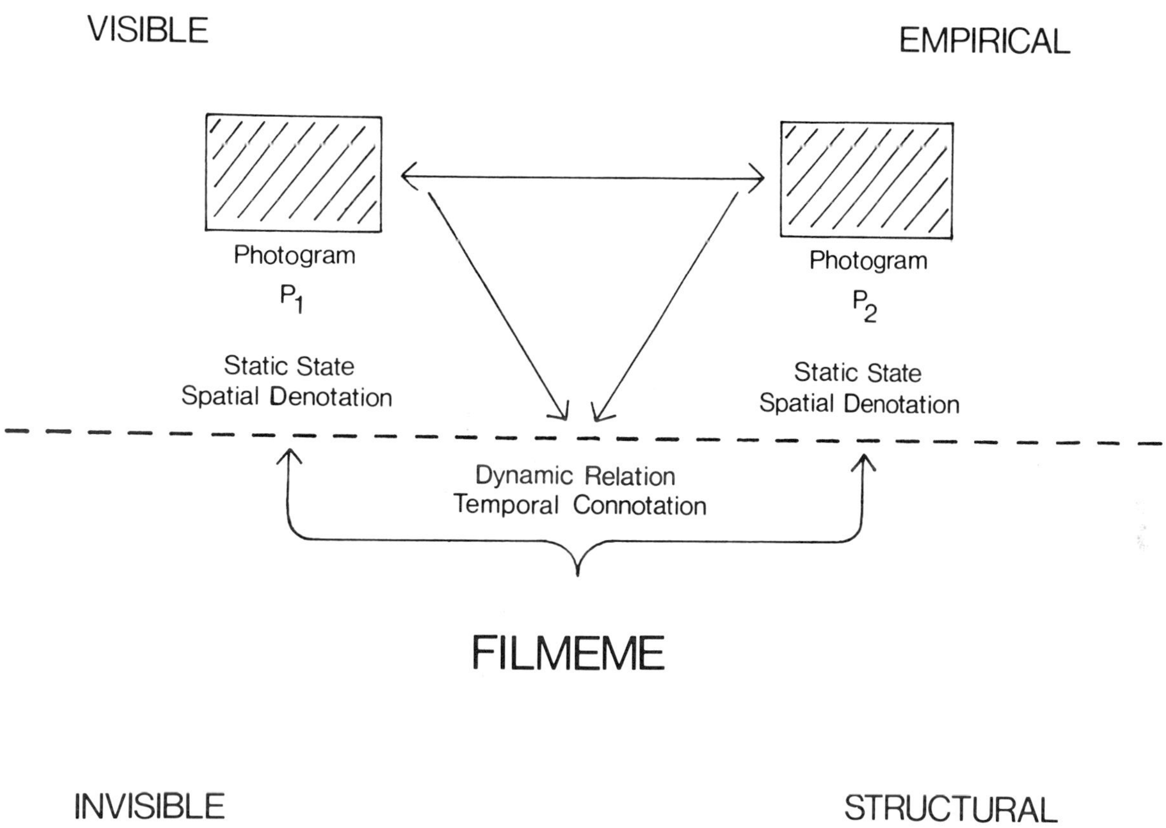

Fig. 3-8: Diagram of Filmeme.

that we are—and fairly "cinemate" to boot, we promptly protest in true Galilean fashion: "Eppur si muove."

This intuitive *cri de coeur* (or is it *cri d'oeil*?) is a call for the rehabilitation of Time within the epistemological context of filmmaking. And filmmaking is, ultimately, Time visualized in guise of Facts.[14] But such factual visualization can occur only in Space, thus bringing about Balasz's composite concept of Zeitraum. To clear such Chronotope from all spatial residuals would be to repeat Eichenbaum's error in the reverse. It is imperative not to map the lines of the *empirical disjunction* between Time and Space but to detect the mode of their *structural conjunction*.

A hint of the route which such investigation ought to take has been already given in the objections to the logically untenable transition from a static root unit *a* to a dynamic set *b.n*. The logic of static photography is impeccable because in it the static unit *a* leads to a static set *b.n*. The logic of cinematography would be equally flawless if only we could demonstrate that in it *dynamic sets b.n. (films) are the outcome of dynamic root units a (filmemes)*. Such a demonstration is not impossible if we recall that, on obvious cybernetic grounds, each photogram has a specific function only within the overall structure of the film. Furthermore, the successiveness of single photograms is conditioned by their related distinctive features serving the global function of the film. This function is to record factual motion and to present it, mimetically or not, as the case may be.[15] But motion is inconceivable without *reference to two points*: it is the progression between them. Consequently, in a specifically cinematic structure, the center of epistemological gravity falls not on a single photogram, but on two *structurally conjoined photograms*, i.e., on the transition between two diachronically discrete, spatial and static (material) photograms which, considered synchronically, function as a synthetonic, temporal and dynamic (ideal) concept. This ideal construct is the elusive filmeme, the missing dynamic root unit *a* from which a dynamic set *b.n*. called film can logically evolve. And if this be so, the images of the trotting horse and the sneezing man (Fig.5 & 6) concern us not as empirically isolated items but as structurally welded couples. They show only two sets of filmemes and not four, as traditional theory would have us believe. The filmeme is a *chronotopic relation*. To some, this definition may sound somewhat esoteric.

But it would have been quite understandable to Braque, who said decisively, "I do not paint things. I paint the relation between things." (Fig.8).

Filmeme: a Logical Dilemma

Once endowed with structural authority, the filmeme lays the methodological foundation stone of a streamlined grammatical system embracing all syntactic structures of the film object. The nucleus which generates them is the Janus-like theoretical blend of concurrent empirical discontinuity and semiotic continuity. Its theoretical distinctive features are relational and opposite, but their contrariness is converted in concordance called for by their parallel convergence within the global structure of the film. But it is precisely this structural accord between static and dynamic states of affairs that disturbs the supporters of the traditional one-photogram filmeme. For them a composite concept like the static/dynamic two-photograms filmeme is untrue because it does not correspond to any "true reality." Such ambivalent "true reality" cannot exist because it infringes the rules of axiomatic non-contradiction:

$$\sim (\exists x)(x \text{ moves} . \sim (x \text{ moves}))$$

This objection has been originally expressed in plain English. The reason for its transcription into the dry graphs of formal logic is not the vain whim to drape commonplace statements in pseudo-erudite vestments: such logical transvestites are soon disrobed. The recourse to formulations alien to facile verbiage is imposed by the very nature of arguments put forward in defense of the allegedly "unreal" two-photograms filmeme. These arguments intend to be strictly and sternly logical and not to turn for a conventional support to objectivist physiology (retinal permanence) or mechanics (perforations-controlled gate flow) and subjectivist psychology (optical delusions). It is devoutly to be wished that such a logical approach to technological occurrences may not be unwelcome in a conference devoted to the unity of the sciences.

The debate centers on the opposition of the "true" reality of the Real and the "untrue" reality of the Reel stemming from the inflexible Aristotelian principle of non-contradiction updated by Quine. But this apparent inflexibility, aided and abetted by behavioral habits mistaken for mental imperatives, rests exclusively on a *particular approach to logical thinking*—on a logic, not on logic. Both Aristotle and Quine see the ordering of thought processes as a *deontic* discipline concerned with what "ought to" be thought and inexorably postulating what "ought not to" be thought. All judgments are to be arrived at according to either/or criteria which tolerate no middle term mediation: categoric acceptances or rejections are the only possible alternatives. In such exclusivist regimentation, filmemes must be one-photogram static shots and, consequently, the Real must take precedence over the Reel.

But the nature of the filmeme and the relation between Reel and Real will appear in quite a different light if the deontic logic of imperative necessity is replaced by a *modal* logic of preferential possibility and the "either/or" way of thinking is superseded by an "if . . . then" mode of reasoning. Proceeding along this alternative path, eminent logicians reach the general statement that, "If there exists a relation R between two entities a and b, then these two entities a and b exist in their own right," formulated as:

$$Rab \rightarrow a;\ Rab \rightarrow b.$$

Conversely, if two entities a and b exist, their relation R exists in its own right. From these universally applicable propositions derives yet another relevant formula:

$$Kap \rightarrow p,$$

in which *K* stands for the predicative relation "knows that," *a* signifies the knowing subject and *p* indicates the knowledge in question. In discursive terms this means, "If I know that an entity exists in such and such a way, then this entity is such like."[16]

This brief incursion into the arid lands of formal logic is fraught with dangers. It may easily disconcert, and even repel, readers unaccustomed to such skeletal presentation of propositions usually clothed in the florid flesh of literary embellishments. Still the risk is worth taking: in its laconic austerity logical formalization is easily the most convincing witness testifying in favor of the sui generis reality of the relational filmeme and—by extension—of the cinema as a whole. And indeed, we are entitled to assert in a model key that as photograms a and b exist, then the filmeme R signifying their relation exists in its own right. But it does so within *its own—and different—mode of existence —i.e., within a different reality.* But, however different, this preferential reality is never untrue or unreal, it is simply *co-real.* And finally, if I (a) know that (K) the filmeme (p) manifests such a mode of co-real existence, then I know also that the global reality in which it exists (the film) must belong to the same logical class as the co-real filmeme.

But such transitive implication does not entail a mere agglomeration of unaltered filmemes within the vaster compound of the film. Already the Stoics had seen in the atoms only an impermanent and transitional state of the universal *hyle* capable of being molded into definable form. Echoing them, Hegel had made us aware that the very thought-act may abolish/transfigure its initial constituents. Such, too, is the fate of the filmemes. They are destined to be swallowed up by the spiralling elaboration of an integrating film grammar. This transformational expansion unfolds according to the same structural rules which regulate the disappearance of phonemes in the more complex (and categorically different) grammatical morphemes brought about by the phoneme's structural self sacrifice. And just as the phoneme is the child of linguistics processing chaotic speech acts, the filmeme is a product of technology processing random sight acts. By this technological molding of the film's syntactic constituents *the Reel makes its first bid to assert its ascendency over the Real.*

Technosyntactic Structure—and Beyond

Hopefully, this lengthy endeavor to legitimize the relational two-shot filmeme has helped us to clear the main methodological hurdle in this techno-logical investigation. Our inquiry can now gather speed and map, without much ado, at least the cardinal nodal points in the generative syntax of the film. Their structural hierarchy is neatly delineated: filmemes aggregate into morphemes, morphemes combine in film phrases and film phrases culminate in film sentences. And because the root unit of this concentric increase in complexity is camera-conditioned, its hierarchically superior derivatives show analogous techno-syntactic symptoms. And, for all the hue and cry of scandalized classicists abhorring the dark satanic machines, it may be quite in order to assert that cinematic syntax is an *anulus technologicus*, a modern reincarnation of the time-honored anulus Platonis.

Like the phonemes in verbal discourse, the filmemes are, in themselves, meaning free, but carry the seed of meanings to come. Pointing the camera at a given target in a given direction is a technosyntactic equivalent of the verbal recourse to demonstrative pronouns. High or low key lighting is parallel to the use of adjectival qualifiers. The quantifiable duration of a screen image acts as a variable intonational stress. Dissolves act (or, to be historically accurate, acted) as punctuation codes. Time and again in its technologically controlled unfolding the film language manifests its meaning-free syntactic coherence.

It is not inconceivable or logically untenable to deliberately limit the cinematic discourse to an aleatoric assemblage of meaning-free technosyntactic filmemes—much in the same way in which Khrutchyonikh constructed his "transmental poems" (zaum) or Schwitters structured his phonematic "Ursonate." Striking examples of such technosyntactic autonomy are to be found in the unsurpassed legacy from the authentic daredevil days of the cinematic avant-garde (Viking Eggelung, Hans Richter, Fernand Leger). But are these apparently "wild" filmemes truly meaningless as their makers had declared them? If semiological analysis has taught us something, it is not to confuse declared subjective intentions with unpremeditated objective occurrences. Seen as the consistent manifestation of an albeit private set of signs, the obsessive adherence to riotous filmemes is far from meaningless. On the contrary, it reveals a subliminal but socio-logically not indefensible rejection of the Cartesian cogito and its corollary —the idolatry of rational communication. If this be so, this filmeme-fixation of the early avant-garde and its illegitimate progeny does not invalidate, but confirms, a basic linguistic law: when "en un certain ordre arranges" (shade of Maurice Denis), *meaning-free filmemes constitute identifiable meaningful entities termed morphemes*. All syntactic structures are subject to precise generative structures.

In the cinema, the birth of meaning is a triumphant by-product of technology. Considered as a source of conceptual connotation (signifie—in Saussure's terminology, signatum—in Jakobson's, denotatum—in Morrice's), the morpheme is inescapably conditioned by its size within the photogram. But while the empirical dimensions of the photogrammic rectangle remain constant[17] the size of the morphemes framed by it can vary from Long through Medium to Close Shots taken from all angles optically achievable within a span of 360 degrees. Such variety is the outcome of privileged factual distances between camera and morpheme and/or appropriate lens specifications. Discursive meanings are generated by technological procedures: *the viewpoint semantizes*. Morphemes are not restitutive reflections of ambient reality. They are constitutive elements of a *technosemantic co-reality*.

Eisenstein was one of the first to sense and systematize the full logical implications of this technosemantic co-reality.[18] To demonstrate the generative power of sizes and angulations he chose to de-construct the portrait of the actress Maria Yermolova by V. Serov now hanging in the Tretyakovskaya Gallery in Moscow—a trail-blazing procedure which still remains unsurpassed in its semilogical perspicacity and calls for a closer scrutiny. Within this misleadingly uncomplicated picture (Fig.9a) Eisenstein isolates, as it were, three potential photograms (Fig.9b, c. and d). The full size portrait in its totality (a) is clearly isomorphic to a *High Angle Long Shot*. Yermolova poses at a slight angle in front of a large wall mirror which reflects the ceiling and the opposite wall. A large portion of the floor is also included in the picture. But this detailed inventory of the visible reality undergoes a notable change if Yermolova's towering figure is cut knee length (b). We are now faced with the analogue of a *Level Medium Long Shot* from which the floor has been eliminated. Because of this exclusion the actress appears to be standing parallel to the back wall dominated by the newly-acquired authority of the mirror. A further, and this time truly dramatic, change occurs when Yermolova's figure is cut at the waist (c). All spatial denotations of (a) are radically subverted in this model of a *Semi-Low Angle Medium Shot Mirror* and flat backwall have vanished and a new spatial depth has appeared by virtue of the deliberate ambiguity between mirror image and real localization (to be achieved in the cinema exclusively by technological procedures: change of lens and angulation). Yet another such change of size and angulation would be responsible for a final analogue of a *Low Angle Close Up (d)*. The accent now falls only on Yermolova's high lit facial expression enhanced by the total neutrality of the background. In this way, perhaps not deliberately and certainly not explicitly, Eisenstein corroborates Victor Shklovski's hypothetical "quantifi-

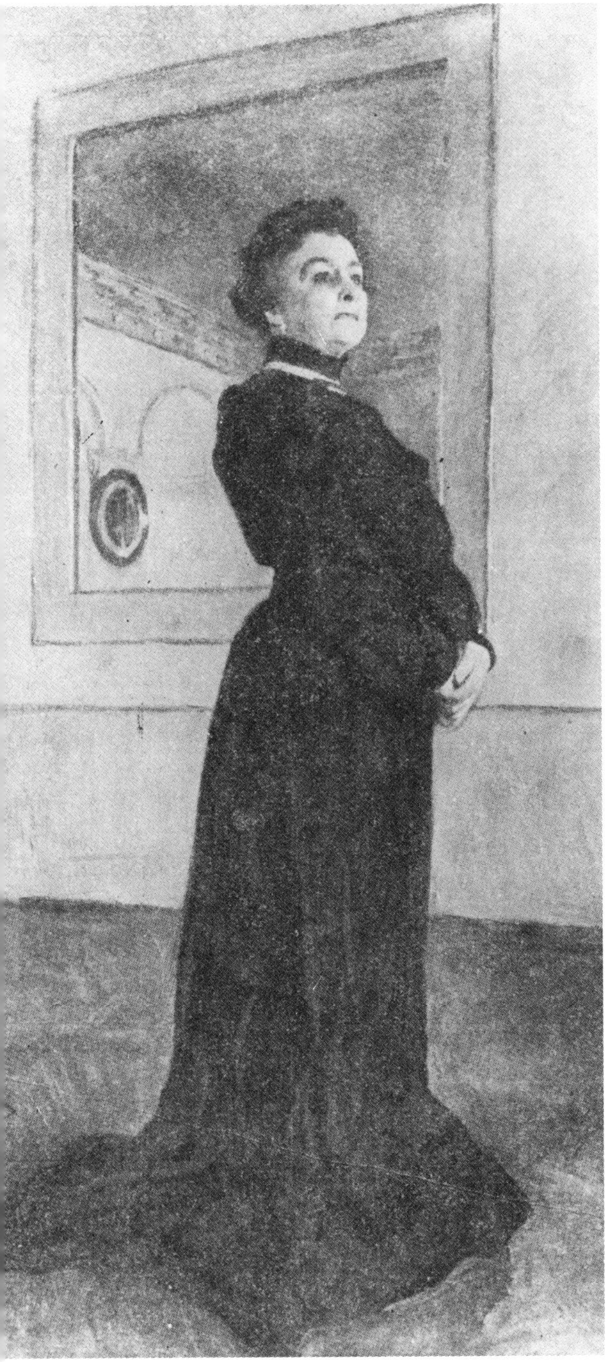

Fig. **3-9A:** V. Serov: Portrait of Maria Yermolova.

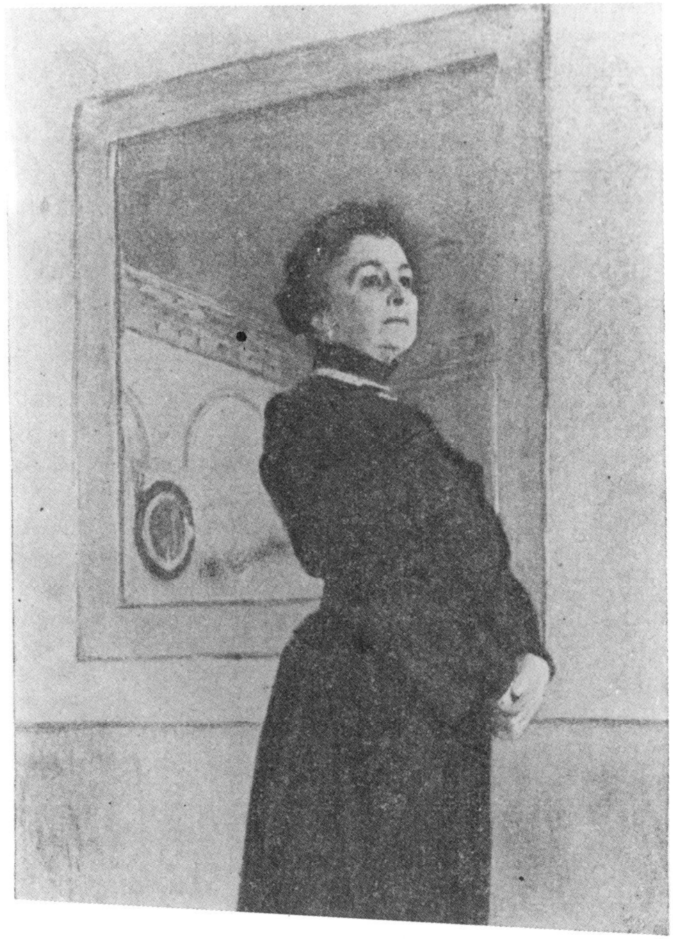

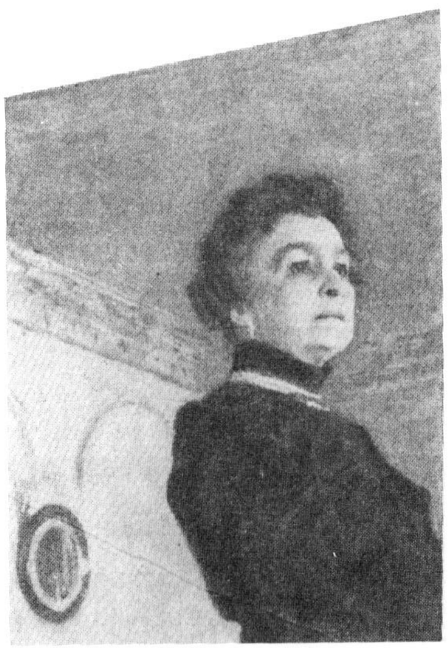

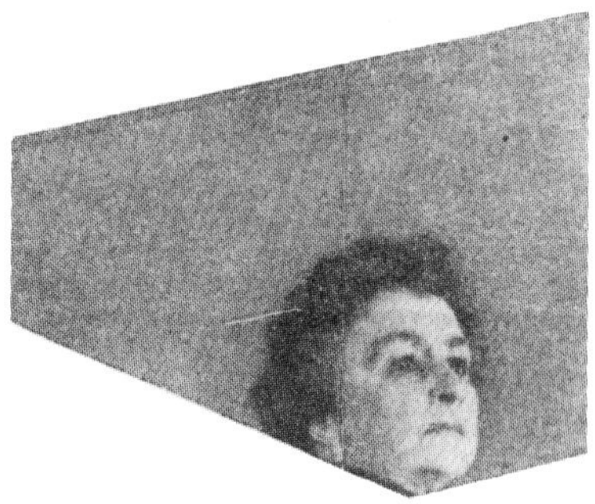

Fig. 3-9B, C, D: Three photographs isolated by Eisenstein from the Portrait of Maria Yermolova.

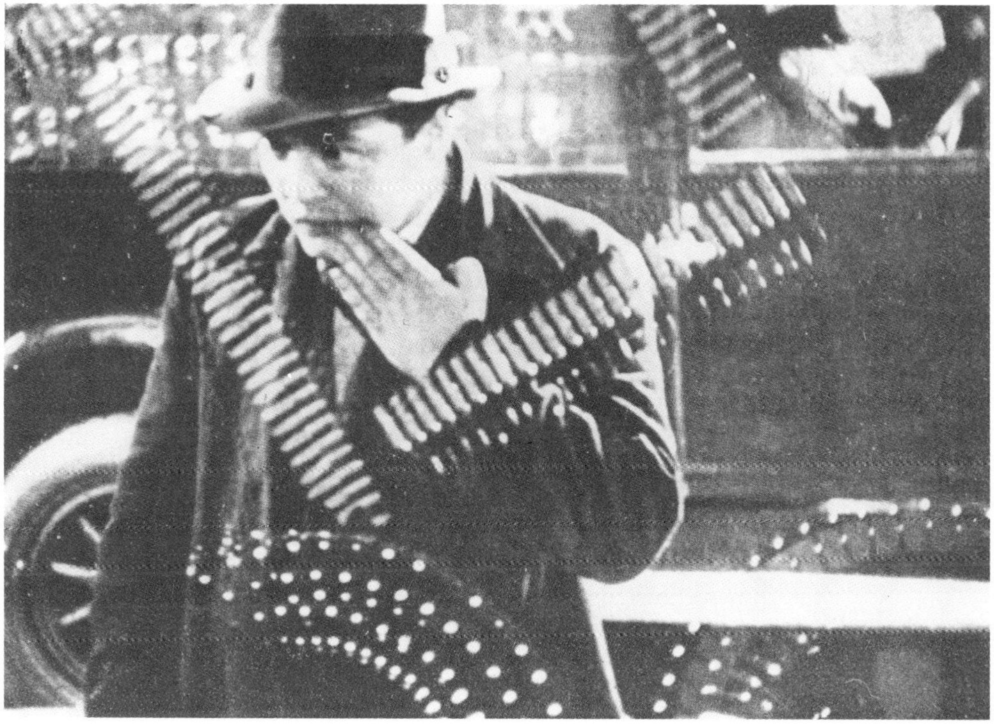

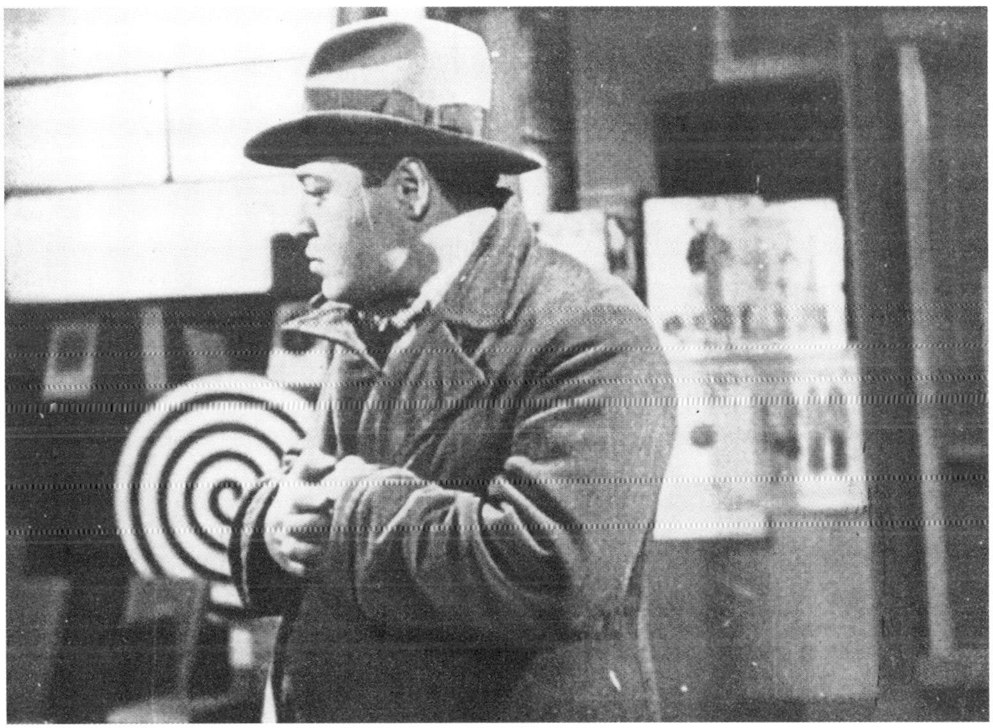

Fig. 3-10A, B: Peter Lorre in Fritz Lang's "M" (1931).

able meanings" (smislovie velichini) in the cinema offering, "formal geometrical solutions not unlike those current in verbal metrics."[19] Finally, Eisenstein concludes his mental deconstruction of Serov's portrait by punctually relating each one of the four imaginary shots to different facets of Yermolova's complex personality described by K.S. Stanislavski in his biography *MY LIFE IN ART*. And, in so doing, he retraces with admirable precision the constitutive potential of technosyntactic entities and their inevitable transmutation in technosemantic morphemes.

It is only fair to point out that Eisenstein uses a somewhat different terminology which, at times, does not do justice to his piercing intuition. But the occasional lexical vacillations do not prevent him from reaching the essential conclusion that, in the cinema, technosyntactic structures and their semantic transformations are *not restitutive, but constitutive.* Because, unlike most pontificating theoreticians, he is also a master practicioner; he knows that this intentionally constituted co-reality of the film is inconceivable outside the structuring code of technology. This code determines the mode of observation and the mode of observation determines the co-reality of the observed.

These self-propulsions rely on the meaningful manner of assembling increasingly more complex technosyntactic structures which manifest their constitutive energheia in each successive transition from morphemes to film phrases, and from film phrases to film sentences. No communicable film discourse would be possible without such an aim-oriented widening of the anulus technologicus. And this widening must be capable of accommodating a coherent, whole, successive chain of visual predications. Hence, the structural function of the specific film script. But to fulfill it without excluding the vivifying presence of the human factor from the actual making of the film, this verbal project must not be mistaken for a cast iron blueprint. This time, in Eisenstein's own words, "film scripts are codes," to be deciphered through a judicious concatenation of film phrases and sentences.

The 1930 script for the classic, "M," written by Thea von Harbov and Fritz Lang, states dispassionately, "The murderer shivers: he has seen in the window pane the reflection of a little girl. As if hypnotized he looks at her moving away." In the film (Fig.10 A & B) Lang first shows Peter Lorre through the shop window: the morbid shock of the psychopath reverberates in the glitter of a multitude of knives displayed in the window. For the second shot the camera is on the pavement outside the shop. The script instructs the child-killer "to stare as if hypnotized." A less sensitive director would have gone straight for a *Big Close Up* of Lorre's famous frantic eyes. But Lang resists the obvious melodramatic temptation. He "decodes" the script's instruction by framing the actor in a Medium Shot, and in profile, thus shifting the center of expressive gravity from the eyes to the hands clasped in the invincible urge to kill. A problematic state of affairs has been neatly predicated in a perfectly balanced film phrase.

In "Alexander Nevsky" (1938) Eisenstein is confronted with a situation which seems tailor-made for spectacular blockbusters. With growing tension, Alexander and his improvised army of peasants and youngsters await the attack of the inhumanly disciplined Teutonic knights advancing across the frozen Ladoga Lake. Yet there is nothing trivial in the long film sentence with which he chooses to convey this intricate succession of phraseological contrasts: leader/army, Russians/Teutons, young/old, ice/clouds . . . (Fig.11). To blend this lexical variety into a thematic unity, Eisenstein relies on a semiotic de-construction of the global sentential context and achieves it through the considerate recourse to different lenses and angulations, privileged determiners of sizes and shapes. Each change in such technosyntactic procedures is a clear cut *visual phrase marker*, as, for instance, the transition from *Low Angle Long Shots* (Alexander) to *High Angle Medium Shots* and *Close Ups* (army). Furthermore, this visual patterning is reinforced by the variable duration of each shot echoed in strict synchrony by the structural features of Prokofiev's score (Fig.11, 12 A & B and Note 20). Musical phrases depend on visual phrase markers, which depend on technosyntactic structures, which are vehicles of technosemantic statements. From this disparate but coordinated assemblage emerges the global meaning of the film embodied in a newly constituted co-reality. Knighted by technology, *the Reel is now peer of the Real*. Perhaps even his liege-lord. And thereby hangs a quarrelsome tale.

Editing: Chord, Accord, Discord

We have seen it happen before our very eyes: the blossoming of technosyntax into stylistic semantics has reached its climax in the *purposeful manner of assembling* and integrating all filmemic and morphemic constituents of the film object. Usually this cohesion-generating procedure is denoted by pragmatically bi-

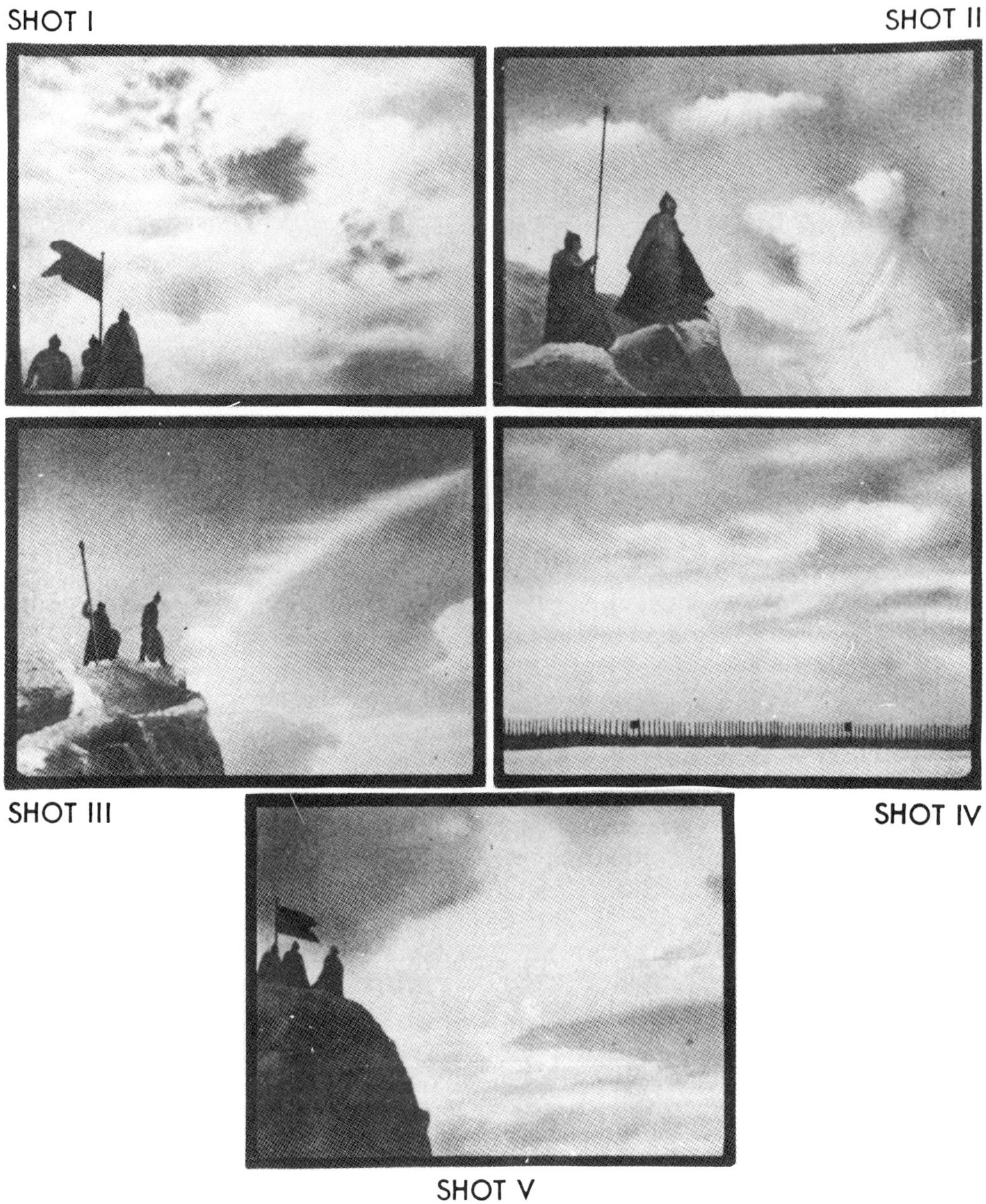

Fig. 3-11: Battle scene in Eisenstein's "Alexander Nevsky" (1938). 12 shots edited together to make the scene.

SHOT VI

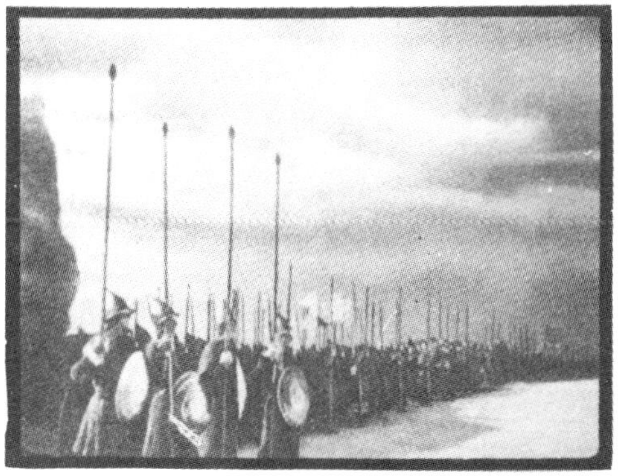

SHOT VII

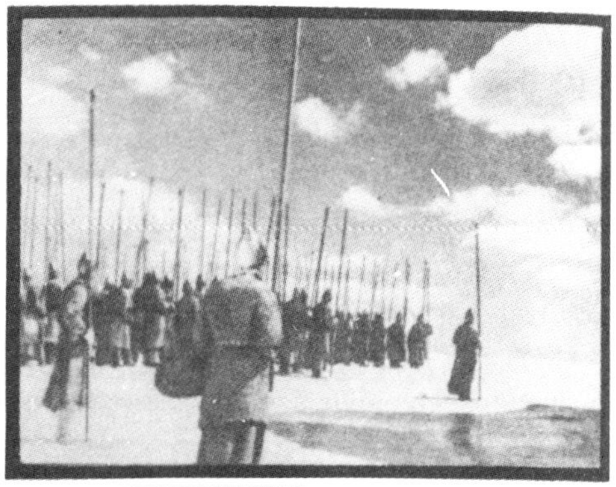

SHOT VIII

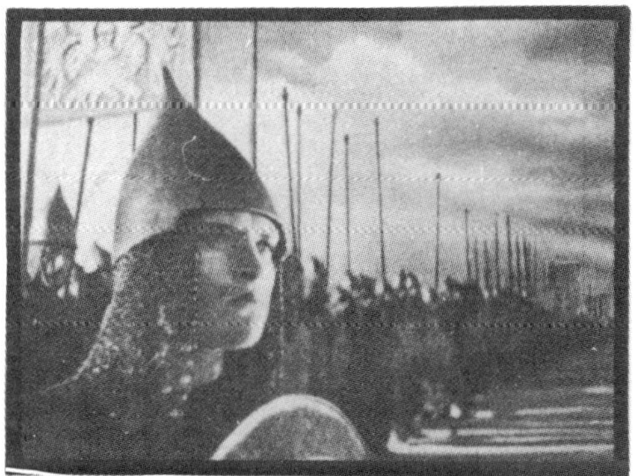

SHOT IX

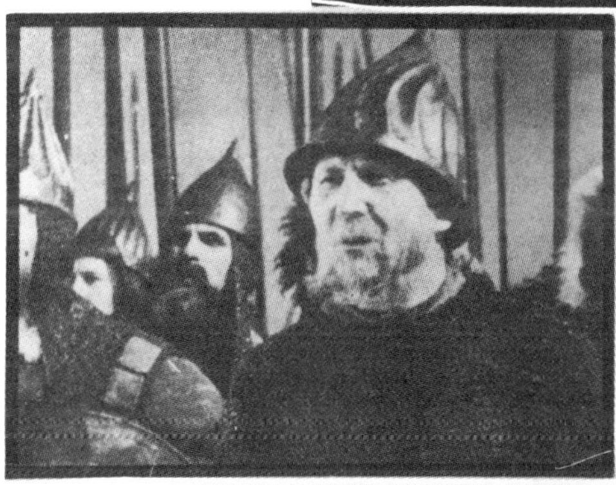

SHOT X

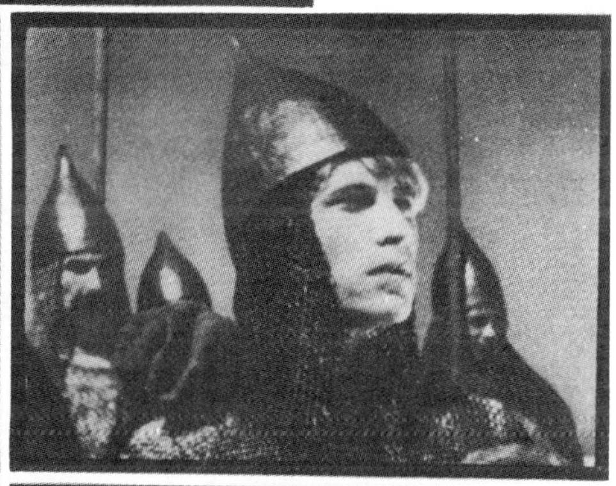

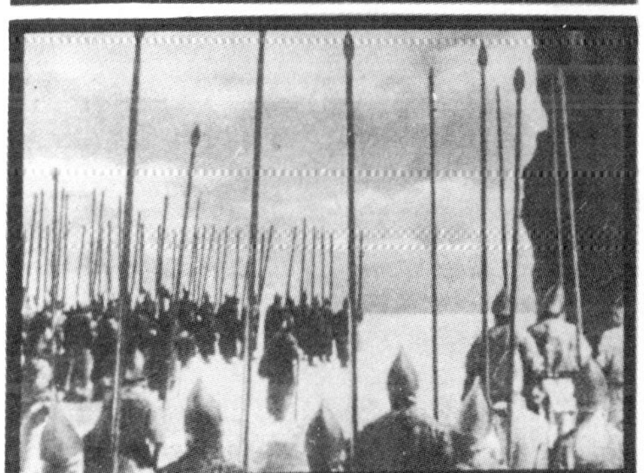

SHOT XI

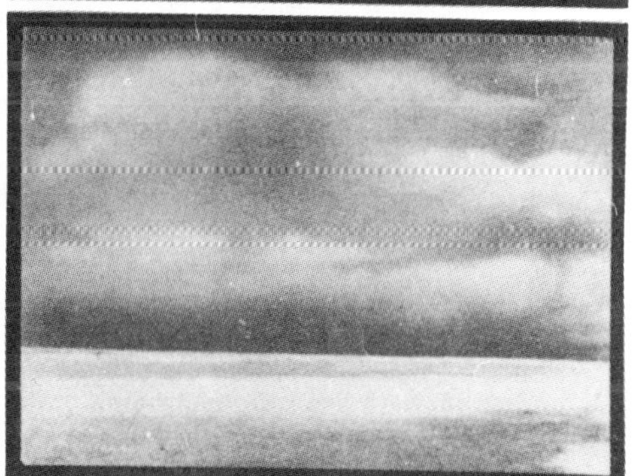

SHOT XII

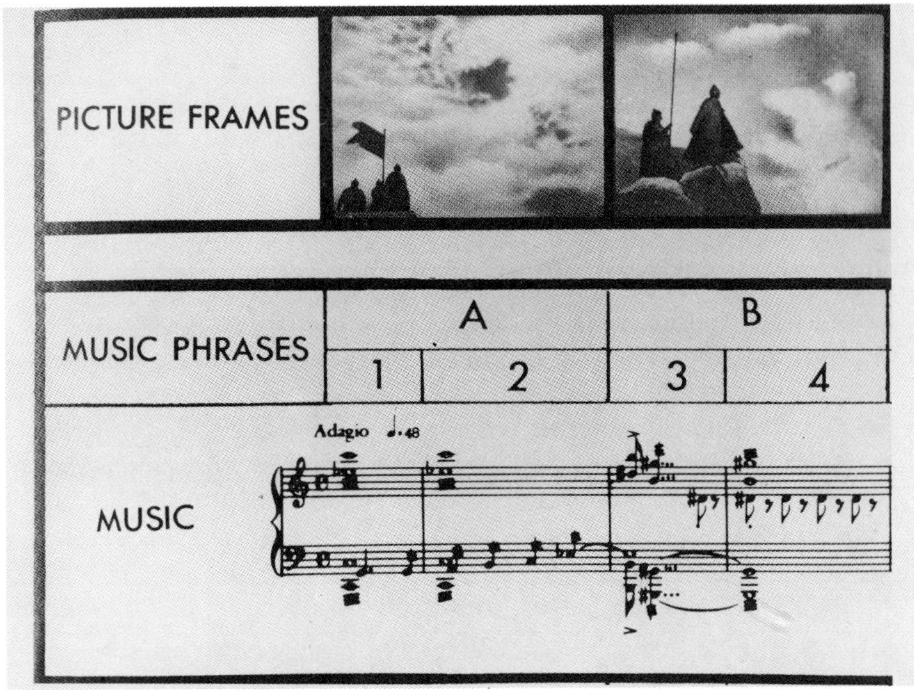

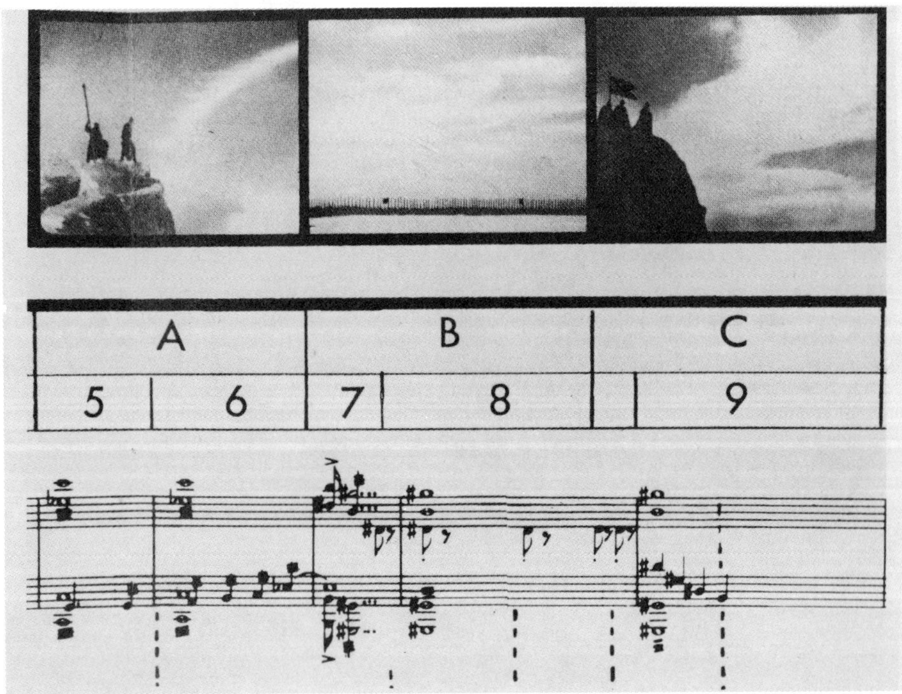

Fig. 3-12A, B: Music score and film frames from the "Alexander Nevsky" battle scene.

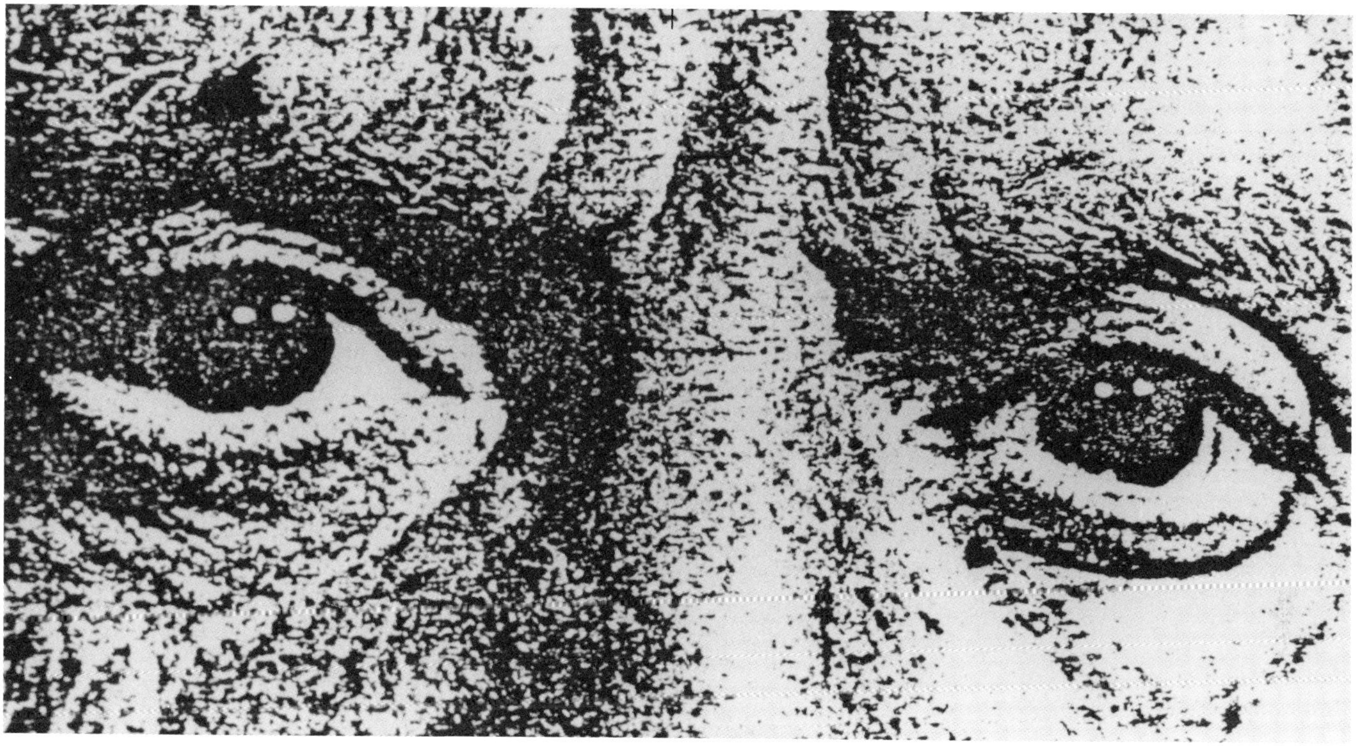

Fig. 3-13: Picasso in H. G. Clouzot's "Le Mystere Picasso" (1956).

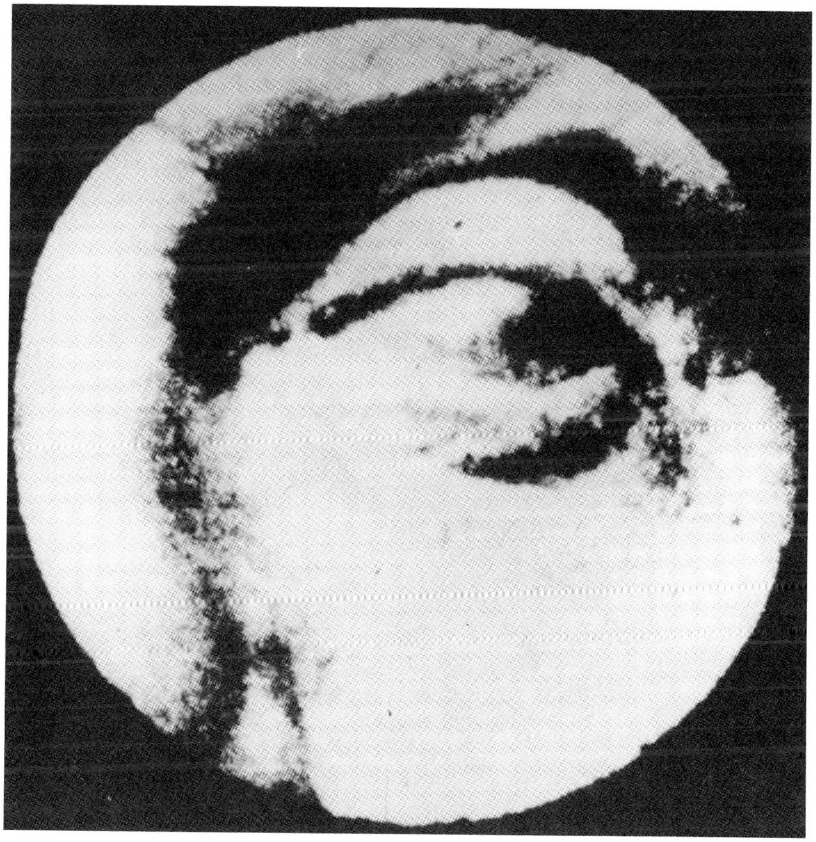

Fig. 3-14: Close up from G. A. Smith's "Grandma's Reading Glass" (1908).

ased descriptions. The elegant Gallic extraction of "montage" does not altogether disguise its assembly line ancestry. The blunt Anglo-Saxon "cutting" points at a mindless mechanical activity redolent of Chaplin's "Modern Times." The authoritative "editing" has an air of self-complacent superiority. A semiologically more accurate definition has been proposed a decade or so ago: *cinegraphy—leaving a trace through motion.*[21] But it has been largely ignored: bad terminological habits die hard. Reluctantly, this inquiry must bow to the power of this micro-Tower of Babel and use interchangeably these various appellations. But such concession to unhampered communication does not alter in the least the steady semiological course of this investigation. Nor does it impede us to realize that all through this investigation we have been not unlike Moliere's character who was startled to learn that all his life, unbeknownst to him, he had been speaking in prose. Without ever mentioning it by name, this inquiry has been relentlessly concerned with what cinegraphy (alias montage, editing or cutting) is and does. The time has come to confess it.

Cinegraphic procedures are supreme manifestations of semiological unity in technological diversity. Each cinegraphic segment (editorial sequence, Montazhni Kusok) is a chronotopic (spatiotemporal) chord conceived in accord with the global structure of the film. Editing is the efficient cause of the purposeful deconstruction of a constant initial empirical reality and its reconstruction in the guise of a variably compounded co-reality. In it, each discrete empirical item becomes a technological phoenix available for annihilation and regeneration through a variety of deliberate permutations in size, shape and duration. These regenerative modalities map the differences which oppose passive reproduction to active productivity, in the Kantian sense of the term. Because of them, the film is not, as Dante has said (naturally in a non-cinematic context), "di natura buona scimmia"—a docile ape of nature. On the contrary, as Klee would have it, the film object "gibt nicht das Sichtbare wieder, sondern macht Sichtbar"—it does not oppose the visible, instead it makes visible.

But what is thus made visible? And does not such a technological epiphany conjure a host of philosophical resonances seldom associated with the lowly reaches of the menial techne? As an intriguing technological co-reality emerges through select editorial strategies, cardinal metaphysical problems loom large on this technological horizon. We have seen that the relation between

Unity and Multiplicity is one of them. By its very essence, editing postulates the isomorphic equivalence between punctual global film morphemes, phrases or sentences and the theoretically infinite possibility of partial lesser entities systematically subsumed in hierarchically higher structural categories. This editorial assumption opens the discussion on the relations between quantitative constants and qualitative variables in the constitutive interplay between part and whole. And, in so doing, brings to the forefront, the epistemological two-way traffic between Identity and Diversity. It may be unusual but it is hardly unreasonable to ask whether cinegraphic technology corroborates Eddington's conviction that all particularized "ordinary objects" have a correlative scientific "double." But if this Doppelganger theory is brought to its logical extreme, must we not accept Pirandello's existential equation of One to No one or Hundred thousand, of rock firm Being to fluid Becoming? In the imaginary projection room of philosophy where Zeno's tortoise is racing Achilles, Heraclitus and Parmenides are fighting for the seat of honor.

Or can it be that such editorial mutations revive Leibnitz's *principium identitatis indiscernibilium*, that there could never be two identical objects? The entire structure of the film object would be then seen in quite a different light and connotations will take the upper hand over denotations. So, if Long and Close Shots are to be equated because they have the same referent, and yet have unmistakeably different connotative resonances, may we assume that their difference is of a *rhetorical* nature?[22] In such case, the film object would attain its persuasive co-reality through an unobtrusive but incessant generation of technologically produced "*figures of sight*" semiotically symmetrical to literary figures of speech. The Big Close Up of Picasso's eyes in Clouzot's film of 1956 is an eloquent synecdoche (Fig.13), while G. A. Smith's purely denotative Big Close Up of mouth in "Grandma's Reading Glass" of 1908, clearly is not (Fig.14). The High Angle Long Shot of Monica Vitti in Michelangelo Antonioni's "L'Avventura" of 1960 connotes metaphorically the girl's aloofness (Fig.15), while a similar High Angle Long Shot in Laurence Olivier's "Hamlet" of 1942 merely denotes—and rather clumsily at that—the interior (!) of Elsinore Castle (Fig.16). And the "Odessa steps" in Eisenstein's "Battleship Potemkin" of 1925, probably the most quoted sequence ever filmed, remains to this day the most limpid demonstration of how the semiotic fragmentation of an initial informa-

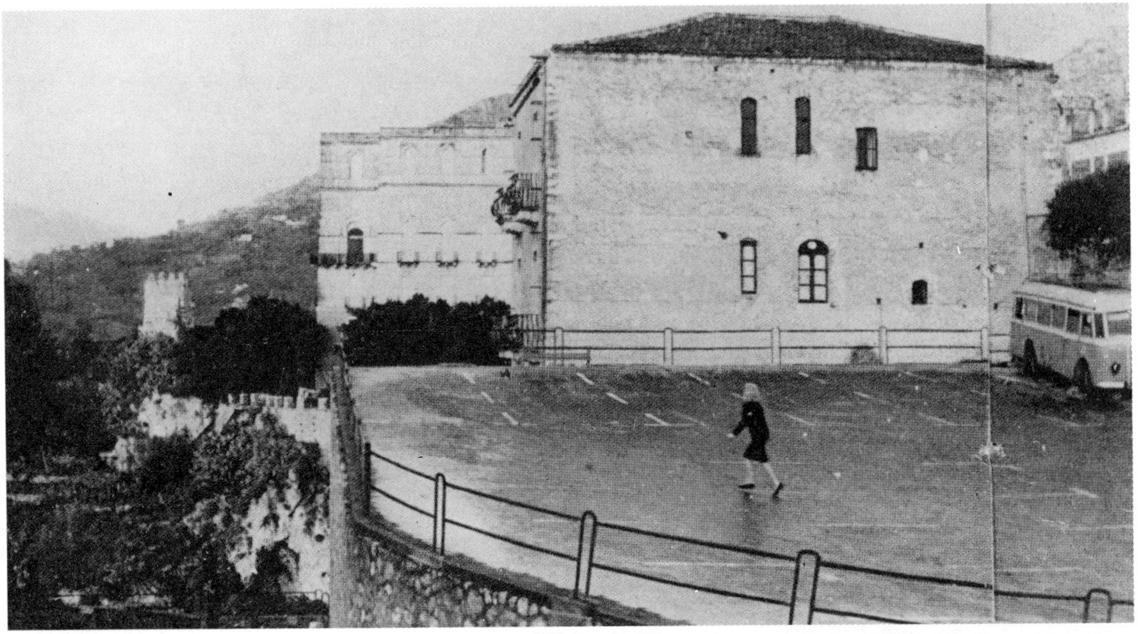

Fig. 3-15: Long shot from Antonioni's "L'Avventura" (1960).

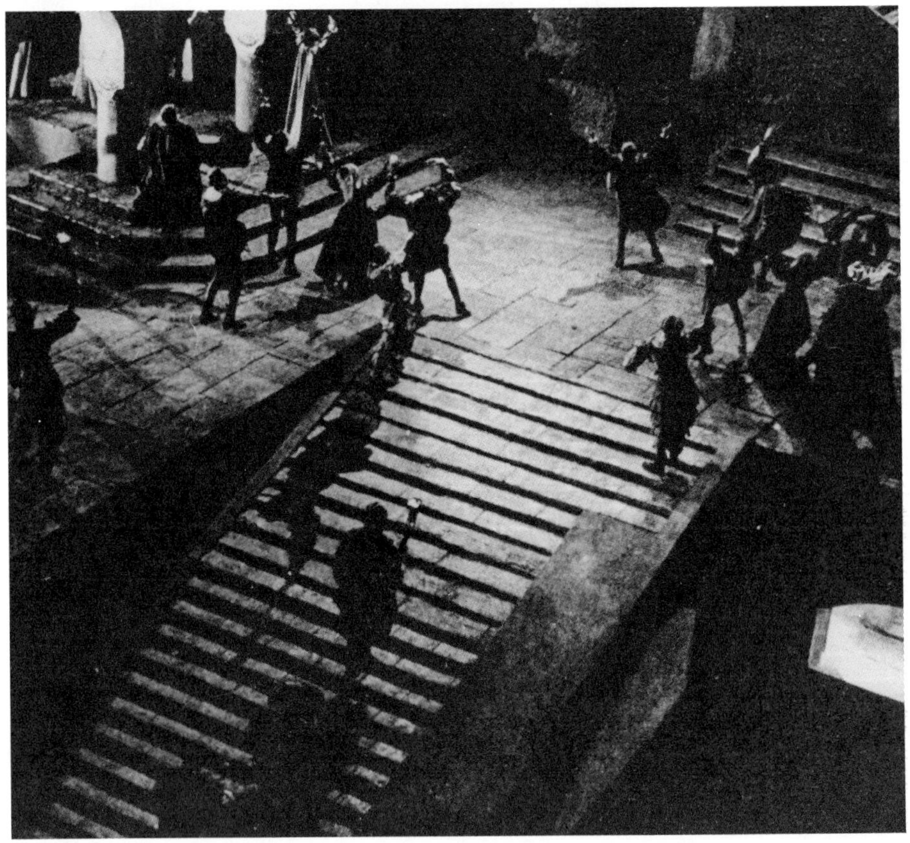

Fig. 3-16: Long shot from Laurence Olivier's "Hamlet" (1942).

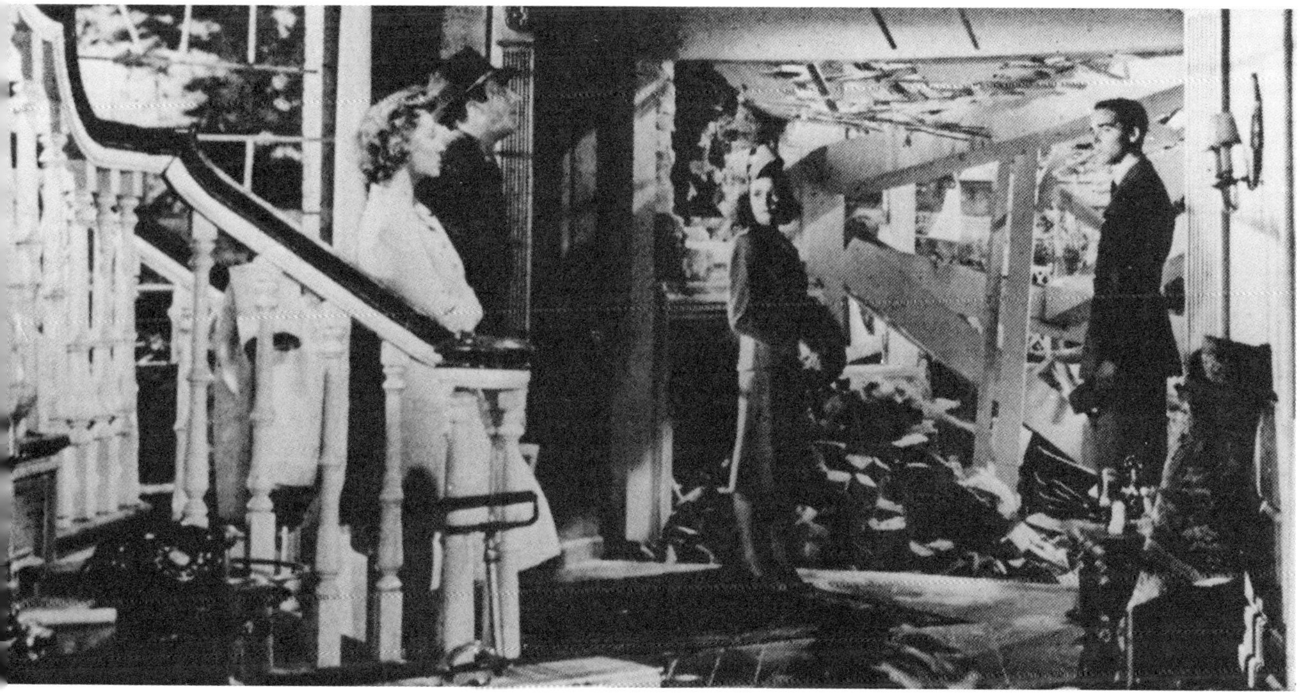

Fig. 3-17: Greer Garson in William Wyler's "Mrs. Miniver" (1942).

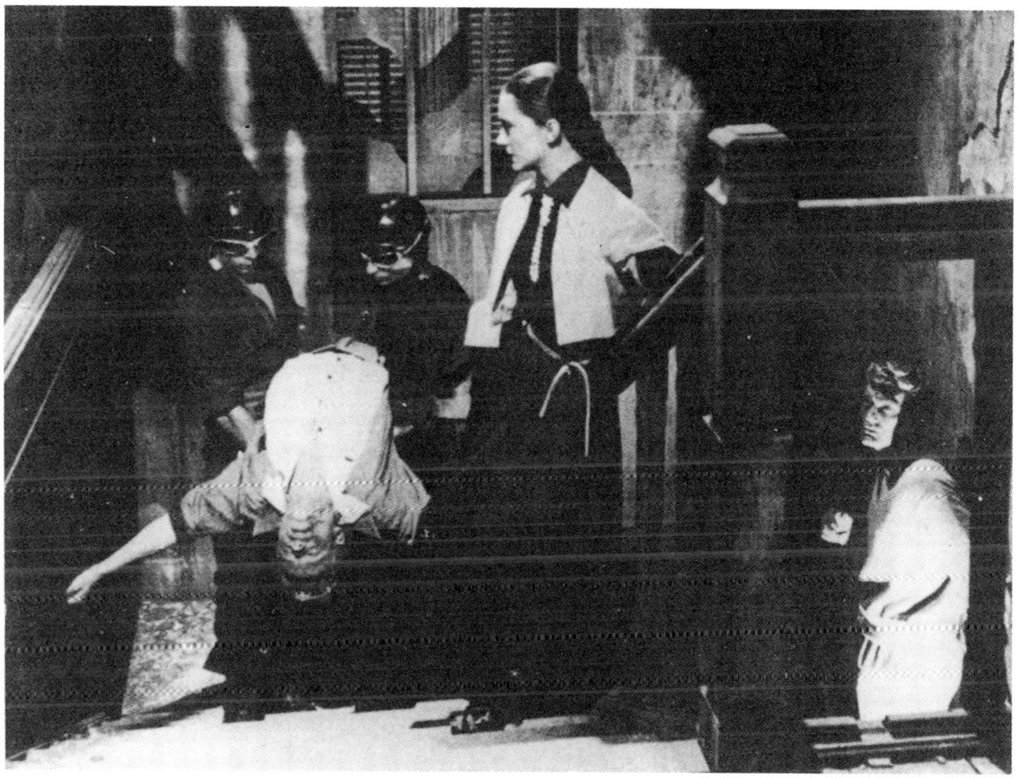

Fig. 3-18: Maria Casares and Jean Marais in Cocteau's "Orphee" (1950).

tive Long Shot can generate a host of metaphorical Close Ups: "each cinegraphic segment (Montazhny Kusok) does not exist as something unrelated to any other entity (bezotnossitelnoye). It is a discrete particularized image of the overall theme with which all such segments are imbued. The juxtaposition of such particularized details within a well-defined montage structure brings to life a generality born out of all particulars and intertwines them in a coherent totality."[23] Eisenstein also envisages the feedback rebound of such emergent totality which retroactively affects both the contiguous cinegraphic fragments and the structural motivation of such contiguity. This leads him to the eminently rhetorical conclusion that in cases of such pertinent contiguities the ultimate totality would emerge as "*a third entity.*"

Unfortunately, he does not analyze in depth this notion of a "third entity" brought about by contiguity. Nonetheless, his intuition is cardinal and it leads the way to a vast new field of visual rhetorics which is the preserve of cinematic co-reality. If semiological fragmentations produce metaphors coralled within the frame of discrete photograms, apposite contiguity of related shots generates by virtue of their editorial vicinity a "third" meaning outside and beyond the frame of the contiguous photograms: a striking technological *metonymy* specific to motion pictures. Such is the seminal Lev Kuleshov effect dating from the early twenties: the same facial expression of the actor Mosjukine was "read" as lust, horror and hunger because of its successive juxtaposition to shots of a scantily dressed girl, a decaying corpse and a steaming soup plate. The pragmatist declaration that we do not believe what we see but we see what we believe is thus placed on firm structural ground. Submitted to strict, semiological scrutiny, each panel of the metonymic dyptich reveals the presence of *semiologically rhyming marked elements*. The semiotic generative power of cinematic metonymies depends on this symmetry of such marked elements in the juxtaposed photograms. Only then the generated (meaning) transcends the twin generators (photograms). In Kuleshov's experiment this marked element is the precisely calculated complementary continuity between the eye line of the actor's gaze in photogram A and the angle from which the alleged target of this gaze has been shot in photogram B. Thus a punctual filmic procedure generates three distinct meanings. The result is far reaching and it proves the crucial relevance of editing for the specificity and creativity of cinematic language. Empirically non-existent occurrences have acquired a reality of their own—a technological co-reality much more expressive than the unstructured chaos of our existential environment. Technological factors act as logical revelation. *And the Reel conclusively asserts its right of precedence over the Real.*

But, it is precisely this right that has been strongly challenged in recent years.[24] It has been asserted that editorial procedures are no more than clumsy counterfeiting techniques which only pervert reality. This reality was said to be perfectly self-evident within the static frame of each single photogram. Any attempt to "add something to the actually visible things and events" was branded as an ethical transgression. Images and lies were proclaimed synonymous. And far and wide resounded the battle cry, "Down with the Image. Back to Reality." This return journey to an honest-to-camera truth entailed a categorical refusal to "quarter reality." In practical terms, this intransigent stand limited cinematic photography to a single viewpoint, and consequently to single, long-lasting, long shots—the so-called "one-shot sequences." And because it was axiomatically postulated that reality "always confesses its meaning" without any need of external coercion, it was prescribed to use exclusively such privileged one-shot sequences and flat, indifferentiated lighting. Only then would films show the truth, the whole truth and nothing but the truth.

In fact, this ethical revivalism was brought about by two technological improvements: panchromatic stock and deep focus lenses capable of recording with equal clarity large shapes in the foregrounds and tiny details in the distant background. Documentarists saw immediately the advantages of these new techniques. In feature films, too, some directors were anxious not to "add anything" to the sights which raw reality had to offer. They quickly seized the possibility to prescribe to their actors lengthy movements backwards and forwards within static one-shot sequences. The outcome was something akin to a crowded mini-stage shot from an immutable panoptic viewpoint (Fig. 17). That did not seem to worry the sworn enemies of "montage monstrosities." They went on proclaiming that the end of editing has brought about the redemption of "reality such as it is."

But is it really so? Could it not be that precisely the much-maligned cinegraphic "vivisection" of reality and the "cosmetic partial shots" provide us ultimately with a much deeper insight into what reality is and means? The entire course of this inquiry points in that direction. And unless it has remained a feeble voice clamoring in the desert of fashion, it should be clear that such reductive theories are utterly untenable. What is more, they disguise as an excess of ethical rectitude what is in fact a default in semiological sensitivity. Their vituperations do not go any further than the recriminations of someone who would be foolish enough to reject Shakespeare's linguistic subtlety because it differs from the verbal poverty of illiterate vagrants. With an unsavory blend of arrogance and ignorance, their scorn for "Images" is nothing more than the illiterate's fear of language heightened by metrical modulations and rhetorical amplifications which "add something" to the speech actually heard in the street. Terrified by such "additions" which they fail to fathom, the opponents of visual strophes and tropes, called "montage," proclaim the dubious virtue of the total lack of articulation. But to restrict cinematic syntax to the sole "one-shot sequence" is no less absurd than to reduce the alphabet to a single letter and discursive morphology—to one type of predication. And, worst of all, it is naive in the extreme to confuse reality with appearance in the belief that apparent "reality always confesses its meaning." Reality never does, unless it is asked the proper questions. The only way to de-mystify appearances is to de-construct them so that their concealed essence is compelled to surface in an epistemologically parallel co-reality—be it literary, scientific or cinematic. One of my productions, "Blow Up," perceptively directed by Antonioni, makes precisely this point. Editorial procedures ("vivisection" of an indifferentiated Long Shot into meaningful Close Ups) chart the quest for occulted truth. And even then, ultimately, reality refuses "to confess."

The "one-shot sequence" theory has obviously little to recommend it. But there is some method in its madness. All its aberrations stem from the initial misunderstanding of the minimal constitutive unit of film objects. If it is posited that this minimal unit is the static single photogram, it does logically follow that the only outcome of such one-shot units would be the one-shot sequence. It also follows that there is no substantial difference between static photography and dynamic cinematography. And, because the camera is motionless, the actors have to move: when the mountain did not go to Mohammed, Mohammed went to it . . . But because the actors have to deambulate within a permanent frame similar to the proscenium arch, the difference between screen and stage is abolished. This is the final blow which the ill-advised "one-shot sequence" delivers to the cinema, particularly to the social resonance of the film event. The essence of authentic editorial segmentation is not "cosmetic," it is logical and semiological. The judicious distribution of Medium and Close Shots lends an additional intelligence to the spectator: *all sizes are semiological markers*. Even in complex Long Shots, perceptive directors find ways of stressing their relevant components: spotted lighting is one of them (Fig. 18).

More evidence of such selective "in frame" clarifications is to be found in a rank and file thriller which I scripted and directed some years ago.[25] The task at hand was to pinpoint certain memories haunting a drunken recluse, admirably portrayed by James Mason. One simple option would have been to shoot the inevitable flashbacks in the recluse's house cluttered by decaying Edwardian bric-a-brac. But the significant objects and events evoking the reasons for his retreat from life would have been lost in that confusion. Furthermore, it seemed worthwhile attempting to show on the screen a purely mental state without a tangible counterpart: the recall of such obsessive objects and events. This was achieved by doubling the established naturalistic "reality" of set and props with a technological co-reality, based on the premise that retention implies oblivion: we remember some things because we forget many others. To signify this state of affairs the entire set was repainted white. In this neutralized interior were left only those objects which had a clear-cut bearing on the character's obsession. These select semiological markers were all dyed black for the occasion and were photographed with a special bifocal lens which subverted the monocular perspective and made it possible to alter their sizes according to their changing significance for the character played by Mason. Popular audiences from widely differing cultures all "saw the point" of this "extended" editing and the film was a box office success.

All this goes to show that there are semiologically conditioned irreplaceable safeguards against possible noises along the communication channel between film and audience. This should hardly surprise us. The spectator, however non-cinemate, is inobtrusively guided by such carefully considered procedures towards the visually meaningful moments of the filmic narrative, much in the same way in which ordinary speakers lay a heavier stress on relevant words in their utterances. It simply is not true that listeners or spectators feel offended by such logical clarification. On the contrary. Because of this support, audiences all over the world are much more relaxed in the cinema than in the theatre: there they are sure "not to miss the point." It is inevitable that it should be so: reality truly "confesses" its depth and span through the technological co-reality of the film.

Notes and References

1. F.H. Bradley, *Appearance and Reality*, London, 1893.
   J.G. Fichte, *Grundlage der Gesamtenwissenschafstlehre*, Jena.
   N. Hartmann, *Moglichkeit und Wirklichkeit*, Berlin, 1938.
   M. Heidegger, *Sein und Zeit*, Halle, 1927.
   C.L. Musatti, *Analisi del concetto di realta empirica*, Castello, 1926.
   N. Schlick, *Les énoncés scientifiques et la réalite du monde extérieur*, Paris, 1934.
   P. Sonderguer, *Realidad inteligible y realidad pura*. Buenos Aires, 1948.
2. Boex-Borel, *Le Pluralisme*, Paris, 1909.
   R. Carnap, "Empiricism, Semantics and Ontology," in *Revue Internationale de Philosophie*, 1950.
   W. James, *A Pluralistic Universe*, London, 1909.
   R.H. Lotze, *Metaphysik*, Leipzig, 1841.
   W.O. Quine, *Ontological Relativity*, Cambridge, 1969.
   D. Spassov, *Edinstvo i mnogobrasie*. Sofia, 1977.
3. P. Carus, "Ostwald's Philosophy" in *The Monist*, 1907.
   A. Kozharov, *Monism i pluralism*, Sofia, 1972.
   F.C.S. Schiller, "Lotze's Monism" in *Philosophical Review*, 1896.
   Chr. Wolf, *Philosophia Prima*, Frankfurt, 1729.
4. L.A. Bowden, *The Oxford Companion to Film*, London, 1976.
5. Y. Lotman, *Semiotika Kino i Problemi Kinoestetike*, Tallin, 1973.
   J. Monaco, *How to Read a Film*, New York, 1977.
   P. Wollen, *Signs and Meanings in the Cinema*, London 1969.
6. J. Dewey, *Psychology as Philosophical Method*, 1969, p. 148.
7. G. Bettetini, *Cinema: lingua e scrittura*, Milano, 1968.
8. Lotman, *Op. cit.*
   Chr. Metz, *Essai sur la signification au Cinema*, Paris, 1968.
9. D. Vallier, "Malevitch et le modele linguistique en peinture," in *Critique*, Paris, 1975.
10. P. Rouve, *From Opteme to Sight Sentence*, Helsinki, 1983.
11. P. Rouve, "Aesthetics of the Cinema," in *British Journal of Aesthetics*, London, 1972.
12. B. Eichenbaum, *Poetica kino*, Moscow, 1927.

13. B. Balasz, *Der Film*, Vienna, 1955.
14. A. Tarkovsky, "Zapechatlionnoye Vleme," in *Voprosi Kino Isskustva* 10, Moscow, 1967.
15. By mimetic time-presentation we refer to the flow through the gate at the standard twenty-four frames per second necessary to reproduce motion at normal speed.
16. G.H. von Wright, *An Essay in Modal Logic*, Amsterdam, 1951.
    J. Hintikka, *Knowledge and Belief*, Cornell, 1967.
17. We have in mind the constance within a chosen aspect–ratio class: Academy aperture (1.33:1), American standard widescreen (1.85:1), European standard widescreen (1.66:1) etc.
18. S.M. Eisenstein, *Manuscript of 1937* (?) published in *Voprosi Kino Isskustva* 7, Moscow, 1963.
19. V. Shklovsky, "Osnovnye zakoni kinokadra," in *Ikh nastoyashchee*, Moscow, 1927.
20. S.M. Eisenstein, "PKVF," in *Izbrannye Statii*, Moscow 1956.
21. P. Rouve, *Aesthetics of the Cinema*, cit.
22. P. Rouve, *Retorica de los hechos de vista*—Course taught at Universitad National Autonoma in 1981/82. Unpublished.
23. S.M. Eisenstein, "Montage 1938," in *Izbrannye Statii*, cit.
24. A. Bazin, *Qu'est-ce le cinema? Ontologie et langage*, Paris, 1958.
25. "Strangers in the House" (1968) with James Mason, Geraldine Chaplin, Bobby Darin, Ian Ogilvy. Written for the screen and directed by Pierre Rouve.

On Pierre Rouve's "Reel To Real: The Cinema As Technological Co-Reality"

Eddie Wolfram

Introduction

Once again, Pierre Rouve does not hesitate to swim against the current and convincingly bare the inconsistencies of the fashionable film theories prevailing, not only in marginal buff-colored publications, but also in many authoritative academic circles. His paper is yet another "structural scrutiny" bearing all the particular distinctive marks of his personal investigative method. This method has already been praised by the late Roman Jakobson, one of the founding fathers of modern linguistics and undoubtedly one of the most powerful analytical minds of our time. Jakobson stresses Rouve's fruitful recourse to "sign theory philosophy, psychology and cultural history," which he defines as "unprecedented," and considers it to be "full of creative stimuli for any interpreter of the arts, both visual and verbal," because "it gloriously opens new vistas for a genuine scientific approach to the cardinal problems of life and creation."

Rouve's perceptive originality (enriched in recent years by an enlightening use of formal logic) has also been most favorably commented on by a number of European scholars, including the eminent French anthropologist Claude Levi-Strauss, who considers Rouve's very singular diagnosis to be "a powerful intellectual spur"; while the leading Italian art historian and former Mayor of

Rome, Giulio Carlo Argan, examines it at some length in a study which defines Rouve's views as fundamental. This acclaim is echoed by Professor Gillo Dorfles of Milan University, a frequent guest lecturer in the U.S.A., who declares Rouve's views to be "epoch making." All of which leaves me only with the more specific task in this paper of seeking out some hypotheses for discussion in this area of uniquely "20th Century ways of seeing."

Mass Audiences and Film Syntax: Rouve's "A Touch of Taxonomy"

Clearly, Rouve is correct to assert that *films as objects* are hardly apprehended either by popular audiences or, for that matter, by most popular critics of the genre as such. Instead of being "read" as some kind of literary potpourri (possibly cooked with proscenium arch and novelittish ingredients), perhaps the way that cinematic grammar nevertheless works on audiences—communicates or mis-communicates—needs to be more closely scrutinized. Certain contemporary directorial and cinegraphic cliches are at least apprehended and accepted by present day generations, even if they are not aesthetically understood or appreciated. The conveyance of changes of time and place as used by every run-of-the-mill 30-second commercial maker today, for example, would probably have been totally bewildering to filmgoers even 25 years ago, yet today they are absorbed and accepted without thought as so much colloquial jargon.

This leads to a number of questions:

(a) Are we now into an age of "cinematic seeing," albeit spawned intuitively through media saturation?

(b) If so, not only how, but what do the children of the cinematic and video age see and understand, having been weaned—from the moment they first opened their eyes—on a daily diet of 8 hours of visually-transmitted information?

(c) Possibly, in relationship to this, we should consider the exploratory work of Max Bunge, Jaus Weinrich and Wolfgang Iser and the "Theory of Reception": what occurs in this multiplicity of transmission→noise←reception?

(d) Rouve's *filmeme* or not, is there any more specific meaning in the "objectness" of the frame-by-frame film reel, than the electronically bombarded video reel or in the "basic floppy disc"? Establishing distinctions or points of concert here might be worth investigating.

The Birth of Co-Reality

Although Paul Klee told us that "a line was a dot going for a walk," and Braque said decisively, "I do not paint things. I paint relations between things" (Rouve p.98), the thought occurs that the exceeding majority of comments towards understanding twentieth century Painting and Sculpture (I am purposely avoiding calling it Art History), conversely to Rouve's assertion about film (usually experienced incorrectly as event first, film as obscured object), has been conceived of as object first and foremost. Thus, until very recently, much entrepreneurial aesthetic discussion (ironically, in view of Rouve's paper) about such "art objects" can be seen to be no more than a closed dialectic endlessly revolving around "Art about Art," hardly ever even hinting towards an understanding of these chronically historically-laden monoliths as vehicles which are also possibly burdened with sociological, psychological and anthropological pertinencies.

Gyorgy Kepes's work (vis: "The man-made object," "The nature of art and motion") springs to mind as relevant to a discussion about how visual art objects are "seen" or "read." Is the *real* as recorded on the *reel* a linear time/space notion of experience as words in a novel? Or Klee's dot going for a walk? Rouve would suggest not. And, therefore, what can we "see" or do, I mean "understand," about "visual dynamism" as in a direct confrontation with the art object (painting) or, indeed, in the still frame? How useful is Rouve's concept of *co-reality* to a better comprehension of how static, physical, visual art is conceived as a fulcrum of form and contour, and how it communicates, or at least affects the audience *or spectator*?

With proven justification, much has been made of the visual influence of painting and sculpture on cinematic language. Most of these influences will be common knowledge, and I need not belabor the point. Interestingly, Rouve uses such cousin relationships with rapier precision himself (Rouve, p.99), but little has been written or clearly established about the influence of cinematic language (co-reality and filmeme) on static visual art. Erroneously, the advent of still photography is accorded by some to have had considerable effect on the evolution of twentieth century painting and sculpture, and pragmatically it cannot be denied that photography made many an artist, portrait, history and narrative redundant. Scientific questioning about the nature of light undoubtedly affected Chez Renoir's artistic vision, but the aesthetics of photography only echoed the aesthetics of painting in terms of compositional, perspectival and adjectival devices, etc., until the inception of the moving picture and the aesthetic language of cinema. It might be worthwhile to ask whether the "photo-montages" of Rodchenko, Heartfield, Hoch and Co. could ever have seen the light of day without Eisenstein? Indeed, would "cubist simultaneity of form and space" have been a gnawing preoccupation for the Parisian Gang of Four? Or would the revolution against the cant of Renaissance perspective have been at all possible by the brothers of Die Brucke without Fritz Lang, wide and deep-angled lenses, extended pan and elevated shot, not to mention three-point directional lighting? Drawing is a way of thinking as well as looking. Has all our way of thus seeing been changed by the camera and lens and the syntactical root unit *filmeme*?

How did the filmeme affect James Joyce, Samuel Beckett, William Burroughs, *et al.*? We must also inquire about cinema East and West, in view of Rouve's aesthetic of film language which pertains to the Western eye, but, perhaps, only presumes for the East? What about Indian films, etc.?

Reel to Real and Reel to Tape

Cineography

In my own somewhat fleeting experience, and the more durable directing careers of some of my friends and associates, the phenomenal influx of young and decisive talents from television to the ranks of filmmakers that occurred in the 1960's might best be accounted for in the emphasis that Rouve puts on what more outmoded persons still describe as film editing. The multi-camera techniques of television drama create a strict routing of exacting pre-planning in terms of pictorial syntax. Here, indeed, directing is a question of instant frame and sequence editing. I venture for your consideration that such video-weaned young lions helped de-atrophize the cinema of Hollywood and brought more focus to bear on the questions of creativity in the director/editor role.

Film on Tape

How is the co-reality of reel to real affected by transfer to tape, and how is tape actually dispensed as to communication, i.e., home video receivers as opposed to theatrical cinema presentation? What, if anything, is lost in the translation? Consider Marshall McLuhan's views (medium as message or medium as massage), etc.

Video Technology

With eyes to the future, how should we regard the video technological hardware explosion with its macro (network presentation, and large screen viewing) and micro (cable and community talking heads) implications? Is cinema to video what painting is to etching plate print? Or is there a video aesthetic language, and how is "reel to real" affecting its pubescent growth?

The Question of Color

A factor not really engaged by Rouve in terms of seeing with meaning is the question of how the optical phenomenon of color should now best be thought of and understood in any visual communication equation. Technicolor is not impressionist invective. Does the dispensation of home color choice control offered by the television receiver imply a degree of "subjective noise" between transmission and reception, with quite devastating, if democratic, implications for any concrete values of meaning?

These questions were prompted by Rouve's incisively original and (more often than not) preempting answers, for which I genuinely and humbly thank him.

Chapter 4

Music and Technology Today

Music and Technology Today

Daniel Charles

> Ohne Kraftwagen, ohne Flugzeug, und ohne Lautsprecher hatten wir Deutschland nicht erobert.
> (We would not have conquered Germany without the automobile, without the airplane, without the loudspeaker.)
>
> —Adolf Hitler[1]

Introduction

If electronic technologies are used for extending the nervous system instead of for extending muscle power, they will give birth in the music field to goods which cease to seem "natural" (in Rousseau's vocabulary) since they consist no more in "bowing, plucking and striking skins or strings."[2] But even if the loss of natural origin (or of manual labor) means the loss of a *certain* virginity, every composer's product tends nowadays to be evaluated in terms of market value, and market value is more and more determined and measured by electronic-technical developments, such as electro-acoustic recording and storage. Therefore, it becomes clearly more and more difficult to proclaim music as an antipode of technology, and to advocate—from the point of view of humanism—the separation of art and technics.

Lewis Mumford, for instance, tells us that:

> Most of the great artists of the last two centuries—and this has been equally true, I think, in music . . . have been in revolt against the machine and have proclaimed the autonomy of the human spirit: its autonomy, its spontaneity, its inexhaustible creativeness.[3]

But he knows very well that:

> Art and technics go together, sometimes influencing each other, sometimes merely having a simultaneous effect upon the worker and the user.[4]

Part I: Storage Techniques and the Calculative Mind

Music and Memory

And since he speaks only of "machine" and "mechanization," i.e., of "the first form in which natural science has emerged" (Hans Jonas), the possibility of a far closer net of relationships between music and technique in the age of information and post-industrial technology remains entirely open. Our purpose here is to explore that field under the three perspectives of storage techniques, telecommunication technique and "live electronic" systems (including some recent uses of the computer).

In his posthumous *The Mathematical Basis of the Arts* (1948), the composer and mathematician Joseph Schillinger—the author of *Electricity, a Liberator of Music*, a book published as early as 1932—gave one of the most impressive manifestos ever written in praise of modernity and of the calculative mind in music. For Schillinger:

> If art implies selectivity, skill, and organization, ascertainable principles must underlie it. Once such principles are discovered and formulated, works of art may be produced by scientific synthesis.[5]

Let us recall also the definition given by Paul Henry Lang of Western music as such. Its "essence," he says, lies in:

> . . . causal relationships achieved by the grouping of many small units into larger ones and finally into a great system of architecture.[6]

According to Christopher Small, when Paul Henry Lang described such an "essence"

> . . . he might equally have been a nineteenth century physicist expounding his essentially mechanistic view of the cosmos.[7]

Schillinger's and Lang's views are clear; they argue in mechanistic terms from their perspective in the age of electricity. For them music and science have in common an analytical, objectifying, logical and somehow abstract methodology, and such a methodology begins with a certain selection and memorization. For instance, we do not recognize, in a bird twittering without repetitions, without order, and without laws, a "real" bird-song. We need periodicity, a certain amount of structure, a balance of meter and sound, etc. But even with the most precise criteria of selectivity, we may go wrong—and that is just the point.

A contemporary philosopher, Don Ihde, has shown how our traditions—our memorizing patterns "hide certain features of musical presence"; and the calculative and all-measuring mind is clearly concerned here. Ihde tells us:

> The multiplicity of sounds competes for my attention. On one side I seem to have little control over their presence . . . But the control I do have is largely psychic. I can choose to attend, to attend in degree, or to diminish my attention with limits. This attention overrides many of the merely *physical* features of auditory presence . . . Not only may I attend to one sound—make it *stand out* in the center of my attention—to the near exclusion or at least relative exclusion of other sounds, but I may concentrate upon particular features of a sound to the relative exclusion of others. Thus the sound of the bird sounds like a musical note with a particular melody. And *for me* the melodic quality of the song is what *stands out*. But note here, . . . what is important for me may not be at all important for the bird. It may be that the signal he is conveying to his peers is actually to be found not in the melodic quality of his *song* but in the barely discerned *clicks* which are included within the total presence of his sounding. In this case, my traditional metaphor of a bird's sounding as *song* may keep me from noting what, to his mind, is precisely the important feature. My *control* over sound is my attention and its selectivity. But this very selectivity is both what *reveals* something about sounds to me and at the same time *conceals* other aspects of sound.[8]

The forgetting of technological concealment is thus typical of the calculative mind.

Music and Notation

"Objective," "calculated" music is at its best when it can be *written*. It may even be said that graphic storage of sounds, and moreover of relationships between sounds, has become—at least in Europe—*music itself* (Werner Kaegi). The meaning of technological concealment becomes evident here. According to Ihde: of the medium itself. As Gould wrote:

> The way in which the performance of music has been changed by its exposure to electronic media is . . . significant. . . . Today, the performer conditioned to think in terms of electronic projection automatically comes to think in terms of an immediacy of reception between himself and the individual listener (represented by the microphone or whatever the electronic conveyance happens to be), and, because of this, a fantastically varied and subtle range of interpretive insinuation is possible.[19]

In other words, the relationship between the listener and the performer becomes closer with the help of new technology, so that the awareness of the "inner workings" of the music will be deeper than on the stage, and the change of attitude required from both the performer and the consumer will lead to the end of the tyranny of "big names" (composers as well as performers). If, even nowadays, "only those works and those names which can be relied upon to *sell*" are "available on records," it is because the microphone is taken only as "a passive extension of our ears"; this very attitude "imposes upon the record industry the conservatism that also dominates the concert hall."[20]

With Gould or, in a different field, with the experiments of composers such as Martin Davorin Jagodic and Costin Miereanu taking the long-playing record itself as a "musical object," we begin to see how the limitations of storage techniques (for instance, the noise of the amplifying apparatus, or the physical dimensions of the record) may give birth to an authentic creation. Even if the magnetic tape "is relieving notation as a form of storage," nothing has been changed in the very fact of storing; but when technology suggests, as if by itself, that storing contains "more" than a mere accumulation, the definition of the work of art and of the task of its guardians begins to oscillate, so that something like a new, yet unknown, but fascinating world seems to appear.

Serialism and Tabula Rasa

In the origin of integral serialism, we find a remarkable *denial of memory*. The fact has been convincingly described by Reginald Smith-Brindle:

> Our minds normally create only out of what memory suggests . . . To avoid this needs deliberately objective reasoning. This was precisely the main reason for the flourishing of integral serialism. It was, in theory at least, a system of composition which obliged composers to think objectively and eliminate memory, so that the musical heritage of the past was blotted out . . . Total organization was, at that time (the early fifties), the only way to create the *tabula rasa* on which completely new edifices could be constructed.[21]

Instead of building on the basis of pitch, i.e. harmony, Arnold Schonberg had sought, in his *Harmonielehre* of 1911, how to relate the pitch-dimension to a larger, more encompassing "proto-parameter," timbre of tone-color. If certain logical laws were already ruling the kingdom of acoustical frequencies, a logic of a higher type was probably involved in the region of timbres, so that the tone-color melodies (*Klangfarbenmelodien*) of the future would allow an absolute sovereignty over the various parameters of the individual sound. We are thus very near—or not far at all—from Schillinger's and Lang's ideal of composition as synthesis. The "serious" musical work must be built—after all the elements of individual sound have been analyzed, i.e., *atomized* —according to the requisites of a higher logic. This is somewhat like a superior number-matrix or a supreme law of proportions, which allows us to compose, to put together in an ordered manner—and not only to organize, i.e., to make juxtapositions.

The difference between Schillinger or Lang and the integral serialist composer is that the last one considers his sound materials as "neutral," not yet polarized or hierarchized from the point of view of classical harmony; his first (and perhaps main) task will be to establish "neutral" relationships between the various sounds (or frequencies) he will use. Negatively, he will have to avoid all the "false" relationships (those remembering the abandoned tonal system); all his calculating power must be put in the service of oblivion, the deconstruction of the past.

One of the most striking examples of such a "forgetting" procedure is to be found in the *Structures of two pianos* by Boulez (1952). The point of departure is an *homage* to Olivier Messiaen's etude for piano, *Mode de valeur et d'intensites* (1949), a piece of particular significance since the author was trying (probably for the first time in the world—with the possible anteriority of a comparable score by Milton Babbitt, 1948) to dissolve the "sound

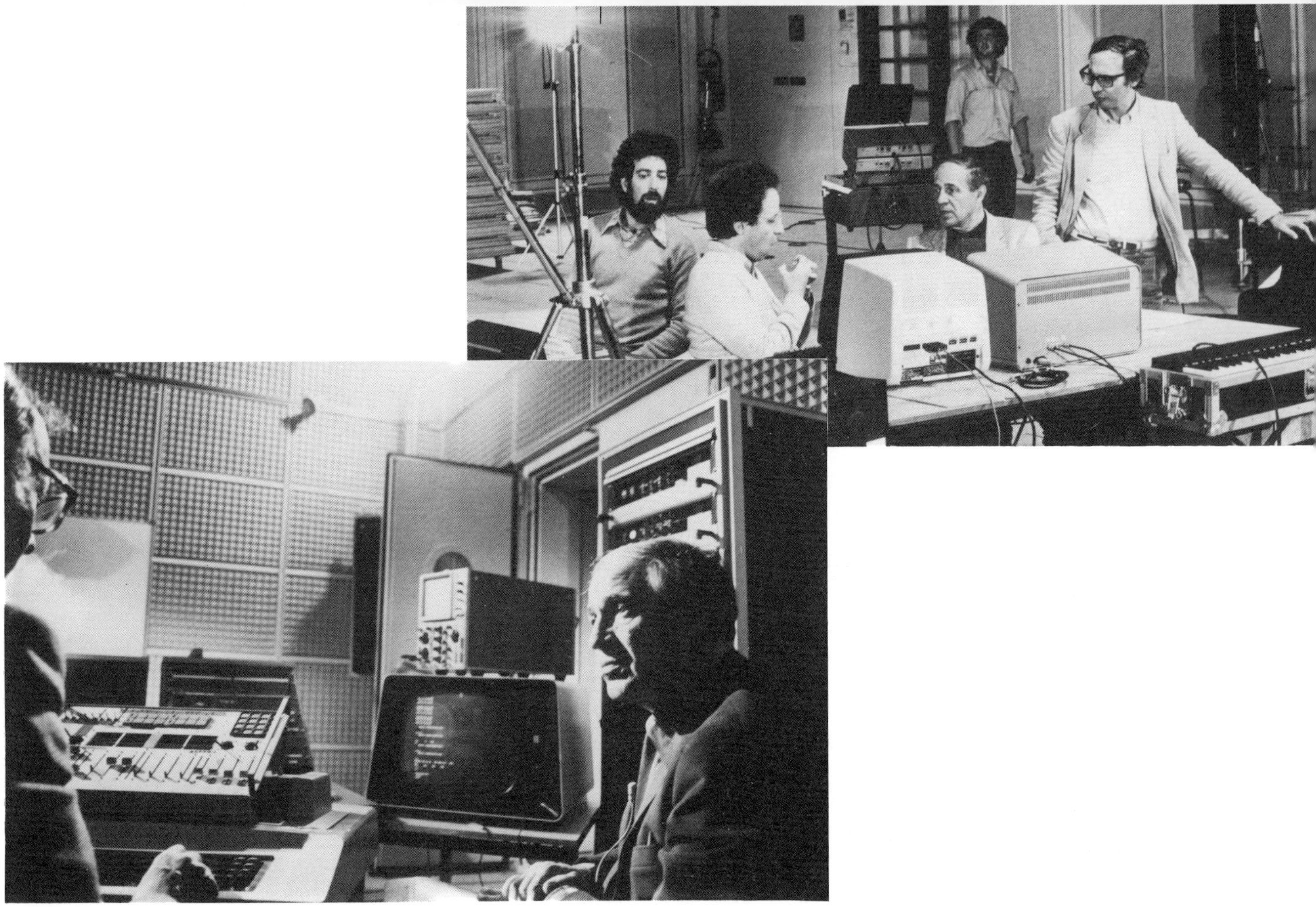

Fig. 4-1: Pierre Boulez with Giuseppi di Giguno, the inventer of the 4X in a Studio at IRCAM, Paris (Photograph by Jean-Pierre Armand).

objects" into an analytical puzzle and build a *mode* with that puzzle. Messiaen had in fact separated the distinct "parameters" of the sound material and subjected the resulting *de-composition* to a quasi-stroboscopic fixity, as if repetition, or the eternal return of the same, was the last possible relationship between such isolated fragments.

> . . . a culture which has a written language is able to *accumulate* knowledge in its records—it makes its words come to *stand* in a preserved written word. But this advantage which reveals things to us about the past, about what happened, about many things, also conceals. Writing "reduces" a living tongue and a "reduction" is a simplification. Thus writing leaves out gestures, tones of voice, the peculiarities of the speaker's style, nuances and innuendos which can be noted only in the spoken word. In short, a reduction to written language preserves only an essential, but bare con-

> ceptuality. Indeed, our sense of "objectivity" is very closely related to the reduction of live speech into an *object* which is the evidence of the written word. (Obviously, also, we are on the verge of a very new reduction in today's records, now no longer merely written, but preserved, made to stand in tapes, videotapes, and phonograph records.)[9]

Schillinger's and Lang's choice of selectivity and control involves the same reductionism in the field of music.

> The introduction of a notation reduced musical forms into noted ones. But just as writing affects a culture, so does notation affect a musical language. One early and formative effect was clearly due to the Greeks who approached their music theoretically and mathematically. *Good sounds* were those which represented certain ideal forms, basically harmonic . . . It also directed what sounds were permissible to a certain extent. (Our notation would be incapable of dealing with a twenty-two tone scale, thus we exclude certain sounds from our musical language. We *conceal* a part of Sound.)[10]

Such a concealment is linked with the analytical-logical methodology praised by constructivist minds. As Ihde shows:

> . . . our attitude has been highly influenced by our theorizing. We compose, put together, our music. In certain periods some theoretical constructions were considered more valuable than others —these rules rather than those ought to be followed. But apart from our folk traditions—which incidentally often were associated with the nonliterate part of society—we did not question the basic theoretical concern with music. Musicology is the meta-physics of music education.[11]

Therefore it is clear that musical memory, as long as it uses classical notation techniques, maintains itself in the sphere of abstract theorizing. It preserves in itself, so to speak, the forgetfulness of concealment. In such a situation, what effects will the emergence of the new electronic technology of storage have?

The Renewal of Storage Techniques

According to Kaegi:

> We must not overlook a second information storer which is no less essential for music: *the musical instrument.* Just as writing is essentially identical with music for the initiated, the instrument is

> identical with music for those simple beings who cannot read music, and who often simply call the instrument *the music.*[12]

In other times the (relative) automation of "classical" instruments was made possible by the use of cylindrical storage; today's punch-cards or punch-tapes are the descendants of barrel-organs and musical boxes. But "serious music," which was identified with musical notation, did not currently use such "popular" devices; the fact that these devices are again in function reveals without doubt something about the new technology of the computer.

The first phase of the development of phonograph and gramophone recording began as early as 1877 (Thomas Edison and Charles Cros), and was completed in 1909 with Arthur Nikisch's recording of Beethoven's Fifth Symphony with the Berlin Philharmonic. It is typified technically by the method of mechanical sound recording, and "artistically by the tendency of having world-famous stars record standard repertoire."[13]

From 1925 to 1927, the second phase introduced—simultaneously with the installation of radio equipment, amplifiers, etc.—electrical recording at the standard speed of 78 r.p.m. on 12-inch records, allowing a duration of 4.5 minutes. After the Second World War, L.P. and High-Fidelity recordings brought only more impetus to the industrial development of the reproduction systems, which of course involves more and more engineering and commercial starring.

It seemed that nothing was changing, however, at the composing and performing levels. The art of "mechanical sound production by means of traditional musical instruments" was being "neatly separated from the merely technical process of electrical sound conversion for purposes of storage and playback," the first area remaining "the strictly sacred domain of the composer and his interpreters," the second being left "up to the sound engineers and technicians."[14] But in reality, musicians like Milhaud, Hindemith and Varese, or Bauhaus artists, had begun during the twenties and the thirties to see the gramophone record "as more than a means of preserving the music of the past."[15] Already in the 1880's or 1890's, someone had "suggested using the phonograph as a musical composing machine by playing favorite airs backward on it."[16] In 1932, conductor Leopold Stokowski thought of the future composer as one who would "create directly into tone, not on paper."[17] In 1964, superstar Glenn Gould gave up the concert stage "in order to become less of a performer and more of a musician."[18] He wanted to concentrate exclusively upon the recording of music, not only to emphasize "the preservative or archival function of recording," to test the "limitless possibilities"

Obviously, Boulez did not want to hold to the repetition of the mode, a vehicle for memory. Quoting the pitch-scale of Messiaen's etude, he integrated the sequence in a *series*, and instead of the immobility of model reiteration, he substituted Schonberg's rhetoric of dodecaphonic variation. Thanks to the permutations, every parameter, serially treated, varies perpetually. In Boulez's work, Messiaen's mode, while furnishing the matrix of the pitch-material, totally disappears at the end of the first page. The purpose and meaning of the variation is thus to obliterate memory. But at the second level, the variation *repeats itself* in such a way that the author re-introduces, quite unwillingly, accumulation. If Messiaen's evanescent score may be compared to a Webernian miniature, Boulez's architecture reminds us of what Adorno had to say about the dialectical process of *Musik and Technik heute*:

> Those moments in Webern where, in Schonberg's words, a whole book is compressed to a sigh, do not allay the suspicions about highly constructed music, which is essentially poor although giving the appearance of great concentration. Every composer today must ask himself about the actual substance of his constructions, and how much musical substance deteriorates through construction.[22]

The Meaning of Constructivism

The ambition of integral serialists was, at a theoretical level, the same as the aim of technological storing processes at a practical level: to transcend accumulation and uniformity in order to produce something radically new. But whereas the direct, practical manipulations of the media could lead a Glenn Gould, for example, to a redefinition of musical communication, the abstract work "on paper" of the integral serialists only made matters worse. The serial scores became quasi—immediately impossible to play, because of the complexity of the structures involved. Quite often, the only way to interpret them was to use tapes; but the results were hardly satisfying, since the conceptions were too abstract from the beginning.

According to the integral series principle, far from directly or "casually" determining the distribution of other parameters (notably of the durations), the note-series led only to proportion patterns which, if applied in general to time-values, dynamics, types of attacks, etc., made everything interchangeable, exactly as, in experimental science, gas particles are permutable. The logical development of the serialist *tabula rasa* was either an affair of statistics, as Xenakis had shown as early as 1955, or the return to "eternity" via the magical use of numbers, i.e., an indifference to what really may happen. Let us insist on this point. The construction of a classical piece was intended to protect the listener "from being lost in time," or "from not knowing where he is in relation to the beginning and the end of the music."[23] As revealed by the "proliferation of clocks, watches and time-checks in our society," the assumption of the reality of linear time had been paralleled with musical "kits," "introductions, perorations, transitions, recapitulation, as well as whole temporal structures such as sonata, rondo, da capo aria and so on."[24] Far from leaving the listener with a music existing only in the present tense, the composer had to lead him "forward to coming events, often in passages which are themselves of no great intrinsic interest," so that "only those elements we call themes (were) existing fully in the musical present."[25]

On the contrary, the postulation of the equivalence of all the moments of the serial development, under the "eternal" perspective of a hidden number-matrix (Stockhausen), could introduce only a greater fixity; the serialists worked as if living time could be "homogeneous" and progress "at the constant rate suggested by the movement of clock hands."[26] Their denial of memory did not contribute to giving to the listener a more subtle awareness of "how time passes," *wie die Zeit vergeht,*[27] (if time "passes" at all . . .), but provided insights on created objects rather than penetrations into creative processes.

The result was the exact opposite from Gould's use of storage facilities as aids to a more refined sense of immediacy; having lost the dimension of time, the "moments" of Stockhausen were explicitly built to *show* eternity. "To fix the infinite," as Boulez claims,[28] means at last only to commit oneself "to the work of preservation of everything that the past has produced"—even if

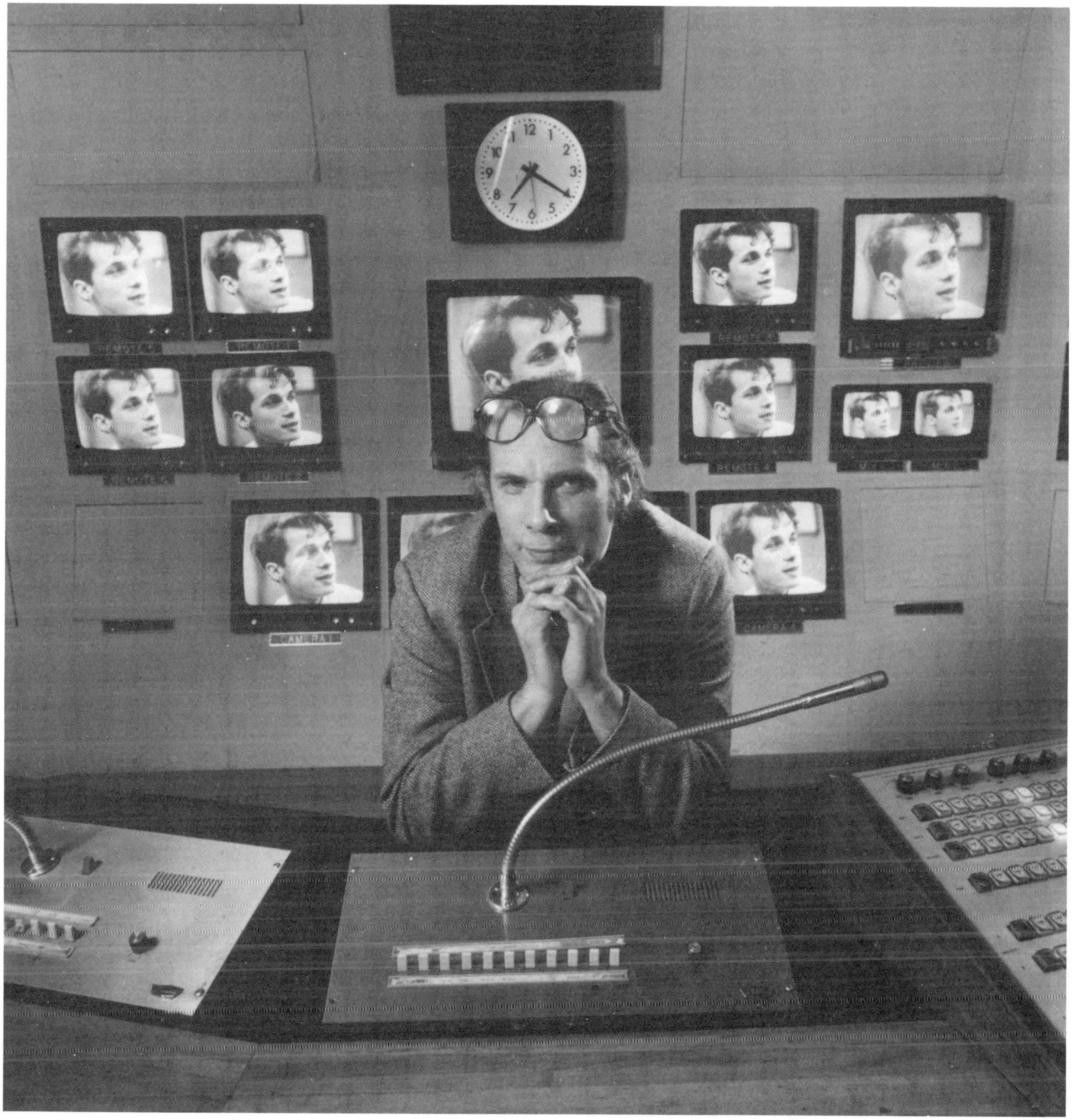

Fig. 4-2: Glen Gould (Courtesy: CBS Records)

that past is an *immediate* one—and if such a "work of preservation" exemplifies eternity, it will gain it exclusively by a total control. We find here, as Christopher Small says:

> . . . a restatement of (the) constant theme of Western music—and Western thinking—the domination of nature by bringing her under conscious control. (That this theme should have been restated so forcefully in new form . . . is perhaps by hindsight not surprising if one considers how the West, and indeed all the industrialized world, after the Second World War, redoubled its efforts towards the scientific and technological domination of the universe, rejecting, on the example of the United States and the German "economic miracle," the opportunity to construct a society on the Webernian model).[29]

What does it mean to say here, from the point of view of technology, a "Webernian model"? It means a model of *meditative* thinking, in opposition to what we have described as "calculative mind"; a thinking which would use technical devices but without dominating them and without being dominated by them. We have already evoked Webern, as seen by Adorno, delivered from the constructivist obsession. Now let us recall Small's appreciation of Webern's *Five Pieces for Orchestra.* They are "equally to be admired as sound constructions and as miniature tone-poems (the third, for example, surely evokes the Alpine landscapes Webern knew and loved so well)."[30] Webern's relations with calculus and constructivist techniques were not the same as those of integral serialists; the new musical technology may help us to listen to Webern in a radically new manner—*poetically.*

Part II: From Telecommunications to the Computer: The Musical Embodiment of Media

The Ambiguity of the Media

If we ask what the meaning is of the uneasiness we have experienced in front of integral serialism, we may say that a too crude or naive technologization of the mental tools which its development involved prevented this music from rejoining the flexibility and refinement of the classical idiom. As Martin Heidegger told us:

> To the calculating mind, space and time appear as parameters for the measurement of nearness and remoteness, and these in turn as static distances . . . Two isolated farmsteads—if any such are left —separated by an hour's walk across the fields, can be the best of

> neighbors, while two townhouses, facing each other across the street or even sharing a common wall, know no neighborhood.
>
> Neighboring nearness, then, . . . does not depend on space and time considered as parameters . . . In order to experience this face-to-face of things with one another . . . we must first rid ourselves of the calculative frame of mind . . . In the succession of "nows" one after the other as elements in parametric time, one "now" is never in open face-to-face encounter with another. In fact, we may not even say that, in this succession, the "now" coming after and the "now" coming before are closed off from each other. For closure, too, is still a manner of facing or excluding something being in face-to-face. But this encounter is as such excluded from the parametric concept of time.[31]

The same can be said about space. The situation created by the emergence of the various media appears thus as problematic; even if they reduce the span of measured time or of measured space between the sender and the receiver of the message, will they allow any authentic relationship between them? How can music be "mediated" at all if it must still include a "true encounter" between sounds?

Ihde's thesis about the "technological embodiment of the media" may suggest some precious answers. First, the media are obviously *nonneutral*. The first transatlantic telephone cable transmitted the sound of the voice, plus a certain amount of noise due to amplifiers, and of course "something had to be done to impede the machinery-produced noise and to enhance the desired sound, that of the voice, if the communication were to occur at all."[32]

But the negativity of such a nonneutral technology can as well be inverted, as in Heidegger's example of the hammer in *Sein und Zeit*; the user of a hammer "does not focus upon or relate *to* the hammer in the work situation at all. Rather, when the hammer is functioning in the use situation, it is the work—that upon which the human using the hammer works—that is experienced." So, "when the technology is good, it becomes semi-transparent with respect to communication" and can be "functionally *forgotten*."[33] We could then hope for a perfectly transparent technology, "one in which the technology was so transparent that it would be invisible and thus *equivalent* to being a nontechnologically mediated situation."[34]

We begin to understand what the meaning of a "true encounter" or a "face-to-face" of men (or things) with one another is by considering the case of a medium which, because it is functioning well, "withdraws"—the telephone, for instance. "I am able to recognize your voice as yours and although your presence to me is *reduced* perceptually to a mere voice, the presence is one in *real space-time*."[35] The "true encounter" happens, because *you* "are" *there* and *now*, "near" me, in my "neighborhood"; the distance, from a geographical point of view, ceases to separate us, and its measure does not really matter. But of course, your presence is not exactly identical to what it would be if you were "entirely" here. You are simultaneously "amplified" *and* "reduced." In other terms, the nonneutrality of the medium, with its inescapable ambiguity, is "transformative of human experience."[36] *The medium opens a world.*

This opening, this trans-formation of daily life, this *performance*, happens at each time and place in a specific way. Moreover:

> . . . every technology has . . . a telos, or weighted center of gravity, which makes it partially selective as to what may be enhanced and what reduced. In the example of the telephone, if all I need is information the telephone seems relatively adequate; but if you are an intimate, while the immediacy of the telephone is better than the more abstract letter, there remains a sense of lack.[37]

We may say that the making-near of various places and times due to the transformative or performing effect of the medium "sounds irreal": that is normal, since the very use of the medium involves "a distance which *reduces* the perceptual or fully lived sense of the phenomenon."[38] But since the transformation *happens*, the medium in itself becomes "part of the total perceptual situation"; and "if at the first level one may detect the irreal dimension of a mediated perception, at a second level one can also claim that this mediation has now been materialized and thus becomes part of reality."[39] The "distance" introduced by the medium does not necessarily disappear; it becomes something else, an "aesthetic stance" which is able to preclude the difference between "perceptual" and "dreamed" experience. For instance, Arthur Clutton-Brock could write, as early as 1924, that "by playing some orchestral record with the thinnest possible needle" he was producing "a delicate, distant kind of fairy music, something not at all like the actual orchestra but *with an original quality of its own*." "This can best be done," he added, "with good but imperfect records, such as the Columbia version of *L'Après-midi*, which crackles and is much cut down, yet makes a faint and beautiful concourse of sounds with a very fine needle."[40]

Consequently, "as the medium approximates perceptual experience . . . it becomes ever more difficult to be a mere observer. Instead, the near distant space-time of media has a center of gravity on the part of the viewer," so that the "aesthetic stance" may very well be brought "into the vicinity of action."[41] Here we may recall Gould's redefinition of the listener, a listener who is no longer anxious about analytical reasonings:

> At the center of the technological debate . . . is a new kind of listener—a listener more participant in the musical experience. The emergence of this mid-twentieth century phenomenon is the greatest achievement of the record industry. For this listener is no longer passively analytical; he is an associate whose taste, preferences, and inclinations even now alter peripherally the experiences to which he gives his attention, and upon whose fuller participation the future of the art of music awaits."[42]

We may also recall John Cage's definition of the composer as the "first listener" of "experimental" music; the attitude of the "ordinary" listener becomes one of activity, since he has to compose what he is hearing; such is the lesson of the "silent piece" *4′33″*, but the other "silent pieces" by Cage, after *0′00″* will involve technology.

Radio and the New Listener

The Berlin State Opera made the first opera broadcast on June 8, 1921 (*Madame Butterfly* at Konigswusterhausen, Germany); ten years later, 200 European and American stations in the world together transmitted Richard Wagner's *Tristan und Isolde*. In the meantime, radio had become an extremely helpful organization for musicians, since it distributed royalties for the diffusion of "light" as well as classical music. The problem of the aesthetic orientation of radio programs emerged clearly at the Conference of the German Program Committee on May 22 and 23, 1929, in Bremen, when conductor Hermann Scherchen spoke about:

> radio's task and possibilities of encouraging public concerts and of creating a community of listeners who understand music, a community in which audiences (would) be educated through the ether imperceptibly to become ideal radio listeners (unmerklich zum idealen Rundfunkhorer).[43]

Still, we wonder nowadays how to manage the programs according to the tastes of the majority; as Kaegi asks, "who, other than the radio, has formed this taste *unmerklich*, imperceptibly?"[44] Obviously, the problem is not one of information. Of course, the amount of exchanged information may be measured if we know the quantity of symbols or rules the emitter and receiver may have in common. But, as Kaegi says, the radio introduces us to a:

> *direct* effect of music on people in the sense that something other than exchanged information is meant. Even the simple example of inexterminable martial music, but also of primitive dance music, shows that information is exchanged without, of course, the recipient being free to decide whether to accept or reject it. Music of this type goes to the feet.[45]

Ever since Plato, we know that this music can very well be used for conditioning people. Even if we can credit radio technology with the preservation of the ephemeral—and with providing at least an aural opening to all the oral stimulation that nowadays risks, at each instant, being definitely lost—radio is essentially concerned with maintaining musical production in the sphere of the market and advertising. Kaegi comments:

> This aspect of music, blown up into worldwide proportions by means of technology, is exploited to a huge extent; there is no escape today from the dangerous and levelling influence of the loudspeaker. (During a recent journey to the Far East . . . I was regaled in Japan, Hong-Kong, the Philippines, Thailand and India with the same music which I cannot escape in Europe).[46]

Now, without forgetting any of the dangers evoked by Kaegi, we must observe that radio in a Post-Renaissance or Post-Modern or "transcendantalist" era (in Leonard Meyer's sense)[47] may also emancipate the listener from a number of (if not all) specific historical and geographical frameworks and allow him to understand the "style of the absence of style" which is so characteristic of experimental music's last trends. What is important then is not the content of the musical message, but, according to McLuhan and Father Ong's thesis, the flavor of the medium.

From Electroacoustic Instruments to the Synthesizer

In the field of substitution of electroacoustic instruments for mechanical (traditional) ones, the ancestor was Dr. Thaddeus Cahill's *Dynamophone* or *Telharmonium* (1906), an electric organ which used as sound sources "dynamos run by alternating currents, weighed 200 tons, and transmitted its music over telephone wires"–a great huge ancestor, about which European composer-pianist Ferruccio Busoni wrote enthusiastic pages in his 1907 "Sketch of a New Aesthetic of Music."[48]

With the development of the radio in the twenties, several electrical instruments due to Mager, Givelet and Coupleux, Martenot, Theremin, etc.—appeared. However, since the constructors all "tried to articulate the parameters in the same way that was usual for traditional mechanical instruments," the new instruments were in fact built "exclusively for an old world."[49]

In a similar vein, when the commercialization of the magnetic tape recorder allowed the manipulation of recorded sounds under Pierre Schaeffer's banner of *musique concrete* (after 1948), the question, unanswered by Schaeffer, was how to deal with the new medium without simply reiterating the old compositional procedures. The "concrete sounds" were already existing ones, recorded in a studio, and subjected to changes of acoustical characteristics—changes of speed, backward play, tape-cutting and editing, reverberation and echo effects, filterings, superpositions, tape-loopings, splicings, multi-track-recording, stereo and quadra-phony—which could lead to very complex (and often unpredictable) effects. But since he advocated the necessity of controlling (in the old sense) all the simultaneous levels of stratification and mixing procedures, Schaeffer reintroduced some "old world" aporias, as, for example, when he divided the "phonogene's" keyboard into twelve steps, or when he developed, in his *Treatise of Musical Objects*, the misleading idea of a "concrete solfeggio." Only after Schaeffer's departure could the Paris studio produce more freely and convincingly, according to the real "center of gravity" (as Ihde says) of the tape medium, without being dominated by the dictates of musical notation and writing.

The elaboration of "electronic music" in the "classical" sense —i.e., the music first composed in the Westdeutscher Rundfunk Studio at Koln (under the direction of Herbert Eimert and Karlheniz Stockhausen) in the beginning of the fifties—depended unmistakably on "old world" habits and prejudgments. Instead of beginning "with prepared sound material, which is molded into its final form by a process of experiment, trial and error, perhaps following unexpected paths to goals which were never foreseen initially," electronic music "was composed like traditional music, being first conceived in the mind of the composer, then written down, and finally realized in sound."[50] In fact, Dr. Eimert was an apostle of orthodox integral serialism, and conceived composition as a slow constructive process leading "from the simple toward the complex." Every parameter of each sound had to be fixed apart from the others, and every sound was elaborated for itself, so that the assemblage of the various sounds in the master tape was the result of a laborious (entirely hand-operated) procedure, which entailed enormous waste of work and time.

The electronic studios of the fifties used sound generators (sine tone, white sound, square wave generators), filtering apparatus, ring modulators and reverberation facilities. Then came more sophisticated devices, such as variable speed tape recorders (the *Telefonbau Normalzeit*, which changed the speed of a rhythmic sequence without altering the pitch), dynamic suppressors and various playback gadgetry.

Very rapidly, the question arose of how to speed up construction of the work. Since what was needed was a way of regulating several parameters of the same signal simultaneously, it was decided to group all the operations accomplished by filters, oscillators, amplifiers, etc., under the same modulation system, the global voltage control which was directly in the operator's hand. The emergence of the "voltage control" principle was the first step toward automation of the studio. The next step was to store the control voltages themselves on tape, and to work directly on them, as if they were audio signals. The programming of control signals could itself be automated, thanks to sequencers or variable function generators (cf. the Institute of Sonology at the University of Utrecht). At that point, the serial control of timbres (and not only pitch, duration or dynamics) seemed to become possible, but the complexity of such tasks led to the computerization of studios (the computing systems being used as memory banks, not directly as composing instruments).

Meanwhile, the invention (by Robert Moog in 1965) of voltage-controlled oscillators and amplifiers and their combination with non-voltage components in the "synthesizer" provided a means of memorizing complex sound-structures and *playing them live without recording and tape-editing.* The studio then ceased to be necessary. As a portable instrument, the synthesizer could be taken on stage. With the tape procedure, the production of sounds (using a number of oscillators and generators), their alteration (by filtering, modulating, reverberating and mixing), and the control of the whole, were successive and (even with automation) time-lasting operations. But with the "synthi" the musician has oscillating sound sources at his disposal, which simultaneously produce a lot of possible alterations by themselves, and also control each other. "Real time" composition makes "music-producing *activity* and its resultant *sounds* . . . simultaneous and inseparable."[51]

It is worth noting that before Moog's (or, on the West Coast, Don Buchla's) invention, all the electronic equipment was first intended to serve non-musical purposes. With the synthesizer, we have "the marketing of the first units designed *exclusively* for the composition and performance of electronic music."[52] Moreover, the "composer working at a synthesizer is, in effect, an instrument-builder as well"[53]—and total improvisation becomes possible as well as total organization; serial imperialism is over.

Computers and Live Electronic Music

Of course, computerization of synthesizers is now available. Computer-control of a synthesizer makes the computer equivalent to a sequencer. Inversely, computer-generated sound results from the composer's instructions, "in and output of *digital* tape, which is then processed onto magnetic recording tape (*analog* tape) to be heard as sound over loud speakers."[54] Thus, the computer itself may be thought of as a "highly versatile synthesizer." It becomes a performing instrument and allows new live musical procedures, instead of being treated (as in the calculative perspective of the integral serialists) as a mere memory or bank of sounds.

Until now, the computer could not give facilities to the musician for "aurally testing" his ideas before the end of the programming process. But at least the composer was not "*limited* by any particular studio set up or any fixed number of modules,"[55] so the preconception of the work was entirely free. By "calling up" the program already existing in the computer's memory and dictating (with punch cards) "the entire course of musical events," the musician could get a total control on what he was doing. Ten years ago it seemed that his aesthetic ought to demand such a control. Let us only recall Elliott Schwartz's formula: "the computer operation naturally attracts composers for whom control is a virtue."[56] Nowadays (in 1983) it can be said that improvisation, "guiding the flow of sonic events while they are in progress,"[57] has become the main concern of the most "scientific" composers (cf. Iannis Zenakis and the "UPIC" experiences). Far from being the "chassee gardée" of rigid serialism, the computer allows the creation of pieces "with" noise, and enlarges the world of chance operations and of live electronic music.

The pioneer in that field was indeed Cage. Elliott Schwartz has given a suggestive view of the *premiere* of the extraordinary piece Cage called *HPSCHD*:

> Six letters are the maximum necessary for computer-coded words, and the six-letter word *HPSCHD* means "harpsichord." Not that anyone witnessing the premiere performance of the Cage-Hiller *HPSCHD* would have doubted that fact. On May 16, 1969, at the University of Illinois, seven live harpsichordists and fifty-one computer-generated sound tapes combined in an assemblage of sound—much of it curiously reminiscent of Mozart—that lasted some five hours. A good deal of the "live" (although amplified) harpsichord sound *was* Mozart, in fact, or made up of possibilities derived from Mozart. Cage and Hiller, in creating *HPSCHD*, had used as one source Mozart's *Introduction to the Composition of Waltzes by Means of Dice*. This late eighteenth-century work had called for a flexible ordering of phrases, although all within the same key scheme and always ending on the same "fixed" measure. A computer program at the University of Illinois had generated a series of possible realizations, and these were played by three of the live harpsichordists. Two other live players performed assorted fragments of music from Mozart to the present (represented by Hiller and Cage), and a sixth player was instructed to perform any Mozart he wished. Each of the fifty-one tapes of computer-generated sound represented a different division of the octave into micro-tones, from a five-tone octave (the minimum) to a fifty-six tone octave—every division, that is, *except* the familiar one of twelve. This last was reserved for the seventh harpsichordist, who

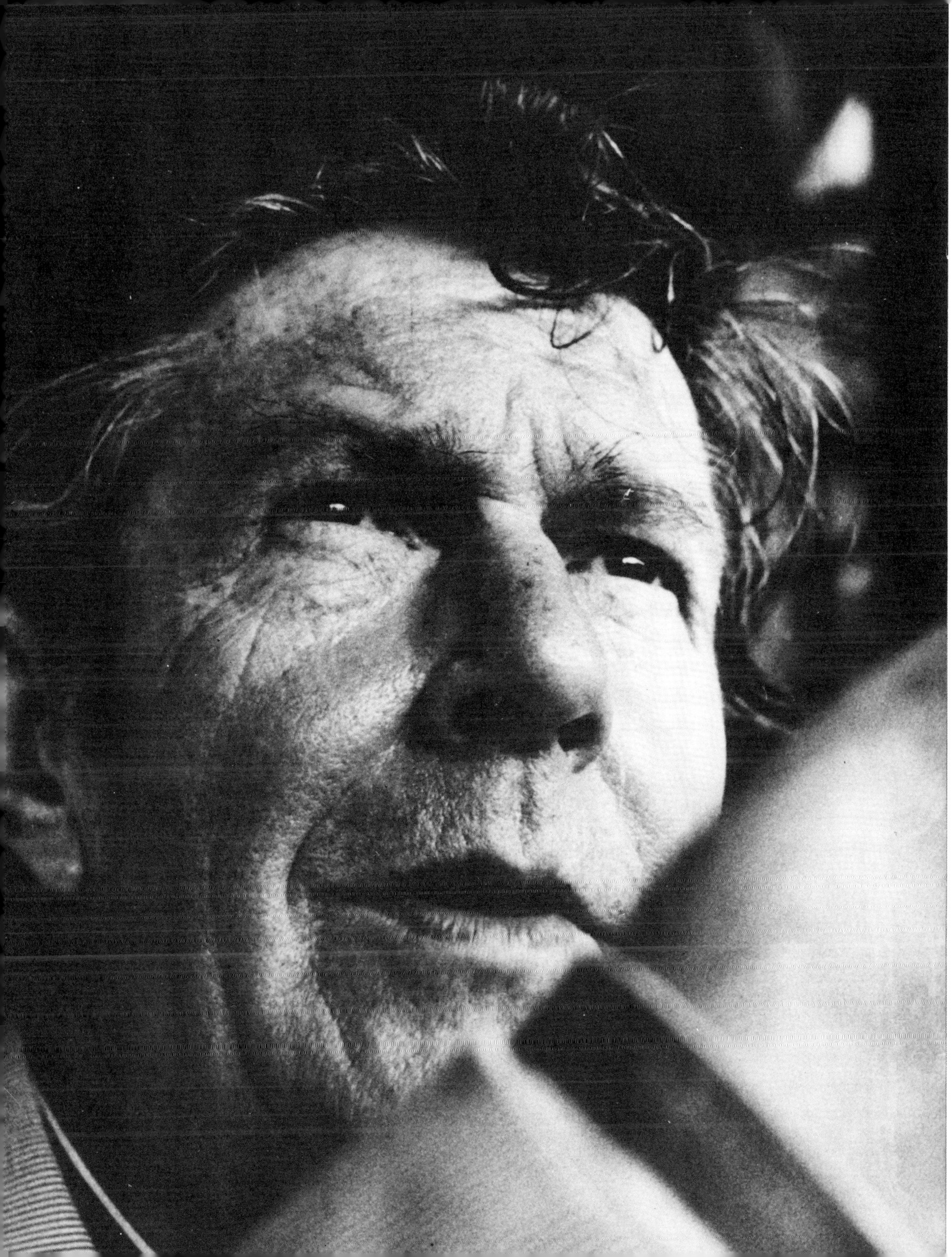

Fig. 4-3: John Cage (Photo supplied by Daniel Charles).

H P S C H D

FOR ANTOINETTE VISCHER

TWENTY-MINUTE SOLOS FOR 1-7 AMPLIFIED HARPSICHORDS AND TAPES FOR 1-51 AMPLIFIED MONAURAL MACHINES TO BE USED IN WHOLE OR IN PART IN ANY COMBINATION WITH OR WITHOUT INTERRUPTIONS, ETC. TO MAKE AN INDETERMINATE CONCERT OF ANY AGREED-UPON LENGTH HAVING 2-58 SEPARATE CHANNELS WITH LOUDSPEAKERS SPACED AROUND THE AUDIENCE.

S O L O I

TRANSCRIPTION BY J.C. OF COMPUTER PRINTOUT FOR SCALE WITH 12 TONES PER OCTAVE. DURATIONS ARE REPRESENTED BY STAFF-LENGTH (5 INCHES = 1 SECOND; ONE PAGE = 20 SECONDS; LARGE NUMBERS AT END OF EVERY THIRD PAGE = ELAPSED TIME IN MINUTES) AND ALSO GIVEN IN MILLISECONDS AT BEGINNINGS AND ENDS OF NOTES AND ORNAMENTS. CHANGES OF AMPLITUDE FACILITATED BY PRESSURE-SENSITIVE CONTROL MEANS DESIGNED AT THE SUGGESTION OF DAVID TUDOR BY HUGH LE CAINE.

JOHN CAGE AND LEJAREN A. HILLER, JR.

UNIVERSITY OF ILLINOIS, URBANA, 1967-68

WITH THE ASSISTANCE OF JAMES CUOMO AND LETITIA SNOW

Fig. 4-4: Original music of John Cage, title page of Solo 1 of HPSCHD (© C.F. Peters Music Publishers, used by permission.)

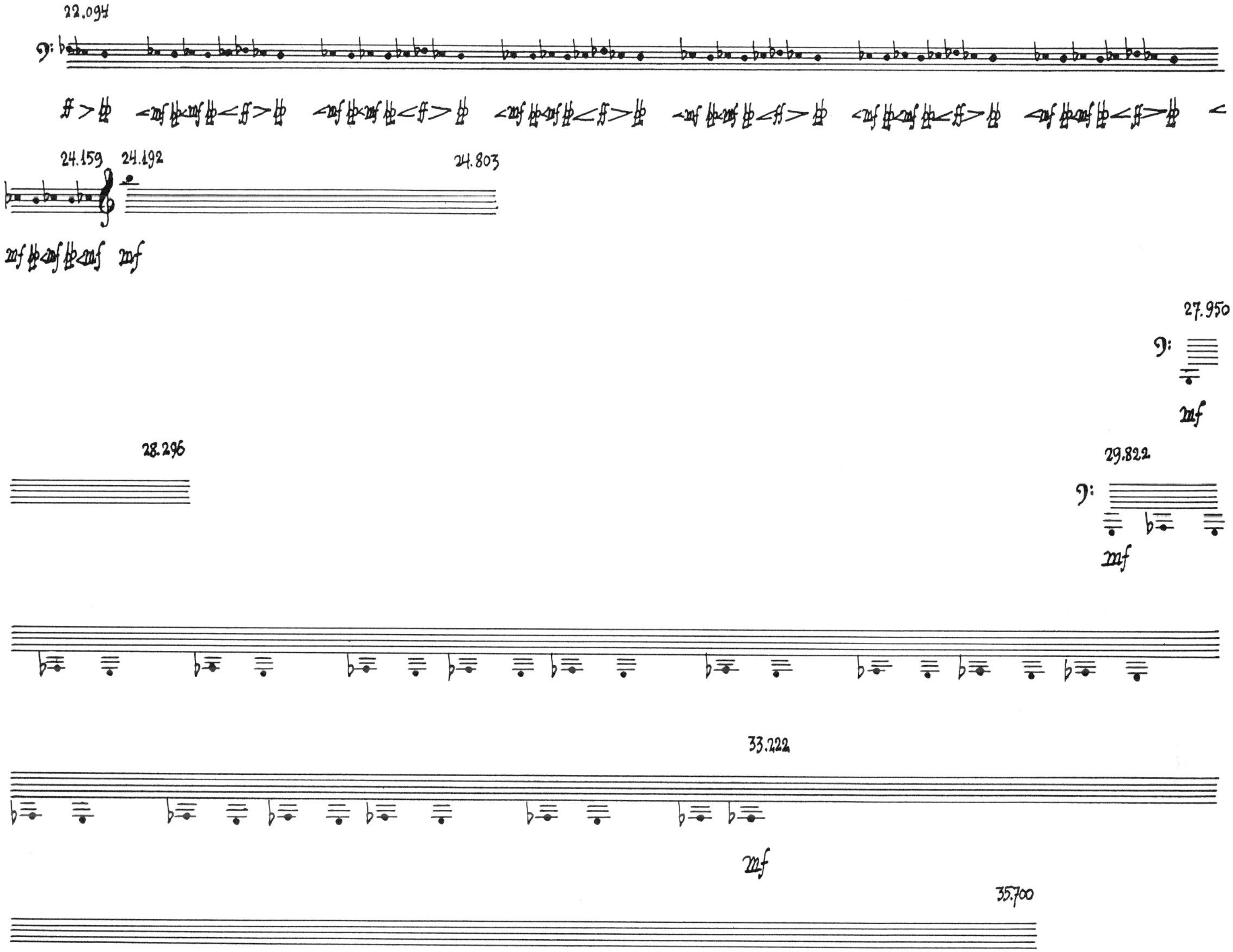

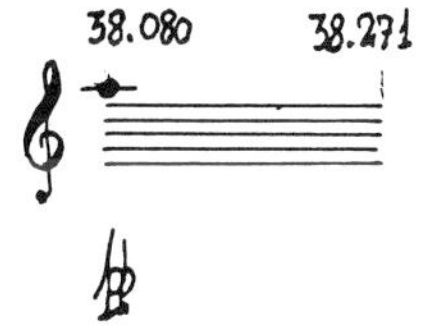

Fig. 4-5: Original music of John Cage, Solo from HPSCHD (© C.F. Peters Music Publishers, used by permission.)

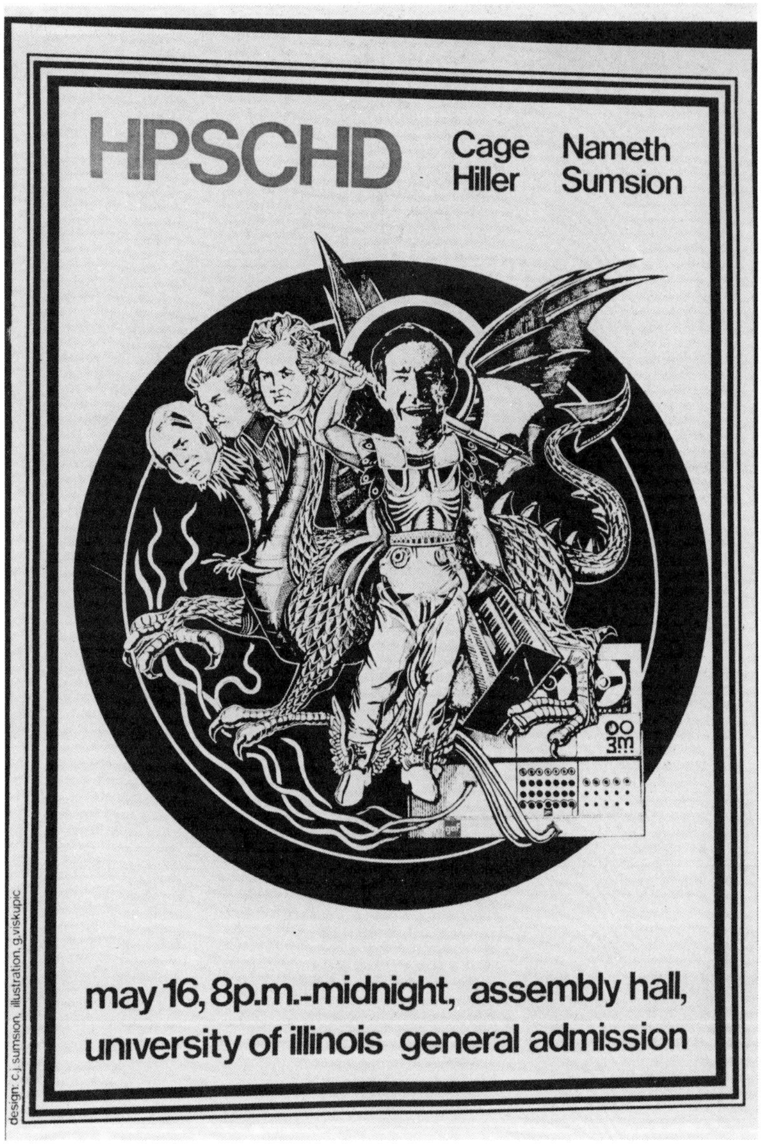

Fig. 4-6: A poster from HPSCHD. The posters were created from random sequences and silk screened (from the collection of David Eisenmann; photo courtesy of Champagne-Urbana *News Gazette*).

Fig. 4-7: Another poster from HPSCHD (from the collection of David Eisenmann; photo courtesy of Champagne-Urbana *News Gazette*).

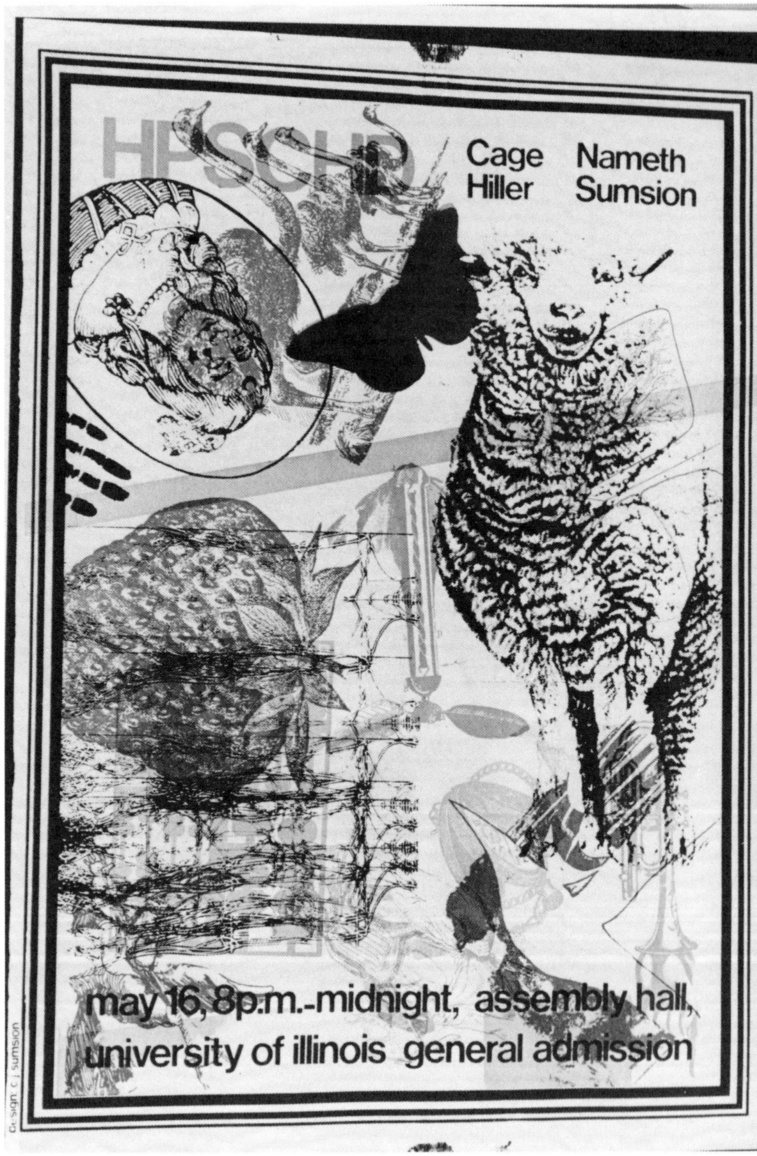

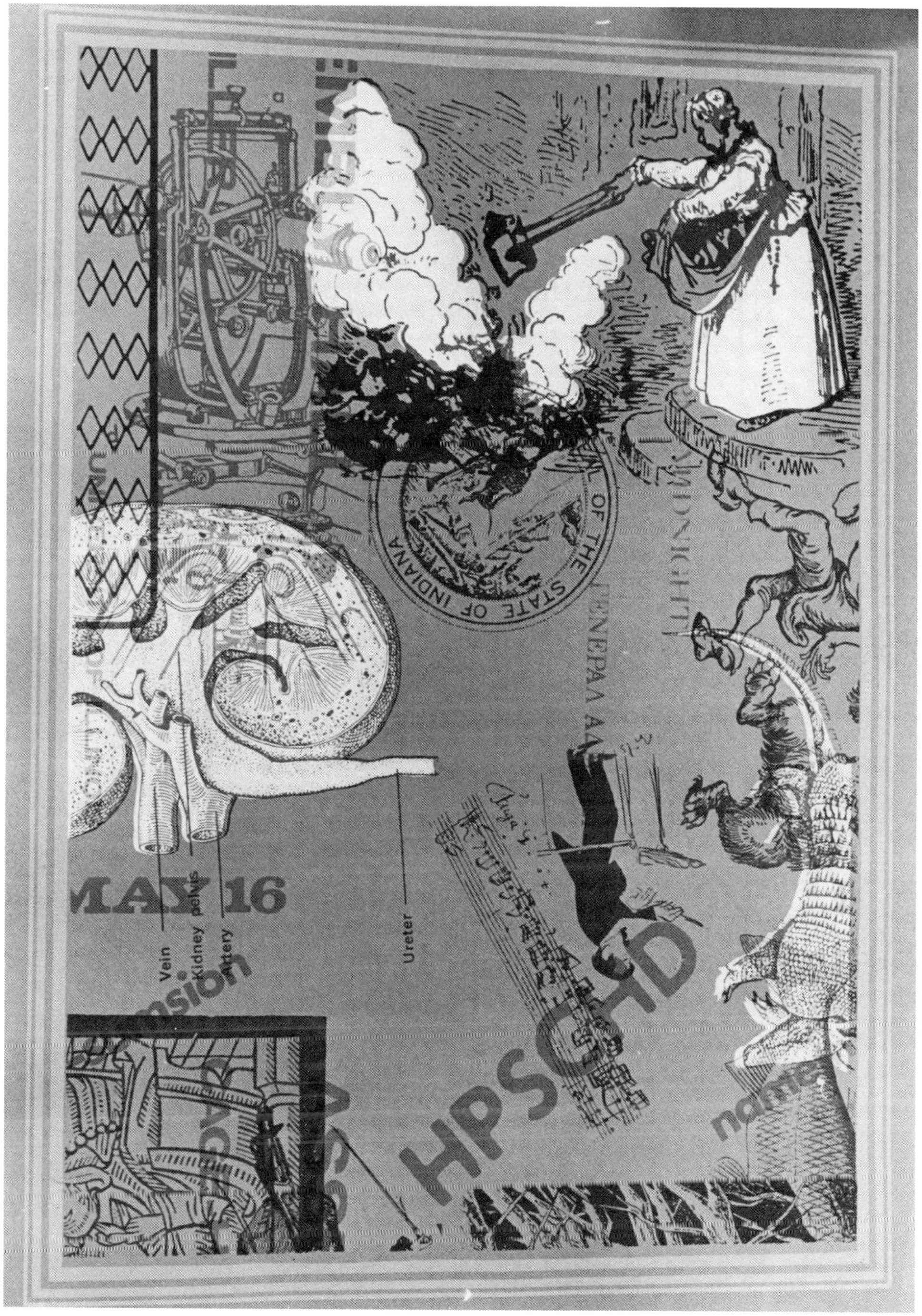

Fig. 4-8: A third poster from HPSCHD (from the collection of David Eisenmann; photo courtesy of Champagne-Urbana *News Gazette.*).

> played a computer-generated "part" written on traditional music paper, designed for a conventional twelve-tone octave tuning. But all this represents no more than a single aspect of the total "performance." Slides from fifty-two projectors flashed on the exterior of the hall, and the interior was a mass of multiple film and slide imagery—a collage of apparently unrelated visual material, from pages of Mozart's music to abstract shapes and scenes from outer space—sifting through a series of colored lights suspended from the ceiling. The audience of several thousand stood, sat, moved about the large hall as they wished, a few occasionally dancing.[58]

The possibility of using a computer without reference to the "total control" ideology (the computer being asked to perform a selection in a set of possibilities) allows for transcending the fixity of the so-called "musical object." In place of such an "object," we are confronted with a *process*, the play of which, as Cage says, is "purposeless."[59] And the very idea of *form*—the "definition" of the work, its obvious limits—becomes flexible. We are in fact plunged into an all-encompassing stream, which unfocuses the individual's attention and opens the listener to a pervading multiplicity of poly-sensory events, since "sounds and non-sounds," as Cage teaches us after Daisetz Suzuki Teitaro, are "interpenetrating without obstruction."[60]

But *HPSCHD* also marked a new step in the field of using electronic material as a part of a live performance. Since *Cartridge Music* (1960), Cage had broken "out of the closed cycle of music committed to tape"[61] by allowing the performance to be "indeterminate of itself." For instance, simply by obeying the score, i.e., the chance operations canvas prepared by the composer, one player may lower to zero the sound volume of another player's activity. In other words, Cage accepted not only the noise of the amplifiers, but the "internal" noise of the score itself, conceived as a camera going wrong. Ironically, in HPSCHD *the nontransparency or nonneutrality of the medium itself* becomes —thanks to the innumerable interferences between the various media involved—the very "subject" of the work. The unfocussing of attention becomes "positively" an activity, a style not only of appreciation but of life, a truly "composing" (and at the same time "de-composing") behavior. For we, the observers, are taken *into* the composition itself, which is viewed, according to Cage's words, "as activity of a sound system, whether made up of

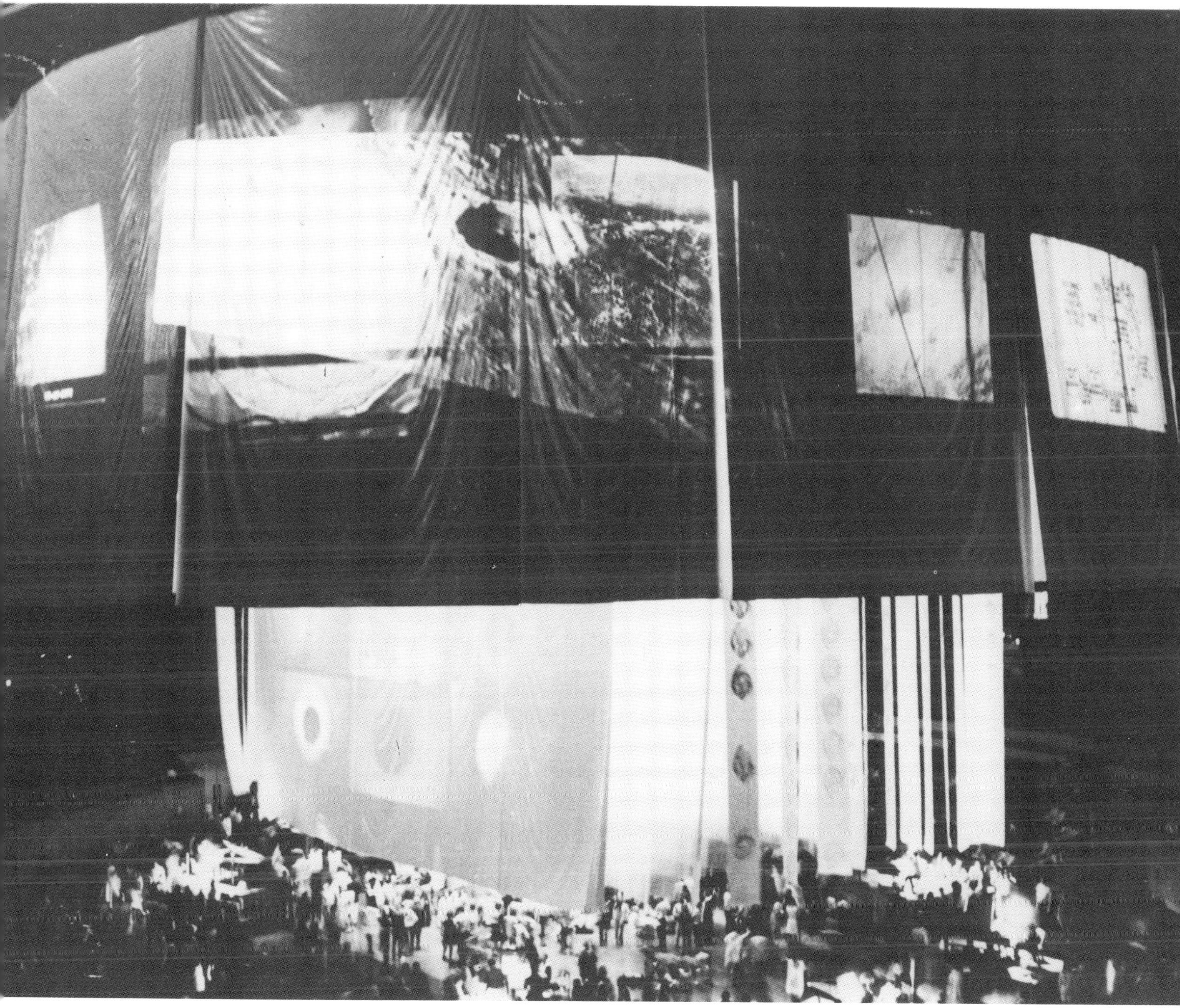

Fig. 4-9: The Assembly Hall at the University of Illinois, Champaign-Urbana, during the performance of HPSCHD. (Courtesy: Assembly Hall, University of Illinois at Champaign-Urbana)

electronic components or of comparable 'components' (scales, intervallic controls, etc.) in the mind of man."[62] We cannot but accept the feedback between the composition and its (multimedia) environment which is at the core of our lives nowadays.

Conclusion: Towards Electronic Meditation

For Cage, technics and art have in common that they are not mere tools, i.e., machines achieving the *material* purpose of producing goods, but *spiritual* items (or *koâns*) leading to the extension of our minds in a global systemics. When the art "object" becomes as disembodied as electricity itself (or when it shares the technological embodiment of the media, according to Ihde's thesis), then art as *process*, or art as *idea*, or art as *concept*, has to be "circulated as information" (even if art always carries *more* than mere information) and not only be (in Kathleen Woodward's words) "tastefully consumed by connoisseurs."[63] "Like white Gaussian noise," the new music, the near-pure circulation of the "most minimally encoded" signals, contains "the epitome of the unexpected" and is thus at the origin of all the arts.[64] So the ultimate criterion of the contemporaneity of any art—not only music—consists in its ability to incorporate and embody environmental elements, whether they are sounding elements or not, i.e., in its ability to behave as an open system.

After Cage, let us now quote Boulez:

> Nothing is based on the *masterpiece*, on the closed cycle, on passive contemplation, on purely aesthetic enjoyment. Music is a way of being in the world, becomes an integral part of existence, is inseparably connected with it; it is an ethical category, no longer merely an aesthetic one.[65]

Unfortunately, such a description does not concern Boulez himself or his music, but non-Western musical traditions as opposed to the Western art tradition. And we have to think seriously about such an opposition.

Cage's celebration of the "electrical sublime" leads to a rhetoric of abundance and utopia which, as James Carey and John Quirk have suggested, risks preventing us from analyzing the real strategy of electric light and power companies. Further, if we compare Cage with those Third World musicians who, like African and Balinese artists, work without any devotion to Western urban culture and independently from the Western concert tradition, we may ask if the real challenge to that tradition is conceivable *inside* that tradition itself. In other words, the challenge coming from "tomorrow's world" is maybe less effective than the one coming from *elsewhere*, if "elsewhere" really means *another world*.

The problems arising in the present-day relationship between music and technology can be solved, it seems, if they involve only matters of programming, of audiophysiology or psychoacoustics, and the like. But solutions are harder to find if the problems concern the very nature of musical life in the different cultures. Even if the old antipathy against everything "technical" no longer operates in our computerizing world, one may have to find some kind of escape from "the leveling influence of the loudspeaker," for instance in the non-Western countries, where European models of what music could be were never relevant.

GUIDE: Fig 4–1

Harry Partch's rejection of European classical music embodies an attitude that maybe leads to a more radical view of "technology made human." Partch said that his work, "parallels much in the attitudes and actions of primitive man," who "found sound-magic in the common materials around him," then "proceeded to make the vehicle, the instrument, as visually beautiful as he could," and finally "evolved the sound-magic and the visual beauty in his everyday words and experiences."[66] Partch built his own instruments for his own pieces, stipulated that they had to play a *visual* as well as a musical role, and insisted on the importance of tuning and maintenance. Such an attitude may not *come* from outside but it clearly *leads* outside of all of the contemporary musical scene.[67]

We may also be reminded of some of the revolutionary ideas of another "unknown" experimental composer, Dane Rudhyar, who treats the whole question of art as no longer an aesthetic matter, but as "inseparable from the question of life itself." (Jose Argüelles) For him, all forms of life are *performers* a term which is to be equated with *transformers*, since *per*–means *through* and *through* is *trans*–, and all performances are *magical*. Therefore:

> . . . if today's magic resides in an electrical generator, and today's magician is the engineer, in another culture the magic may be in the raindance ritual, and the magician the dancer, the shaman or medicine man who brings the rain.[68]

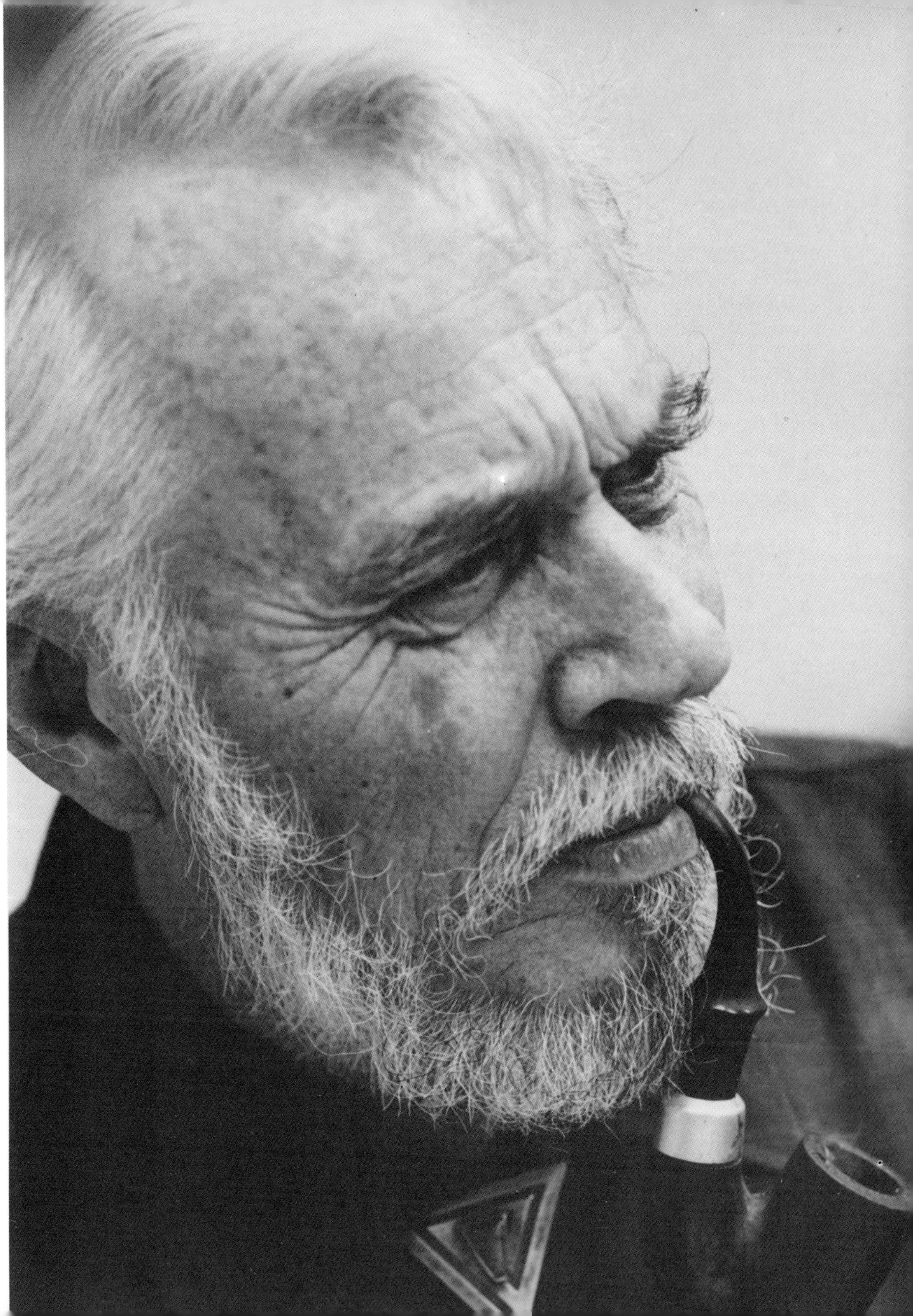

Fig. 4-10: Harry Partch (Courtesy: CBS Records)

Dieter Schnebel wrote in 1972:

> The New Music has enriched itself with much that is exotic, sometimes from as far afield as Eastern Asia . . . What is nevertheless still lacking amidst this more universalized music is a World Music as such. A prime requisite for this would be a real effort to experience the strangeness and peculiarities of the many kinds of exotic music, and that means thoroughly to know them. Only when one can truly see the Other as such, . . . can it truly be combined with one's Own. Certainly, as long as there is so little movement in this direction in the sphere of political and social relationships, the prospects of World Music itself are slim. It can proclaim the *harmonia mundi* only when this has been accomplished, and even then not in unison, but only in an as-yet unheard polyphony.[69]

The very term of "polyphony"—not "unison" shows the difficulty of syncretism. As Peter Michael Hamel says:

> Meditative music will never stand much chance in the concert hall or the Pop-event . . . Thus the sort of things that happened at concerts by the guitarist John McLaughlin or the group *Santana* was that the stars, photographed in white with their guru on the record sleeves, along with quotations of holy words about cosmic consciousness and divine grace, produced on the stage an unholy racket, a mindless magic, because the audiences were impatient and created an atmosphere that was far from spiritual.[70]

But this same Peter Michael Hamel, having asked the eighty-year-old Indian musical philosopher Jaidev Thakur Singh if electronic music could ever be "spiritual" reports that the old man, after hearing some of this music,

> . . . replied that this music in the hands of a *conscious* person could quite well have a spiritual effect, and he could easily imagine a synthesizer-*tambura*. In fact it would be much better tuned than the mechanical Indian instrument.[71]

Now let us listen to La Monte Young or Alvin Lucier or Max Neuhaus or Walter Marchetti; they play perfectly well-tuned electronic *tambura*. And, in the words of Dieter Schnebel, their drones stand *between* "polyphony" and "unison."

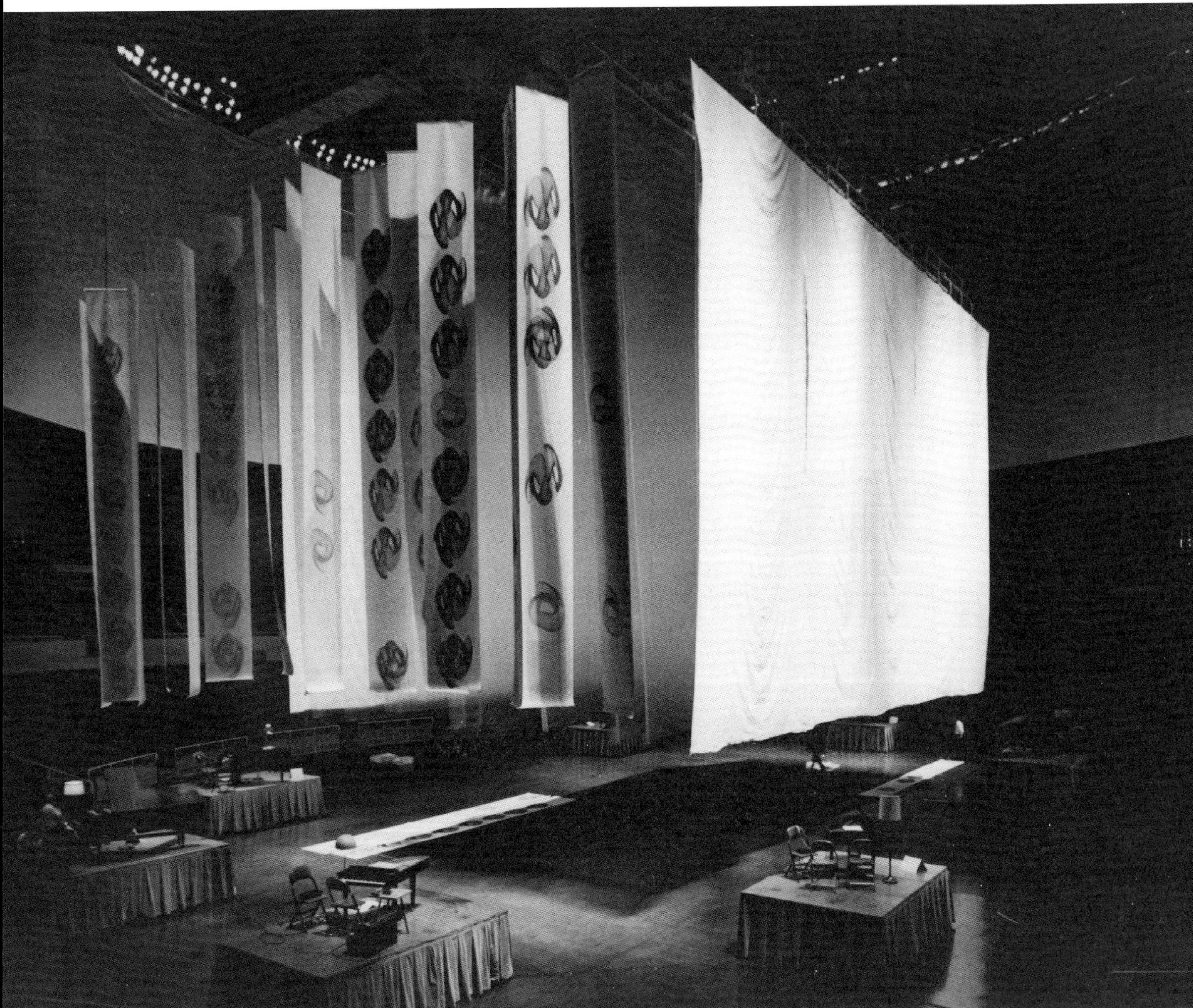

Fig. 4-11: The Assembly Hall at the University of Illinois, Champaign-Urbana, set up for HPSCHD performance. (Courtesy: Assembly Hall, University of Illinois at Champaign-Urbana)

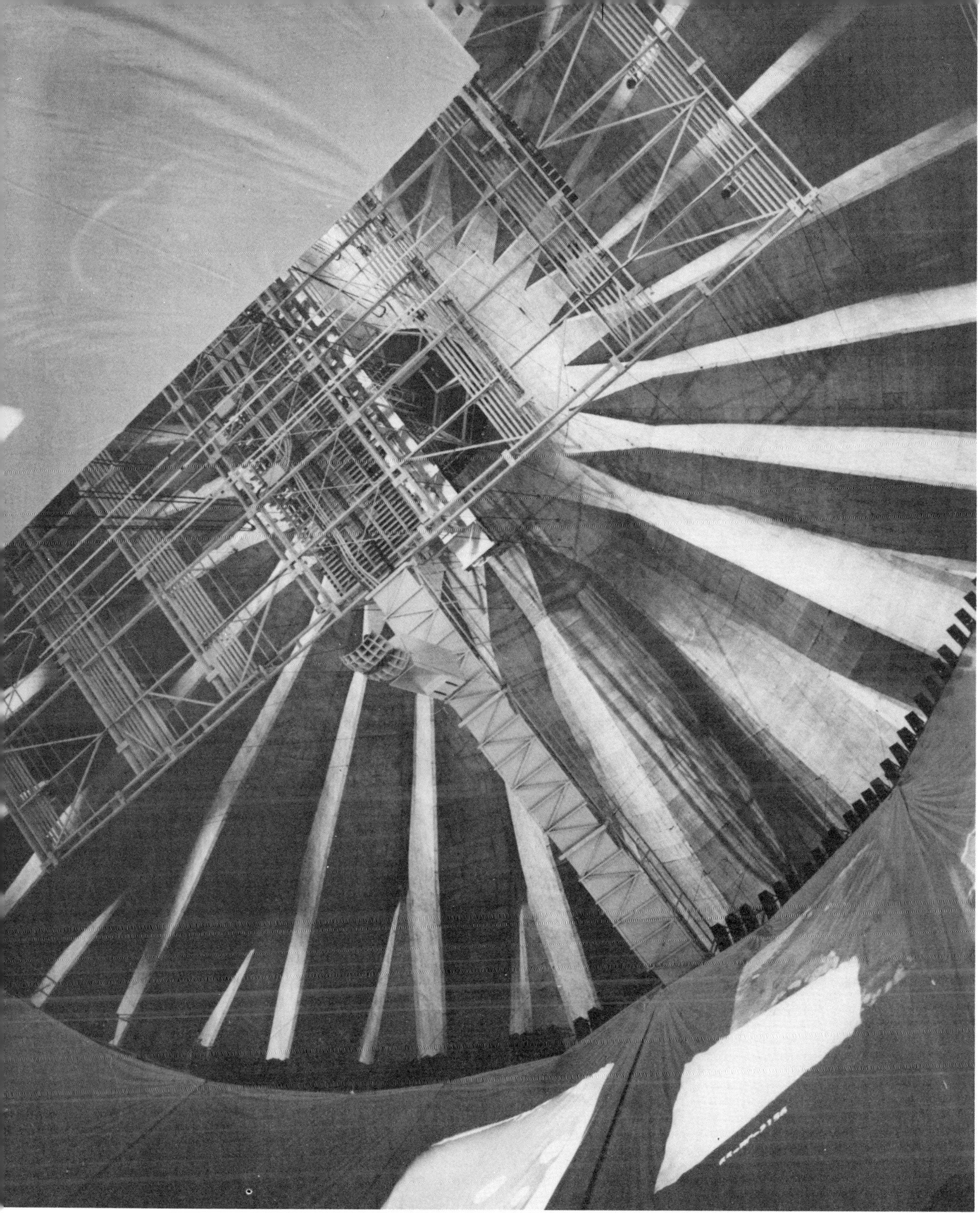

Fig. 4-12: Looking upward toward the ceiling at the Assembly Hall, University of Illinois, Champaign, Illinois, showing curtains for HPSCHD performance. (Courtesy: Assembly Hall, University of Illinois, Champaign-Urbana)

Footnotes

1. *Manual of the German Radio*, quoted by Werner Kaegi in "Music and Technology in the Europe of 1970," *Proceedings of the UNESCO Meeting at Stockholm on "The Composer in the Technology Era,"* June 8-12, 1970.
2. Werner Kaegi, *op. cit.*, p. 2.
3. Lewis Mumford, *Art and Technics* (New York: Harcourt, Brace and World, 1963), p. 7; quoted by Kathleen Woodward, "Art and Technics: John Cage, Electronics, and World Improvement," in Kathleen Woodward, ed., *The Myths of Information: Technology and Post-Industrial Culture* (Madison, Wisconsin: Coda Press, 1980), p. 173.
4. Mumford, *loc. cit.*, p. 58; quoted in Woodward, *op. cit.*, p. 173-174, note 9.
5. Quoted by Reginald Smith-Brindle, *The New Music* (Oxford: University Press, 1975), p. 22.
6. Quoted by Christopher Small, *Music-Society-Education* (London: Calder, 1980), p. 82.
7. Small, *op. cit.*, p. 82.
8. Don Ihde, *Existential Technics* (Albany, N.Y.: SUNY Press, 1983), pp. 100-101.
9. *Ibid.*, pp. 104-195.
10. *Ibid.*
11. *Ibid.*
12. Kaegi, *op. cit.*, p. 3.
13. *Ibid.*
14. *Ibid.*
15. *Ibid.*
16. Roland Gelatt, *The Fabulous Phonograph* (Philadelphia and New York: Lippincott, 1954, 1955), p. 52; quoted by Geoffrey Payzant, *Glenn Gould, Music and Mind* (Toronto: Van Nostrand Reinhold, 1978), p. 41.
17. Quoted by Werner Kaegi, *op. cit.*, p. 4.
18. Quoted by Geoffrey Payzant, *op. cit.*, p. 21.
19. *Ibid.*, p. 44.
20. Payzant, *op. cit.*, p. 40.
21. Reginald Smith-Brindle, *op. cit.*, pp. 22-23.
22. Theodor W. Adorno, "Technique, Technology and Music Today," English translation, *Ordini* (Rome, 1959); quoted by Smith-Brindle, *op. cit.*, p. 50.
23. Small, *op. cit.*, p. 88.
24. *Ibid.*
25. *Ibid.*, pp. 88-89.
26. *Ibid.*, p. 89.
27. From the title of Stockhausen's well-known commentary on his works *Zeitmasse* and *Gruppen fur Drei Orchester*; English version, "How Time Passes," in Die Reihe, vol. 3, *Musical Craftsmanship* (Bryn Mawr, Pa.: Theodore Presser, 1959), pp. 10-40.

28. Quoted by Small, *op. cit.*, p. 121.
29. Small, *op. cit.*, p. 121.
30. *Ibid.*, p. 118.
31. Martin Heidegger, *On the Way to Language*, trans. by Peter D. Hertz (New York: Harper and Row, 1971), pp. 102-104.
32. Don Ihde, *op. cit.*, p. 49.
33. *Ibid.*, p. 50.
34. *Ibid.*, p. 51.
35. *Ibid.*, p. 55.
36. *Ibid.*, p. 56.
37. *Ibid.*
38. *Ibid.*, p. 61.
39. *Ibid.*, p. 59.
40. Quoted by Geoffrey Payzant, *op. cit.*, p. 42.
41. Don Ihde, *op. cit.*, p. 62.
42. Quoted by Geoffrey Payzant, *op. cit.*, p. 26.
43. Quoted in Kaegi, *op. cit.*, p. 6.
44. Kaegi, *op. cit.*, p. 6.
45. *Ibid.*, p. 13.
46. *Ibid.*
47. Leonard Meyer, *Music, the Arts, and Ideas* (Chicago: The University of Chicago Press, 1967), pp. 158-169).
48. Cf. Elliott Schwartz, *Electronic Music* (New York: Praeger, 1973), p. 39.
49. Kaegi, *op. cit.*, p. 7.
50. Smith-Brindle, *op. cit.*, p. 104.
51. Elliott Schwartz, *op. cit.*, p. 70.
52. *Ibid.*
53. *Ibid.*, p. 78.
54. *Ibid.*, p. 89.
55. *Ibid.*, p. 92.
56. *Ibid.*, p. 95.
57. *Ibid.*, p. 96.
58. *Ibid.*, pp. 102-103.
59. John Cage, *Silence* (Middletown, Conn.: Wesleyan Univ. Press, 1961), p. 12.
60. John Cage, *For the Birds* (in conversation with Daniel Charles) (New York & London: Marion Boyars, 1981), p. 91.
61. Michael Nyman, *Experimental Music: Cage and Beyond* (London: Studio Vista, 1974), p. 75.
62. John Cage, quoted by M. Nyman, *op. cit.*, p. 77.
63. Kathleen Woodward, *op. cit.*, p. 176.
64. *Ibid.*, p. XXIV.

65. Pierre Boulez, "Alea," trans. David Noakes and Paul Jacobs, *Perspectives of New Music*, Vol. 3 No. 1, Fall-Winter 1964, pp. 42-53.
66. Quoted by Christopher Small, *op. cit.*, p. 154.
67. Cf. Christopher Small, *op. cit.*, pp. 154-159.
68. Jose A. Arguelles, *The Transformative Vision* (Boulder Colorado: Shambala, 1975), p. 229. Cf. Dane Rudhyar's last book, *The Magic of Tone and the Art of Music* (Boulder, Colorado: Shambala, 1982).
69. Quoted by Peter Michael Hamel, *Through Music to Self* (Boulder, Colorado: Shambala, 1978), trans. by Peter Lemesurier, pp. 153-154.
70. Peter Michael Hamel, *op. cit.*, p. 153.
71. *Ibid.*, p. 138.

Concerning Daniel Charles's "Music and Technology Today"

Jose Maceda

There are two different topics that can be discussed concerning the historical process of interpenetration of music and technology—the process so well described by Professor Charles—as that process developed from the 1950's to the 1980's. The first topic concerns the dissemination of existing music through storage and communication techniques; and the second topic considers the creation of a new music using live electronic systems including the computer.

Storage, reproduction and dissemination techniques for music affect the whole musical world. In Europe, America, Africa and Asia all kinds of music (folk, pop, reggae, classical, avant garde, etc.) may be heard essentially through one dimension—that of the loudspeaker. Of course, this instrument varies greatly in quality, size and make. Nevertheless, when one speaks of the medium as the message, the loudspeaker becomes a rather restricted medium in comparison to original sounds of gongs, bamboo and human voices. Moreover, the facility of this medium as used in exchange for the original musical instrument reduces listeners to mere recipients of a market commodity, for it takes more than mere listening through an electronic device to understand the inner meaning and significance of a music detached from its original environment and cultural setting. Techniques of communication are indeed a great help for the dissemination of a music culture, but they need to be supplemented by human contact—by participation in the actual environment of the music or by a live performance that would show hidden refinements of behavior and performance styles unseen in the reproductions.

When we consider music composition and the use of live electronic systems and computers in fashioning a new aesthetics of music, such a music—like the computer it uses and serves —can influence a lifestyle more profoundly than the reproduction of existing musical genres can bring about a new world of musical thought. In Charles' paper, some items may be chosen as having something to say in reference to musical technology of the computer:

— the obliteration of memory through accumulation and repetition in serial music as well as in computer music.
— the domination of nature by a conscious control of it—"to fix the infinite . . . to commit oneself to the work of preservation of anything that the past produced."
— the importance of flavor rather than content of a musical meaning.
— the computer allows us to transcend the musical object, the sound itself, and be confronted by a process which is without a purpose.
— "technics and art . . . use machines as spiritual items extending our minds to global systemics." The art object as process, idea or concept in the electromagnetic medium is circulated as information.

Having these as subjects for discussion, I shall now talk of music creation in relation to technology and the computer.

To be sure, following the original concept and aims of technology, it is conceivable to create a world of sound—using silicon chips, transistors, circuits and loudspeakers replacing "natural" musical instruments—making electronic tools an extension of the human mind and body; this follows, however, the same logical-linear thinking, organization and efficiency of parts that a technological society incorporates. Airplane flights, travel schedules, data banks, weather forecasts, communications and a host of other human activities make up a physical counterpart of that world of electronic sound. Both worlds are truly a machinery of moving parts that must constantly be checked, maintained, replenished or replaced. Learning, research and discovery are a part of that maintenance job, an important and necessary tool of the whole machinery. Altogether, the unceasing activity of all these parts results in a great tension inherent in continuous movements and great speeds. Activity and tension go together.

Is it possible to have all of humanity attain such a high state of technology and tension, and then maintain and improve on it indefinitely? One may get tired of music coming out of loudspeakers just as one may become dissatisfied eating chemically treated food from cans and packages. Are we not talking of a computerized society which, in appearance, controls the technology it runs, but which, in another sense, becomes a servant of that technology because the human being is forced to behave like a machine which serves that society? In such a world there is little room for imprecision, uncertainty and doubt, qualities experienced by human beings in a contemplation of things unknown, especially in the spiritual world. Other human attributes like love, kindness and devotion are spurned or given a secondary consideration in favor of reason, efficiency and accuracy, things necessary for a controlled world of matter, but more difficult to apply to human emotions. A material, technological world of comfort, well-being and personal freedom is a view of life, a product of thinking, that is relatively new, and very different from old human traditions in Asia, America, Africa and even ancient Greece, from which a source of that thinking may have been drawn.

In this other world — which has been on this earth for hundreds if not thousands of years, with a learning that may be measured more by its experience in time than by an acquisition of wealth in tools and technology — a great bulk of humanity pursues a life preoccupied less with material well-being than with an accommodation of life with nature, involving a deep spirituality manifest in traditional feasts, ceremonies and ritual practices.

If we meet technology halfway and accept its contribution to human welfare, we may perhaps view technology with another purpose and sense of value—we will see it less as a tool to save time, human effort and to provide for a monetary profit than as a means of employing many men to produce a product, a machinery that admits, rather than rejects, people. A situation of one million people reading one million pages of a text distributes a knowledge of that text to human beings who can not only feed back that same information, but who can also make countless permutations concerning it. The choices of interpretation of that information are much larger than what a computer can muster. More important than quantity of choices is the capacity of each person to generate other constructs covering things unrelated to the original instruction of ferreting out specific information. The human being is a richer source of creative knowledge, so much richer that the computer cannot compete; this reminds us that there is a basic difference between a simple human being and a very well-informed machine.

In the India of the early thirties, statesman Jawaharlal Nehru preferred the employment of hundreds of workers, in a work of pier hands unloading big ships of their cargo, to the use of a machine that would do the job in hours, but that would deprive them of their daily means of subsistence. In rural Asia, farmers work in groups in the yearly planting and harvesting of rice, a work procedure that distributes the products of their labor equitably amongst themselves. In some Asian villages, other teams similarly work together to build a new house, prepare a rice field, construct or repair a road, and so on.

Similarly in music, a melodic phrase is played with the coordination of groups of performers. The Tiruray in Mindanao use a set of five gongs, each one played by one person executing a rhythm of his own; and jointly, the different rhythms produce a melody. The Kalinga of Luzon make a music of flat gongs using the same principle of coordination between six people playing on six gongs in decreasing size and rising pitch. While the first four players play one rhythmic pattern, each delayed in entry by one beat, thereby producing a rising succession of tones, a fifth player provides a drone, and a sixth plays an improvisation on tonal colors.

In Sarawak Borneo, four or five suspended gongs of different beats produce an ensemble of rhythmic colors, impossible to achieve with only one performer. An exchange of musical parts, *kotekan*, in the Javanese and Balinese gamelan illustrates a very sophisticated work of cooperation which is different from the cooperation in a harmonic process of music making in Western orchestral music.

Contrary to the work principle prevailing in technology and the computer, which takes away jobs of people much faster than new jobs can be provided for them, a plan of work employing, rather than eliminating, people to do the job may conceivably be applied in a search for another concept of the machine. The idea of profit would, of course, be completely changed. It would not be a profit of one man or of one giant multi-national enterprise— a profit pattern that brings a disequilibrium in the lives of thousands, if not millions, of people. Instead, the profit would be shared and divided among the very producers of the product. Taken on a global scale, and involving thousands of enterprises, profits would be shared and diffused not only among the immediate employees but also among people who buy the products of the enterprise. The shares for each person—the cost to be paid by the buyer, and the pay for each employee—would be relatively small, but the discrepancy between profits and costs would also be small, allowing for a more restrained economy.

But economic principles of reproduction are not the only source of ideas that could contribute to a new musical thinking and that could be reflected in another use of technology. Such ideas could come from an appreciation of village thinking, especially of a built-in equilibrium between a material and spiritual side of human beings, an equilibrium which modern technology cannot come face to face with.

The danger of an engineering process of ever-improving technology while eliminating man as employee, is that, as one refines technology one is lured into a world of sophisticated circuits that has a built-in efficiency with comfort as its goal and a world with set limits of time. Man himself becomes a part of that time. A telephone rings so many times a day; a computer accomplishes several tasks all within measurable time; and man follows along with his actions and his thoughts also measurable in time—time to go for a vacation, finish a paper or get lost in a theater—in a perfectly controlled world of technology. The contents of the papers he writes, the music he listens to, and the telephone and the computer itself become items, "bits" in a wholly computerized, technological world he has built. He literally becomes a part of his own man-made machine.

A Note on HPSCHD

Thomas P. Parkinson[1]

The Assembly Hall of the University of Illinois at Champaign-Urbana was deeply involved in every step of the production of John Cage and Lejaren Hiller's HPSCHD. We helped locate and transport the seven harpsichords—although some of these were located by Cage himself. We figured out how to construct and handle the big screens. We used huge sheets of plastic, such as contractors use to wrap buildings. These were suspended from the catwalk, 150 feet in diameter, of the dome of the Assembly Hall. Inside that circle we suspended several other sheets of plastic together with a number of long narrow ribbons. On the floor below were seven small stages on each of which John Cage placed a harpsichordist with floor lamp and other trimmings.

The Assembly Hall staff handled the advertising, publicity and promotion for the event. We also handled the lighting, audio and other stagecraft. The Minnesota Mining and Manufacturing Company (St. Paul, Minnesota) lent us 52 Wollensak 1520 AV tape recorders; Shure Brothers Company (Evanston, Illinois) provided pre-amplifiers, microphones and solophones; amplifiers were provided by Ancha Electronics; headphones came from the School of Music of the University of Illinois; loudspeakers were provided by the Public Functions Department of the University of Illinois Physical Plant; 80 Rotomatic 707 AQ slide projectors were provided courtesy of Sawyers Slide Projectors, GAF Corporation (New York City); and the screen material of white opaque polyethylene came from the Union Carbide Company. Funding was provided by grants from the University of Illinois Research Board and from Thorne Music Fund, Inc. Coordination of the production was handled by Jack McKenzie, and the event was produced by The School of Music of the University of Illinois and the Assembly Hall.

Punch cards were used to generate the program on the computers, the Illiac II and the IBM 7094. Computer programming was done by Lejaren Hiller, John Cage, and Laetitia Snow, and the computer-created sounds were generated on tape with the assistance of James Cuomo, Paul Krabbe and James Stroud. The sound system and recording apparatus were constructed by Jaap Spek and George Ritscher, and tape editing was done by James Stroud. The images for HPSCHD were designed by Ronald Nameth and Calvin Sumison.

The performance used 7 harpsichords, 8 movie projectors, 52 tape recorders, and 64 slide projectors. The sound system comprised 7 pre-amplifiers, 59 power amplifiers and 59 loudspeakers. There were a 340-foot circular screen and 11 rectangular screens. The software for the production consisted of 631 pages of music manuscript, 208 computer-generated tapes, 6,400 slides (mostly from NASA), and 40 motion picture films. Silk screened posters, streamers and smocks were made before and during the event.

Cage expected the production to be attended as one would a Chinese Opera: people come in for a while, leave, then return for more. In practice, we never noticed anyone leave. The floor was always jammed with people. In many cases they remained prone for the evening.

Stories about this event were carried in many newspapers and newsmagazines, including the *Chicago Daily News* (May 10, 1969), the *Chicago Sun-Times* (Sunday, May 11, 1969), the *Decatur* (Illinois) *Sunday Herald and Review* (May 11, 1969), the *Chicago Tribune* (Sunday, May 11, 1969, and Sunday, May 18, 1969), *The New York Times* (May 25, 1969), *Time* (May 30, 1969), and *National Geographic*. In the intervening years since the production of HPSCHD we have received numerous requests for information and photographs of this event.[2]

The arena in which HPSCHD took place was opened in 1963, and was designed by Harrison and Ambormovitz, who designed the United Nations, Lincoln Center, Radio City, and others. The Assembly Hall seats from 3,600 to 17,000 people, and is used for ice shows, circuses, rock music, Broadway shows, country music, basketball, and major ballet.

Notes

1. (Editor's note.) The author is the Director of the Assembly Hall of the University of Illinois at Champaign-Urbana. This account of the HPSCHD event was written in June, 1985, especially for the present volume. Mr. Parkinson did not attend the conference in which the other papers in this book were presented, but he provided information and photographs of HPSCHD at the editor's request, as a special supplement to the paper of Professor Charles.

2. The editors acknowledge the information about the event and the photographs received from David Eisenmann of Champaign, Illinois. Mr. Eisenmann produced silk screened posters and T-shirts during the HPSCHD event, wrote an article about John Cage that appeared in the *Champaign-Urbana News Gazette Weekend* (Friday, February 22, 1985), and provided the photographs of the silk screened posters that appear here.

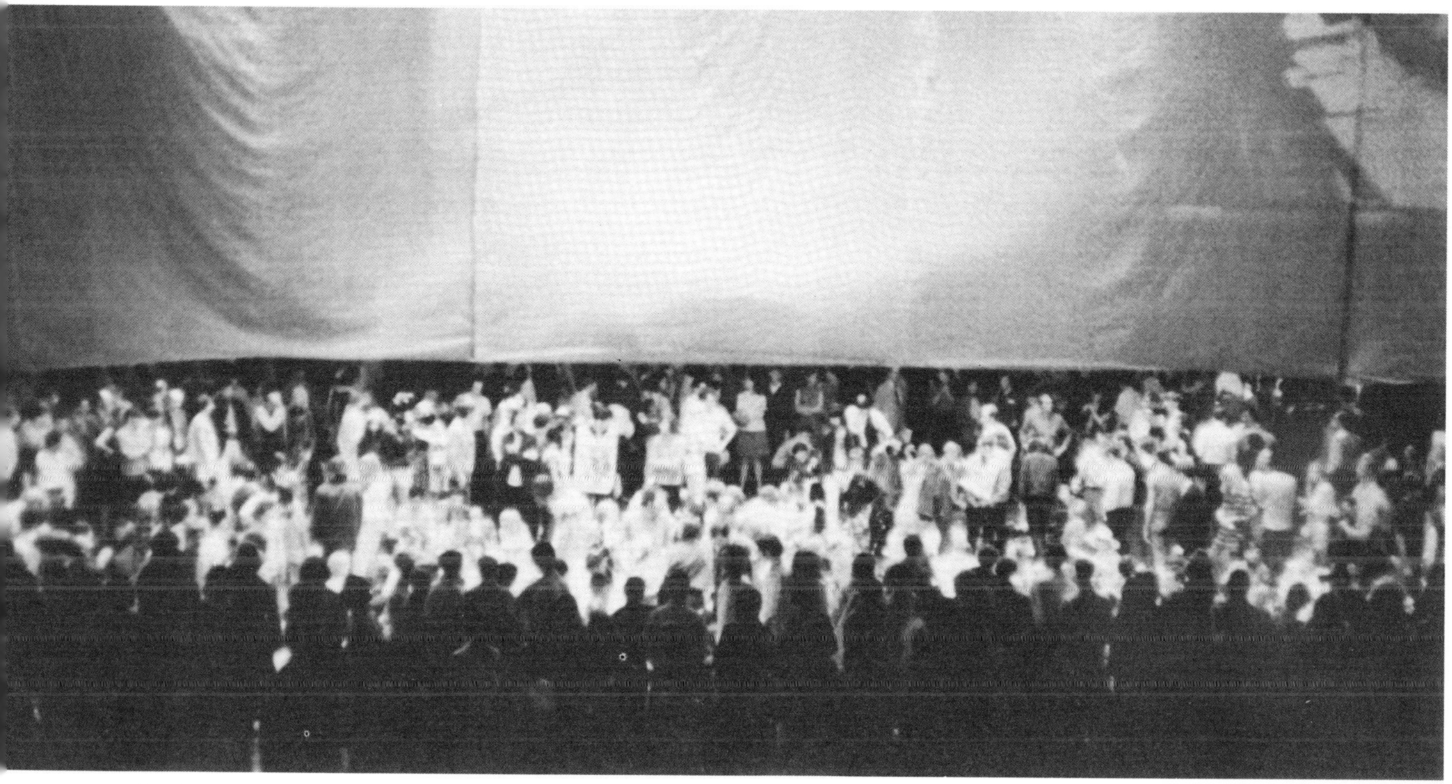

Fig. 4-13A: Assembly Hall, University of Illinois, Champaign, Illinois.

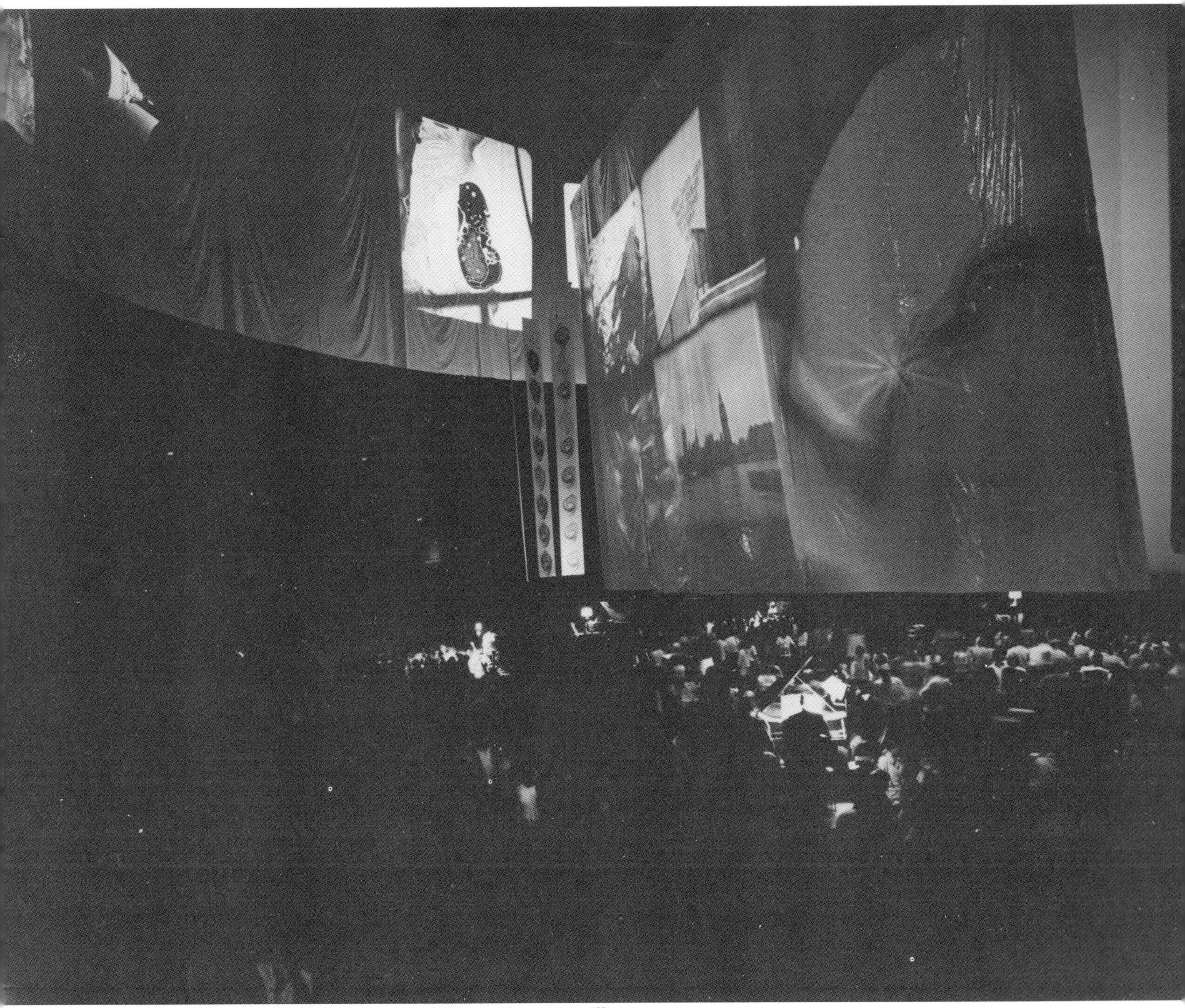

Fig. 4-13B: Assembly Hall, University of Illinois, Champaign, Illinois.

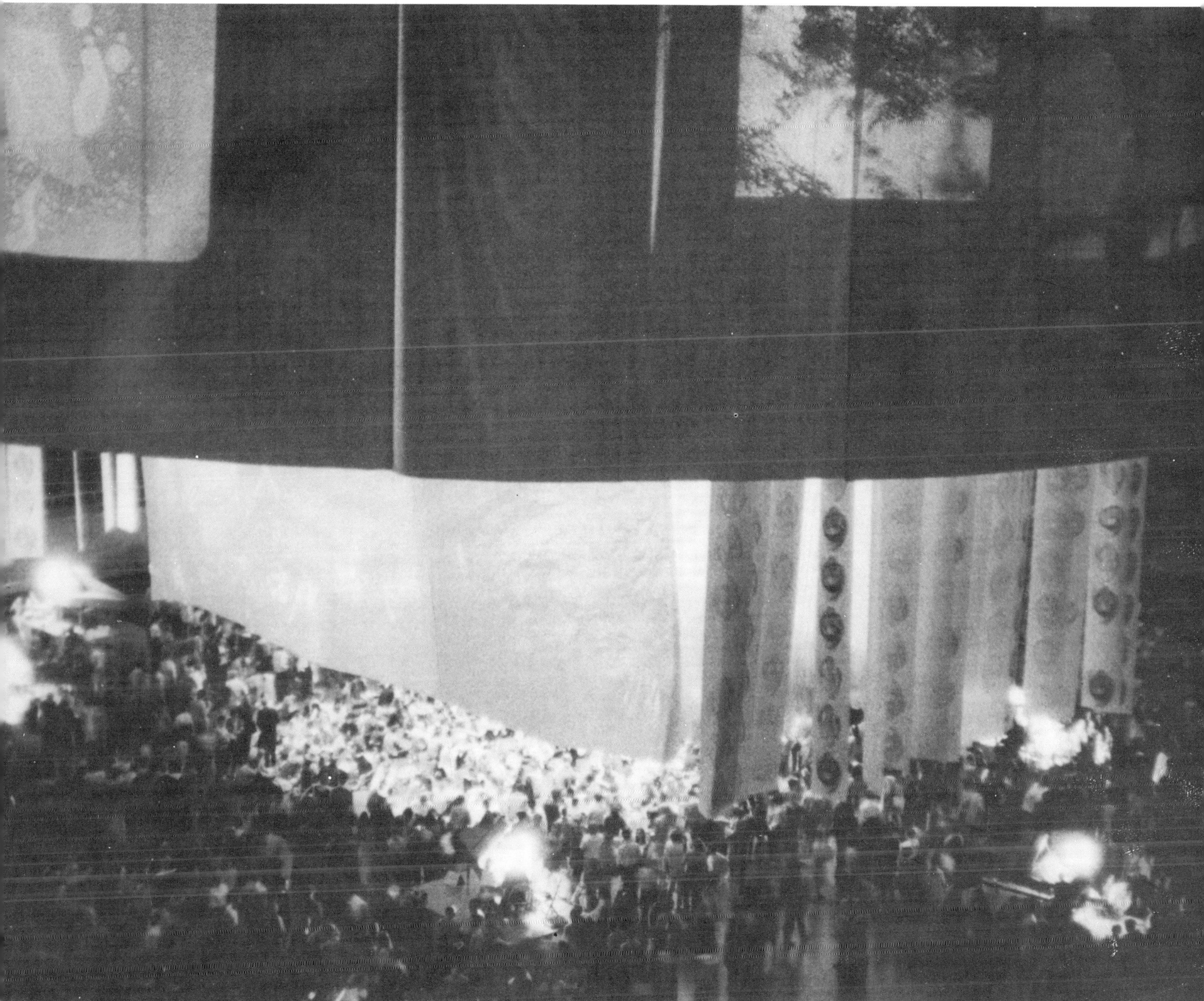

Fig. 4-14: Assembly Hall, University of Illinois, Champaign, Illinois.

Chapter 5

Television And Art: Television As Art

Television And Art: Television As Art

Robert Stephane

Introduction: The Difficulty of a Changing Relationship

Art has never been a more changing concept than today. Some even see it as a dissolved concept. But notwithstanding this extreme position, art is ingrained into a pattern in our societies and our consciousness, a pattern that is quickly changing. At the same time, art-producing technology is exploding. Computer art is only one example of this technical trend.

The scope of what is accepted as art has developed both diachronically and synchronically, and the field of what is considered to be legitimate art has been enlarged by the inclusion of different types of reality. I think for instance of a movement, developed during the forties, which brings together all the creative arts associated with interior design in the field of series fabrication and standardization. At the same time, it concerns industrial design in general. It is a strange reappearance of the "ready mades" that Marcel Duchamp has signed since 1913 ("Le Porte-Bouteille"); current daily objects are taken as such and called "works of art" based on a simple decision by the artist.

In avant-garde work (experimental work), research concerning difference as a tool of assertion has led smoothly to the dissolution of any thread of the notion of art. It has gotten to the point that we have seen the appearance of so-called conceptual art, which is in itself a question put beforehand to art—its self-analysis and its self-criticism. As is known, conceptual art does not mean the creation of a new repertory of forms, as was done earlier by cubism, abstract art and pop art. Conceptual artists tend to think of artistic phenomena along completely new norms, so that any further recuperation is made impossible. The American painter Joseph Kosuth takes the suppression of any connotation (the perfectly neutral thing) to be the supreme aim; he has, for instance, exhibited objects like a chair and, at the same time, the most banal photography of that chair, and—still at the same time—the definition of the chair taken from the dictionary. In that way he thinks that he keeps the viewer away from aesthetic emotion. That seems to him to be the primary condition for all discovery of the very nature of art. In the same spirit, the artist known as Ben has exhibited himself during two weeks in a showroom in London in 1962.

Television, in a similar way, is going through a strongly transformational period. The development of television was so strong and pervasive that it overrode our mental landscape in such a way that, for some, an event does not exist outside of its TV representation. It is here that we get to one of the first differences which makes TV another medium; the necessity of the picture tends to make this picture into the event itself. Here we must recognize a very important cultural change: the problem of scientific truth cannot be put, if we follow Ilya Prigogine, completely outside of *a certain construction* of truth and reality.

An event is not created by itself, but by its audiovisual production. It is through television that a very new, very contemporary (and very old at the same time) philosophical truth seems to become extremely obvious. Events are created through their representation and not through their occurrence. This takes place in political life as well as in daily life. In a certain way, television is becoming a tool which produces so-called cultural truth.

One could add two more elements of the definition of television which increase its influence:

1. Television shows the same thing to millions of people at the same time. It gives to what is shown, whatever it is, a credibility of its own and, without doubt, an unequalled persuasion power.

2. It touches people at home.

To conclude these remarks, let us say that at least television is a certification tool for reality, and sometimes even a tool of reality creation.

This brings us to consideration of the nature of a changing technology and the changing of the gatekeeping process.

1. Television now receives new inputs. New systems of gatekeeping rule, and will later decide on the inclusion or the censorship of certain types of news and artistic messages coming from the creation milieu to the reception units.

2. The television screen is changing in such a way that it now creates, and will recreate, old relationships between viewers and pictures. On another side, the interface between the sphere of the sender and that of the receiver will diversify and become more complex. One can even say that the interface is becoming interactive in a certain way. That phenomenon is completely original and new; it takes the shape of visiotelephony, videotex, phone participation, Qube-type systems and so on.

Television and Art

The TV Terminal: Changing Limits

The nature of the beast is changing. It is evolving in two directions:

1. *Miniaturization*: The pocket TV set will permit us to look at it in the same way as we search for the time on our watch. Miniaturization will put a stop to the set's being immobile and precisely located. Just as one walks around with a Walkman, listening to Mozart, one will be able to move around looking at the news of the world.

2. *Enlargement and greater sophistication for sound and picture*: As one can notice, the wide screen TV projector is extremely attractive to the audience; in public spaces where normal TV sets and new wide screen projectors are working simultaneously, people get together in front of the wide screen, despite the lower quality of the picture. Is it only the attraction of the new, or is it because it looks more like "true cinema"? Whatever the reason, the fast-approaching industrialization of the production of extra-thin, wide screen TV

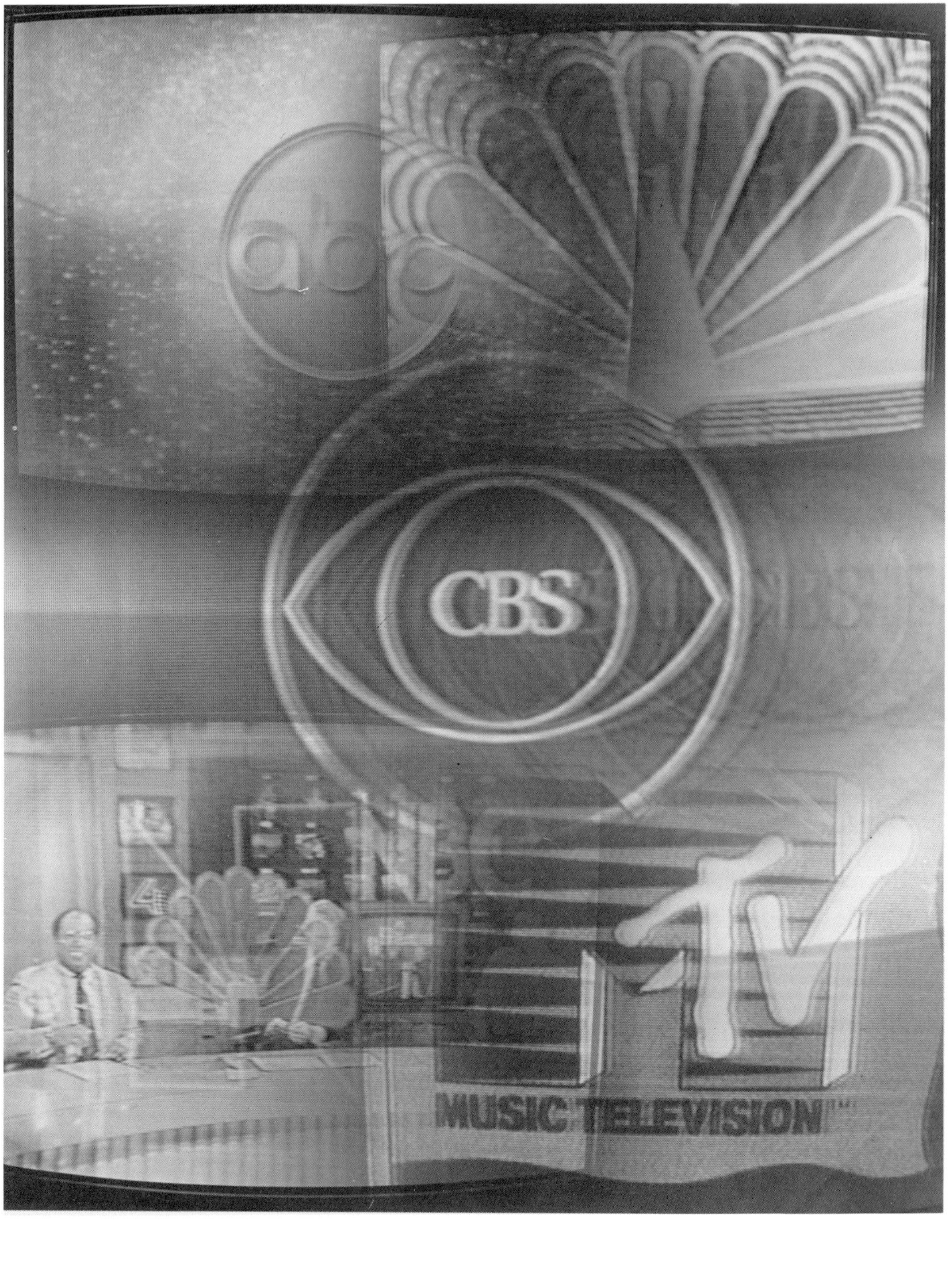

Fig. 5-1: Photomontage of TV logos (Photos and montage by Lloyd Eby).

projectors will permit us very soon to have large screen television at home.

As a matter of fact, we start from a situation of information scarcity with the traditional TV set. It is in that sense that Marshall McLuhan defined television as a *cold medium*. In other terms, one could say that the world through television is never pure spectacularization (picture), nor pure transparency (reality). It is re-presentation. Is it possible for this re-presentation to be trustworthy? We, of course, know that it is not. High fidelity of vision does not exist on a television screen; the picture is grossly structured by the lines of the frame. The picture goes through arbitrary and large variations of light and contrast. It can be heavily distorted. One must take into account that the audience seems to get accustomed to the crudeness of the TV picture without insisting too much on getting any better fidelity.

The notion of visual high fidelity has a value above all in the field of art re-presentation. If television is a cold medium with low visual fidelity, it is at the same time a tactile medium in as much as it is an excellent place for close ups. TV shows things in a very intimate way. It makes us recognize morphologies, clothing and postures. This paradox finally suggests the idea that television is more a tactile medium than a visual one. This has often been underlined by McLuhan. We know the extent to which he contradicted the so-called "idées recues." In commenting on this concept of tactility for television, Jean Collet said, "a good television picture is a body to be touched. The very fact that it is there on the screen is more linked with flesh than with anything else. The true depth is not in the picture; it is between picture and sound, between what is shown and what is said."

From those characteristics of the medium (cool, without fidelity and tactile) derive the problems—the transmission difficulties—that space arts go through when shown by TV. The space arts are what, since 1968, are now preferably called environment arts. The word 'environment' tends today to replace all the concepts describing space-organizing techniques. In the first place it means urbanism and architecture. It is concerned with different disciplines which tend to create the daily milieu of life: landscape conception, urban tools, design and so on. Diversification of those disciplines is a relatively recent phenomenon. If we take architecture first, we can see that, as an environmental art, it means an activity which is artistic even if it is at the same time different from other things. TV representation of architecture creates some problems of its own. Initially, architecture is summarized into a design and a map. The Italian architect and theoretician Bruno Zevi writes about this: "It is this representation method (that art historians use at the same time as photography) which is the principal cause of our lack of spatial education. An architectural construction is not the sum of the length, the breadth and the height of all its different elements. It is at the same time the measurement of all dimensions of the inside spaces in which men walk and live."

A design—a map—represents architectural reality in a very unsatisfactory way. The TV screen, because of its characteristics that we defined earlier (coolness, low fidelity and tactility), shows the same type of defects. If we take another example—home design—we discover other types of problems. Interior design is a subchapter of architecture; it addresses itself to everything which is inside of private homes or any architectural place. It is connected with color, furniture and coverings. It's domain is the finite. It is clear here that television can sometimes adequately render interior design as well as furniture creation. Furniture design and creation has become an important space art. If, in earlier times, furniture as an art concerned only a privileged minority, it has now evolved into a real social need. As a consequence, it has become a mass cultural phenomenon (one remembers how peasant furniture created an interest for the Early American style).

One can still think of landscape creation or the art of gardening, and here the example of Versailles gives us precisely the possibility of understanding the difficult inside/outside relationship which links architecture and garden creation. So too, because of the fact that it is not expressed only in finite objects or isolated buildings, architecture is linked with the notion of continuity relationships between different buildings. It, of course, makes us feel concretely how difficult it is for all these elements to be taken into account by television itself. A question of the same type can be put concerning other aspects of daily life, whether they be telephone booths, street lamps, benches, or all sorts of elements of urban life today, and especially one aspect of the street space that some theoreticians call *signaletics*. Kevin Lynch, in his book *L'Image de la Cité*, studies the signaletics of

Boston, Jersey City and Los Angeles. He introduces an interesting concept: the readability criterion for a city. It seems obvious that the television set can hardly give a significant account of a town's signaletics. But, on the other hand, it can usefully be used as a medium taking part in the signaletics structure of the city and can contribute in organizing the readability of this city.

Television entertains another possible relationship with city structure. In a clear or unclear way, it can provide new material for the city's appearance. Eventually, large moving screens where texts and patterns can develop will become elements of today's city architecture. The architects who conceived the Pompidou Center in Paris must, I assume, have been thinking about this when they designed the building: the building front can be considered to be a huge picture wall. If it is sometimes a problem for television to represent architecture, things can be seen the other way around; applied architecture tries to represent television as one of its own elements.

We have seen how difficult it is for the plastic and environmental arts to be handled in a literal way through the TV set. But there are other ways out. The art documentary has often been conceived apart from any exclusive concern with filming an achieved monument, sculpture or painting.

Relationships Between Television and Theater (Space/Time Art)

I would like to point out two or three elements concerning this topic, without giving an exhaustive discussion of it.

When a TV camera shows us close-ups when broadcasting a theater show, the director shows us what the viewer on location (in the theater) probably never sees: an exaggerated expression by the actor, an expression which normally doesn't work when seen from so close up. The televiewer gets a privileged look. But if the TV production makes us see what a viewer in the theater doesn't see, it excludes, at the same time, what that very viewer can see; TV really misses the movement of the actors, their disposition and their very theatrical space. In this sense, the theater play is as an ensemble, distorted and fragmented. Here television can present itself as an added art to the art of the theater. There are two possible cases; either the aesthetic impact of the TV medium is erased as much as possible and the TV aesthetics are subordinated to those of the theater or, on the contrary, the TV concept takes the lead. These two cases or trends represent two different attitudes for television. It is, of course, clear that the second trend means that other signs, which are properly TV signs, are added to the general meaning of the play.

This leads some to believe that the only possible way that theater can be used in television is in terms of commentary about theater, and nothing else. Otherwise, in the other cases, broadcasting must be highly referential and highly redundant. I think, for instance, of the series produced by the BBC on Shakespeare's most important plays, and of the series produced by French television about Moliere. At such moments, television gets the recognition of the cultivated class which it otherwise so rarely obtains.

Music as an art of time is more easily broadcast, at least in its traditional content. But here, too, it very often happens that the picture somehow distorts audio-space perception. Certain music lovers prefer a total immersion in the sound, detesting the aspect of distraction that is brought because of the TV picture. Others will admit the viewer's being put visually into the concert itself —with its particular mood, rites and so on—only as a possible commentary to the audio. It should be mentioned here that more and more special images seem now to be invented specially in order to promote music; we now have a rapid development of the phenomenon of video-clips, which attempt to present a visual representation for popular songs and music. At the very beginning those video-clips tended only to show the music as it was being performed. But now they increasingly encompass a very different dimension; they present a certain number of pictures which are structured into a very tight scenario, so that in a very short period of time a large number of aesthetic and symbolic elements are brought together.

Just as with music, cinema is—despite elite prejudices—very easily transmitted by television. Of course, a part of the intelligentsia has always maintained that a film on TV is not really a film anymore. This idea is very widespread. The small size of the TV screen and the uncertain quality of the colors seem to create a situation in which films are distorted. To be sure, cinema as an aesthetic discourse is destroyed under these circumstances. But other people hold, to the contrary, the view that a film can survive aesthetically on television if the film is itself strong enough. Stated simply, what happens is that something which is essential in a film or in a theatrical presentation becomes peripheral when the piece is presented on TV, and conversely, something which is otherwise peripheral becomes central on TV. A third point of view holds that certain theater-like films which emphasize acting and close-ups (such as, in the instance of French culture, the films of Pagnol) take on a special eloquence when presented on television.

We all know that one aspect of miniaturization will be that television language will become more and more like radio in that it will emphasize live programming. But there is also a contrary trend which purports that the development of high-definition television (with a picture made up of more than a thousand lines) will reinvent and more efficiently install what can be called home cinema. This development will mean, of course, that everything that has been said before is a debate that has been superceded.

Video Art on Television

Since the beginning of production of video art in the middle sixtics, there has been a great deal of discussion about the possibility of its being broadcast on national television. At the very beginning of video art a number of television stations helped start such production by giving their facilities to so-called artists in residence. This was the case in the United States, notably in New York and Boston with WNET and WGBH. It happened in Europe with W.D.R. in Koln and with R.T.B.F.-Liege in Liege. But after a few years, only some stations continued broadcasting video art. More and more, the more interesting elements of video art were transferred into other types of production, production that found its way into television; this was especially the case with video-clips, as mentioned above.

Gatekeeping in the Process of Change

In *purely commercial systems*, the gatekeeping process tends to exclude from mass communication what is considered to be art production (except in its more populist manifestations as in folk and rock music, etc.). Those systems tend to search for success on the most vast scale (largest number of viewers). Consequently, in most cases, they drop whatever could lead to a decrease in ratings, which is primarily highbrow and middlebrow culture.

In *public systems* art has been included in the structure to a certain extent. This is the model of the BBC, which, in its paternalistic wisdom, thinks that at least some art has to be brought to the masses. The BBC has always considered the mass audience a student to be taught, and thought that it had to give a certain cultural background to those masses.

We tend now to shy away from those purely commercial and paternalistic systems and tend toward a new type of selection which might be called a free market view. Increasingly, viewers will have more time, more leisure and the desire to choose for themselves how to use that leisure time. There is a growing desire for more than one program or one similar kind of programming (against the monopolistic attitude in Western Europe); one obvious evolutionary change is specialization of channels. This change, of course, takes place against a background of the theories of the early sixties for radio of "thousands of stations"; this meant simply that through the constraints of the market, a situation should develop in which the choice of programs would be enlarged to such an extent that any minority could find what it wants. For example, one could imagine a situation in which the public service station (whether the PBS/CPB type or the European type) would create a specialized network which would have as its only mission giving the audience a knowledge of the cultural heritage of a community, in order to understand creativity better and promote its development. Such a network should have creative persons from all disciplines participating in the creation of programs; it should organize permanent links with all the most important cultural institutions of the community. In that way, musicians, writers, painters and so on would have a means of having their work become known through television. A concept of that sort did take shape in certain initiatives that were offered through cable, such as Cable-arts, for instance. In a liberally oriented society, this proves to be a difficult concept to institute, but as a concept it still represents an interesting goal.

This type of specialization is only one example. Another trend which is obvious today is a change in the nature of programs. We are moving more and more toward what has been called peri-television; the television set is increasingly becoming a multipurpose home terminal which brings us teletext, videotext, teleconferences, telemessages, videodiscs, videocassetes and home computers. Both the nature of the pictures and their uses are undergoing vast changes. At the same time there is a drastic evolution of the supply networks. Cable TV (which diminishes the difference between television and the telephone, and which creates a situation where it is possible to subscribe to a network or to a station as if it were a newspaper), and direct broadcast satellites (which eliminate all the defects of traditional broadcasting and

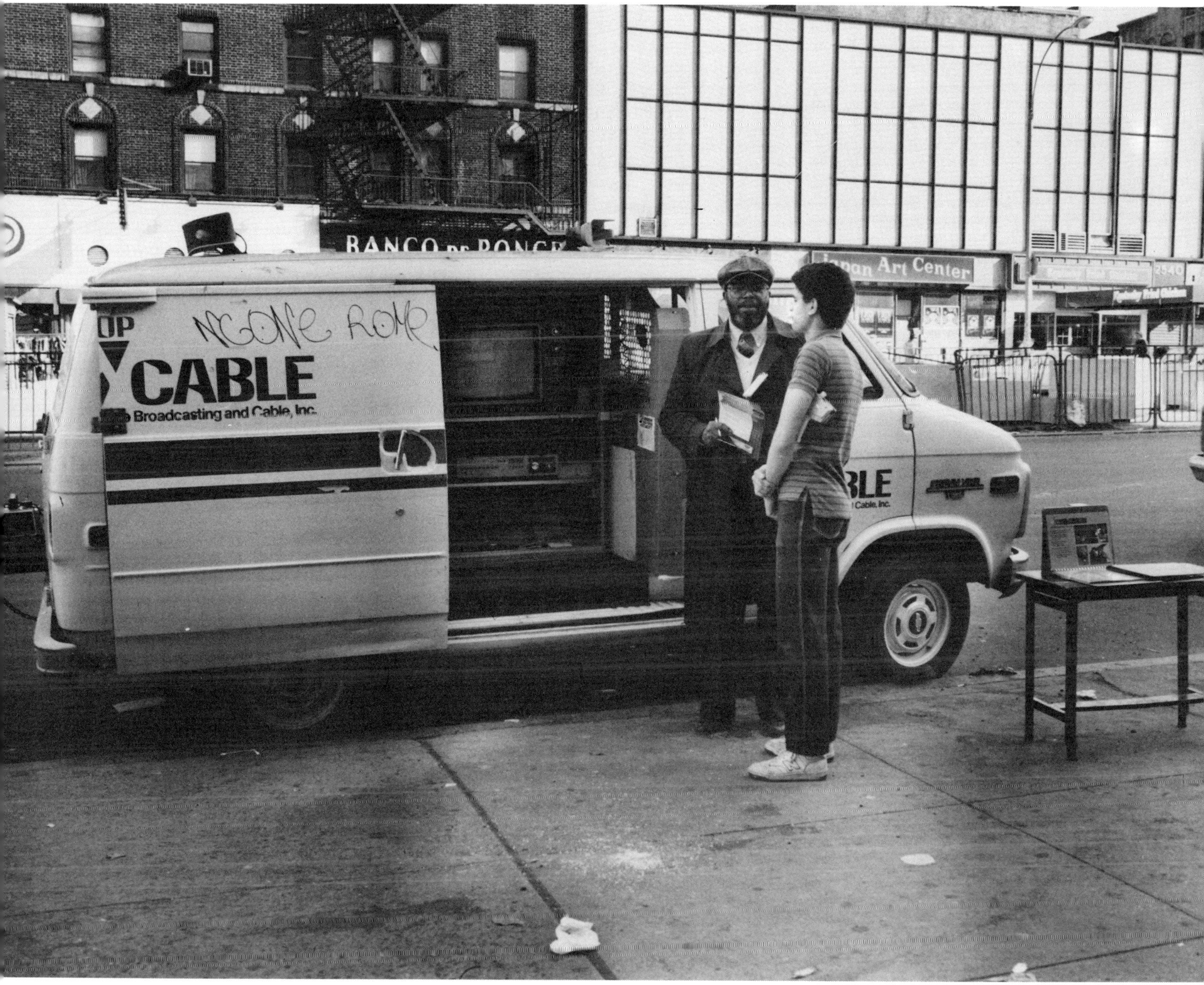

Fig. 5-2: Soliciting cable TV subscribers on a Manhattan street corner (Photograph by Lloyd Eby).

make it possible to receive broadcasts from every place on the planet twenty-four hours a day) give strong impetus to these transformations. In the Netherlands, for example, direct cable-casting has been accepted at night after the end of the official TV programs. In other situations, as in the United States, laws have been passed that required cable stations to open free access channels.

Some time ago NASA lent a satellite for an experiment (Artists TV Network) that helped create a communications network between centers of creation of modern art in the world and enabled them to present live broadcasts of their exchanges. At the last Biennial in Paris, the American participation consisted of the sending of an exhibition by telephone. This used the so-called slowscan process by which a picture was sent line by line through a simple telephone wire, and then at the end was converted into pictures. For a month and a half, a new picture was sent every day—and pictures came from all regions of the U.S.A.—to the Museum of Modern Art in Paris, where the picture was photographed and exhibited; this process represented the realization of an old idea of Nam June Paik, the inventor of video art.

The aforementioned event was a transnational one, but the exchange can, of course, take place on a more local level. All these networks create a dynamic which makes it easy to transfer into reality the famous phrase, "think globally, act locally." It is no longer utopian to think that, by means of all these changes which we have discussed, we will find ourselves enjoying the same state of free choice with our TV set that now exists for anyone entering a bookstore or visiting a newsstand.

By means of all these processes, television is becoming more and more oriented toward what can be called the publishing model. In the publishing model broadcasting can be understood as a social institution which is, for the first time, a *distribution entity.* Television encompasses all the intermediary tasks that occur between audiovisual production by the creators and the viewing of the product by the client. In a more restricted sense, the publishing model can also mean that through the development of text transmission, one can easily imagine a situation in which the distribution of newspapers will shift toward the notion of telecopying at home. New concepts like TV cassette collections and personal video libraries are now becoming realities. With diminishing prices and wider availability, collections of video

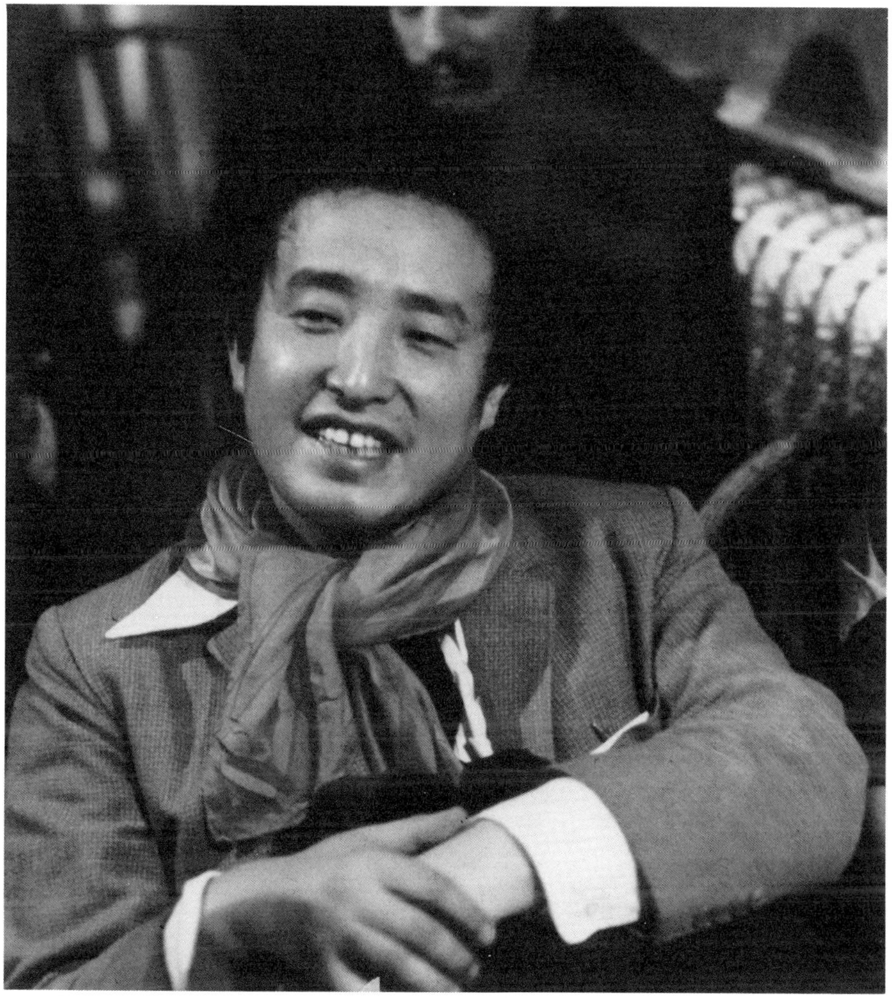

Fig. 5-3: Nam June Paik, the inventor of video art, at Millennium Film Workshop, New York City, November, 1971 (Photograph by Lloyd Eby).

tapes and videodiscs are becoming a more and more banal aspect of daily life.

We are, consequently, now entering a transition period. We will still have a certain number of explicit and centralized gatekeepers (the concept of networks and stations means this) with their positive aspect as well as their aspect of censorship. But, on the other hand, we will also have a more hidden and implicit gatekeeping by means of the marketplace, where, in theory, everything is available to everyone. So the strongest trend seems to be that we will come to have more and more private channels, whereas, up to now, the system was completely—or quasi-completely—public. Adversely, where few commercial systems prevail, we will see art-oriented and culture-oriented commercial networks appearing, starting from renewed public service institutions. I tend to believe that we are going to go through a period of commercialization for public systems and ennobling for commercial ones. It seems that we will never have any pure system anymore, but that the two systems will tend to correct one another in a dialectical way.

Television As Art

If art is the symbolic structuring of elements of reality by use of a certain technology for the enrichment, greater comprehension and aesthetic pleasure of an audience, then television is art in different ways.

We have seen that, compared with records and radio, television presents us with a much wider field because it uses pictures. All the arts can be handled through television, even if problems occur. Painting, sculpture, architecture, cinema, theater, music and video art are all shown in one way or another on television. In that sense, television is a very strong aesthetic weapon. Earlier we analyzed its role as a distribution agent, and we looked quickly at its artistic role and its aesthetic nature. This aesthetic is specific. It can be seen within the old traditional models of broadcasting, and it will take on another aspect within the publishing model that is now being developed in the industrial countries.

The Art of Programming (The Metaprogram)

Television programming is a flow of products organized in a particular way. The programming process itself conveys a certain number of social, moral and aesthetic values. Whether they exist or not, all these variables (the standardized length for programs, the order of presentation of the different programs, the moments of broadcasting) are different from one station to another and from one country to another, even if certain cultural and sociological constraints tend to remain the same. *This metaprogramming itself can be considered as an art.* TV can be analyzed in another way; it is a specific version of an audiovisual language, which defines itself as a certain mode of combination of talk, noises, music, written sequences and moving pictures.

It seems most important to discuss first one key aspect of TV language: the screen or the frame. The fact that the screen is

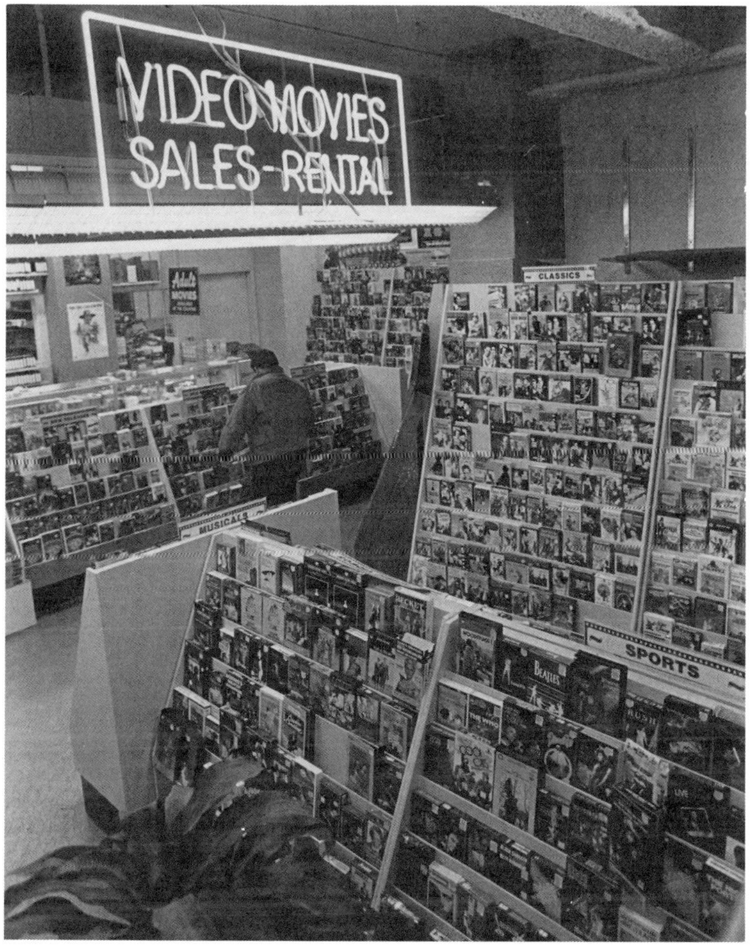

Fig. 5-4 A & B: Videotape sales and rentals (Photographs by Lloyd Eby).

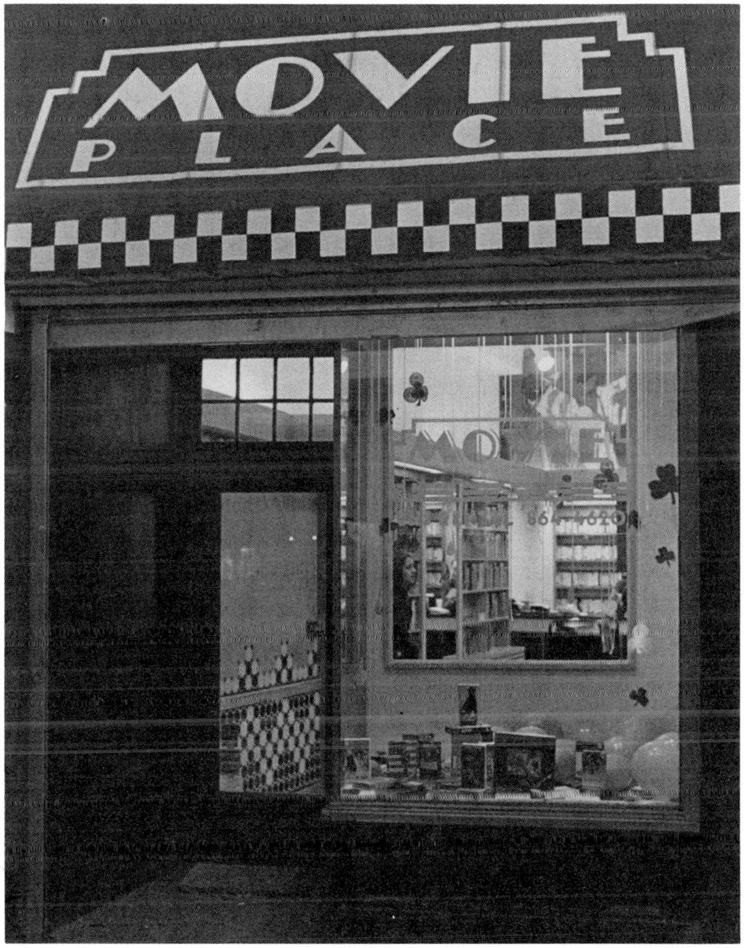

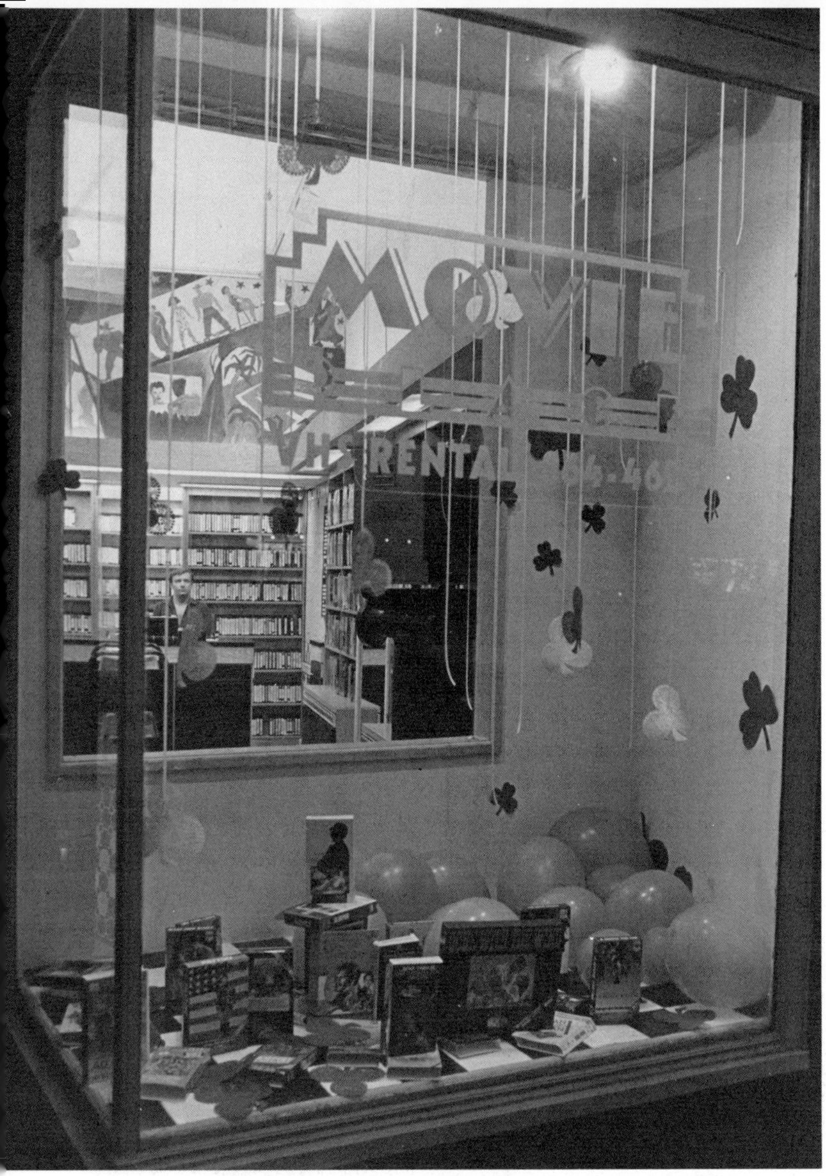

Fig. 5-4 C & D: Videotape sales and rentals (Photographs by Lloyd Eby).

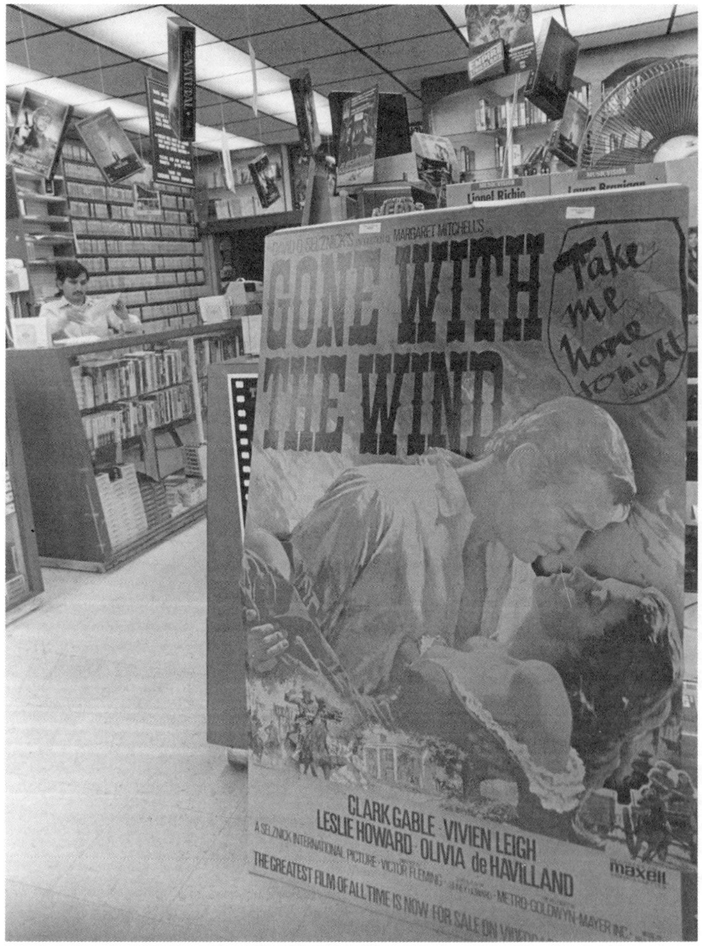

YOUR CLEAR, CONCISE GUIDE TO PRIME-TIME VIEWING

	8 p.m.	8:30 p.m.	9 p.m.	9:30 p.m.	10 p.m.	10:30 p.m.	11 p.m.	11:30 p.m.
Ch.2	Detective in the House		Dallas		Falcon Crest		Local News	College Basketball
Ch.4	Codename: Foxfire		V		Miami Vice		Local News	Tonight
Ch.5	P.M. Magazine	Rituals	Merv Griffin		10 O'Clock News		Taxi	Charlie's Angels
Ch.7	Webster	Mr. Belvedere	Benson	Off the Rack	Matt Houston		Local News	Nightline
Ch.9	NBA Basketball: New York Knicks at Philadelphia 76ers (cont'd)				Billy Graham Crusade		Sergeant Bilko	Burns & Allen
Ch.11	MOVIE: The Quiet Man ★★★½ John Wayne, Maureen O'Hara					INN News	Odd Couple	Honeymooners
Ch.13	Washington Week	Wall $treet Week	Great Performances: Rigoletto					MOVIE
Ch.21	National Geographic		MOVIE: Take a Letter Darling ★★½ Rosalind Russell, Fred MacMurray				Business Report	Latenight America
Ch.25	On the Job	Books	WNYE Presents		Off the air			
Ch.31	Masterpiece Theatre		The Visit		Gold Lust		Nippon TV	Off the air
Ch.41	Charytin		Sexcitante		Dancin' Days	24 Horas		Ojoreja
Ch.47	Amigos Siempre Amigos		Natalia de Ocho a Nueve		Amor Gitano		Informador	Italian Movie
CABLE/PAY-TV PROGRAMS								
A&E	Fanny By Gaslight		Sounds Magnificent				Showcase	Ashley Putnam
ESPN	Col. B'ball: DePaul vs. Syr. (cont'd)		NCAA Tourn.	College Basketball: Southeast Regional. Dayton vs. Villanova				SportsCenter
HBO	MOVIE: An Unmarried Woman ★★★½ Jill Clayburgh, Alan Bates				Boxing: Larry Holmes vs. David Bey			Maximum Security
LIFE	Regis Philbin's Lifestyles		America Talks Back		Good Sex! With Dr. Ruth Westheimer		Smart Money	
MAX	MOVIE: The Dark Crystal ★★★ Puppets			MOVIE: Stacy's Knights ★½ Andra Millian, Kevin Costner				MOVIE
MSG	Boxing (cont'd)				Boxing (cont'd)			Untouchables
SC	Boxing: Bobby Czyz vs. David Todt				Stanley Cup	Inside Racing	Aqueduct Racing	Off the air
SHO	MOVIE: The Blue Lagoon ★★ Brooke Shields, Christopher Atkins				MOVIE: Sahara ★½ Brooke Shields, Lambert Wilson			
TBS	MOVIE: Scalawag ★★ Kirk Douglas, Mark Lester				MOVIE: Eyes Behind the Stars ★★ Martin Balsam, Robert Hoffman			
TMC	MOVIE: The Octagon ★★½ Chuck Norris, Lee Van Cleef				MOVIE: Hot Dog...The Movie ★★ David Naughton, Patrick Houser			
USA	Wrestling TNT		From Hawaii with Love				Night Flight	
WHT	MOVIE: Max Dugan Returns (cont'd)		MOVIE: The Buddy System ★★½ Richard Dreyfuss, Susan Sarandon				MOVIE: A Scent of Heather	

Fig. 5-5: A typical evening program schedule for a major city (New York), comparing the programs available on network TV, local TV and cable TV.

small encourages everything which is homely, intimate and familiar. A close-up is handled much better by television than by cinema. It would, of course, be mistaken to build too general an aesthetic law on that fact; wide shots are indeed used very often on television. This happens all the time when TV broadcasts cinema films in its own language, television should, perhaps, reconsider the nature of its own screen and the scales of the different sequences. For instance, wide angle shots could be used in a very precise perspective when some visual rest is needed or when some distance is to be created between the audience and the show (Verfremdungs Effekt). *In any case, TV should increasingly see itself as the medium of intimacy.*

Next, we should consider another specific characteristic of TV: the syntax of its own language. This takes the shape of a combination of elements whose intention and positioning are rather different than those, for instance, of cinema. That syntax is easily understandable through the movement of the camera, slow motion, transparency, etc.

1. Camera movement: television, to put it briefly, is the place for optical travelling or zoom. The TV camera's focus can go towards or away from the object it wants to describe without the cameraman having to move the camera itself. Furthermore, panoramic shots are often replaced, especially in live broadcasts, by cutting from camera to camera. Instead of having a horizontal or vertical movement from one point to another, the director switches from one camera to another.

2. In television, slow motion has gained a function that is more than simply aesthetic or demonstrative. It is used to attract the viewer's attention to a special event. It is especially used that way in sports broadcasting.

3. Through the techniques of transparency, we are able to see, at the same time, characters being filmed and action taking place behind them. As opposed to cinema, the transparency technique is always obvious in television. The viewer knows very well that the character who is presented in close-up (the newsman, for example) is broadcast live, while whatever is seen behind him comes from another setting. As with slow motion, superimposed pictures are used more for an expressive or informative purpose than for an aesthetic one. The results of a vote, for instance, can be shown in written form (superimposed on the screen) at the same time as the vote is announced orally by a newscaster, or some other person.

Those considerations lead us to an important and special remark about TV aesthetics. As opposed to cinema, which tends to hide most of its technique (the technique is not visible on the cinema screen), television tends to be quite willing to show its own instruments or machinery. This unmasking is rather discreet, but it exists. You can see it explicitly when journalists are shown with microphones in their hands, or when one camera is shown filming another, and so on. We can say that television aesthetics tend toward realism. Another case is the use of transparency, as discussed above, to show a reality rather than to create it. This occurs even if we know, as stated earlier, that it is not that simple.

Two final remarks will finish my quick survey of what makes television syntax different.

Can one say that television itself is the art of live programming? This is, of course, something that differentiates television from records and cinema, which are canned products, and from radio, which is blind. This particular function of TV brings us to the notions of suspense and improvisation. It is noteworthy that those phenomena rarely find a place anywhere else in art, except in theater happenings and in rhetoric.

The second extraordinary specificity of television that we want to underline is its ubiquity; its techniques permit it to present life from different places at the same time. For instance, live broadcasts can show—successively—pictures taken in very different and/or very distant places. At the same time, it is possible to insert many different realities jointly into the same picture on the screen, realities very different in their genre, their original place, or their time. This can be done visually by superimposed picture, or by division of the screen into many sections. The techniques that permit this function of ubiquity strongly enhance the realistic character of television.

The Art of Interstice

In the process of building images by a network or station, there has developed a new type of art specifically linked with television. Logos, jingles, studio design, masters of ceremony style, and so on all tend to convey a certain number of aesthetic values. This is sometimes asserted explicitly and officially, as when the French network *Antenne 2* asked the designer Folon to prepare all its transition sequences. It leads to a point where Bob Wilson used these interstices ("in-betweens") as the space for his own TV creation called *Video 50*. He produced a series of 50 very short episodes, each lasting roughly 30 seconds. These episodes are a kind of very condensed fiction. They are edited in continuity, following a very precise and mastered order, which tends to avoid break-outs or excessive repetition. Wilson himself says that these episodes can be intertwined between programs and used, for instance, as interludes or, in case of program breakdown, as a way of diminishing the anguish of programmers or viewers. Wilson invented these spots by pushing the logic of television language to its very end. At the same time, all of those 50 spots turn away from the technology of Western narration to rediscover a mosaic poetry which is fragmented and without narrative steps.

The rhetoric of these TV interstices confirms the aesthetic status of the medium and its language. Another element which very often plays a central role in creativity and in art on television is advertising, even though it is supposed to be external to the broadcast media. TV advertising has such great importance that very few events can lead to its suppression. When General de Gaulle died, for example, all the scheduled programs on French TV were changed or suppressed; the only thing which remained stable was advertising. This advertising was so important because it served as the backbone of the network programs. What is striking in some of this advertising is its perfection. Because the ads are so short, a very high degree of invention is compressed into the rhetoric of advertising. For sure, the best advertising film will show our time in a more clear way than many feature films. The anthropologist of the future will know much about our fantasies, hopes and fears through studying our advertising. It would be very interesting to study why and how advertising can be so reflective of our current situation, as well as such a vivid place for creation. So far, advertising has influenced even artists who deal with the more advanced forms of culture; the American video artist Joan Logue, for instance, has produced a certain number of "advertising messages" for the Pompidou Center in Paris, done with the collaboration of artists, in which the work of these artists is presented in 30-second spots. This demonstrates how vital the development of a certain conception of space and meaning has been even in very foreign fields.

Mainstream Television As Art

Most intellectuals tend to reject current mainstream television. The tradition of opposition to mass culture is as old in the United States as elsewhere. In fact, since its origin, mass media has been a machinery of massification about which, for good reasons, intellectuals were to a certain extent skeptical. But the same mass media simultaneously carried largely unnoticed potentialities for bringing about invention, critique and language modernization. The denigration of popular mass media by the intellectuals (or the shyness of intellectuals in front of popular media) can be explained differently. Their reaction is perhaps one further aspect of an archaic fear of pictures. This debate is old indeed, at least as old as the notion of "representation." Everyone remembers the classical anathemas against theater representation, such as Rousseau, for instance, published in 1758 in his "Lettre sur les Spectacles." Additional analysis of this problem would be not only useful, but even necessary.

It is high time also that we develop an aesthetic analysis of current TV series. "Dallas," and "Dynasty," as noteworthy examples, convey societal, ethical and aesthetic norms. Recent studies (both essay-type and empirical studies) tend to confirm that TV soap operas transmit important values in a hidden way (see J.M. Piemme, "La Propagande inavouee"). That type of thinking (the thinking represented in TV soap operas and other popular TV series) depends fundamentally on homogenization, but the trends mentioned earlier create a situation of *potential diversity*.

Conclusion As An Opening

For a very long time, mistrust has dominated the relationship between TV, on the one hand, and artists, writers, cinema directors and other intellectuals on the other. It is increasingly clear—and the most inventive of intellectuals do not ignore it anymore—that the audiovisual media (specifically TV and video) could be transformed into a very important instrument of aesthetic pleasure, culture and development. More and more painters, writers and video artists take up the modern challenge of allowing pictures and sound to penetrate their own particular language. But very often artists and intellectuals say both yes and no to television. For instance, Federico Fellini accuses television of having corrupted the relationship between the viewer and the cinematographic picture, but, at the same time, he recognizes that TV can obtain what he calls "an extraordinary style." Bob Wilson accepts the language of TV functioning, but pushes it to its very end in such a way that takes TV pictures away from their ordinary and traditional effects. Current TV hybridization by video art or by computer and teletext technology is beginning more and more to develop new languages and new expressions. Even video games, after a crude first period that is now coming to an end, are perhaps going to use very complex systems of interaction that will bring to TV sets of the future potentialities for viewer creation through interaction with a gigantic electronic mega-system. As Robert Wangermee and Holde Lhoest wrote, television might still exist after television, but it will be another television that could be quickly called hyper-, para- or peri-television.

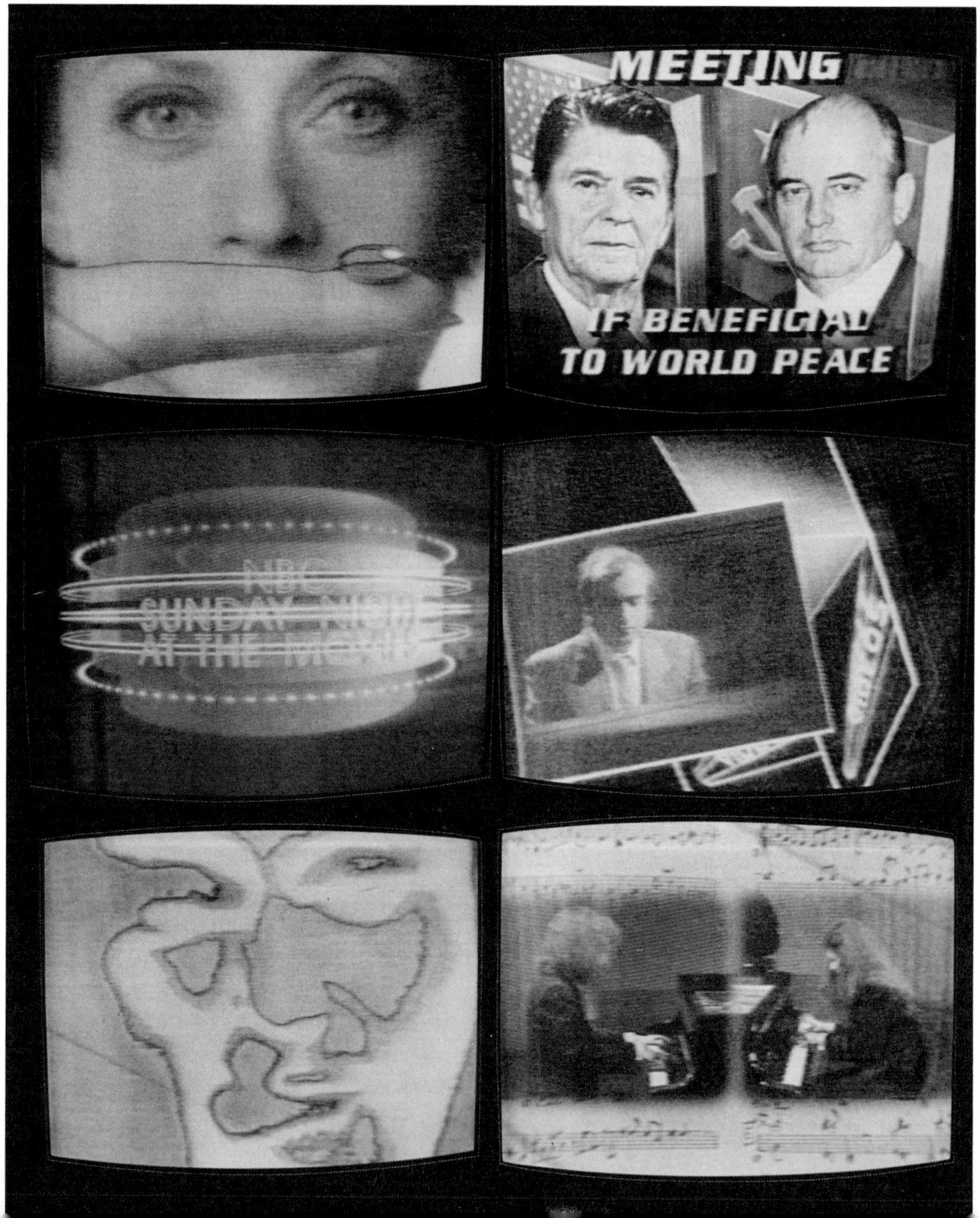

Fig. 5-6: Innovative or computer techniques used in standard broadcast and commercial TV (Photo composite by Lloyd Eby).

On Robert Stephane's "Television And Art: Television As Art"

Woo Hyun Won

Robert Stephane has treated his topic comprehensively, and I have little controversy with what he has presented. It does seem that he is inclined toward a descriptive and linear approach to his topic. I would like to supplement his work by discussing some additional things: audience needs and their gratification, media logic and media culture relating to art.

Before describing what kind of animal television is, we must make clear that the distinctiveness of television depends primarily on how audiences define it, and on what the production sector shows on it. Mr. Stephane seems to mention only the audience's decision-making role, but I will make up for that by describing a hypothesis of audience needs and their gratifications, and of television logic and its culture.

We consider television as an art on the basis of the social origins of audiences' expectations and gratifications. Audiences tend to strive towards an assessment of television as an art, in their related terms rather than in technological, ideological, or other more or less elitist terms. The convergence of various foci on the question of television as an art indicates that there is a clear agenda here—partly methodological and partly theoretical—for a discussion of the right approach for answering the given topic, as follows.

a) The audience actively conceives television as an important part of mass media use for the directed goal. Even if it allows me to think that television exposure often has a casual origin, patterns of media use are shaped by more or less definite expectations of what certain kinds of content, including art, have to offer to the audience member.

b) In the television gatekeeping process much initiative in linking need gratification and media choice lies mainly in the audience member. The categorization of art coincides with how the television competes with other sources of need satisfaction. The needs served by television constitute but a segment of the wider range of human needs, and the degree to which they can be adequately met through television consumption certainly varies.

c) Value judgments about the cultural significance of television should be suspended while audience orientation is explored on the audience's own terms.

d) Methodologically speaking, many of the goals of television need as art can be derived from data supplied by individual audience members themselves—that is, people are sufficiently self-aware to be able to report their interests and motives in television art programs.

Secondly, it should be considered that television has a particular logic of its own, even if television takes for itself all intermediary tasks between audiovisual production by the creators and viewing of the products by the client. And it seems to be true that television logic becomes a way of seeing and of interpreting social affairs. In general terms, television logic consists of a form of communication; the process through which media presents and transmits information. Elements of this form actually include the various media and the format used by television. The format in television consists, in part, of how material is organized, in which style it is presented, the focus or emphasis on particular characteristics of behavior, and the grammar of television communication. The format becomes a framework, as a perspective that is used to present as well as to interpret phenomena.

Audiences for major media have become so familiar with different formats that they automatically know when art is presented on television. Thus the logic of television formats has become so taken for granted by both communicator and receiver that it has been overlooked as an important factor in understanding television.

Thirdly, modern communication media like Cable TV are a technology, and as such carry a connotation of rationality. Both television management and the audience are consequently oriented toward a rational means and type of communication, a rapid dissemination of vital information at relatively low cost. This practical approach to television is a logic in itself. This rational, practical character of media logic further leads to dependability—in other words audiences want information that is relevant to their desires.

In sum, these characteristics become a form of television program content that is consistent with the modern scientific manner in which contemporary society operates. On the other hand, the style in which the technology is used may promote affective mood response, as in art content. Audiences have come to expect that advanced television technology will produce an art-style program.

In an information society, television and other media are the dominant forces to which other institutions conform. Social reality is, in some sense, constituted, recognized and celebrated with television. In contemporary society the logic of television provides the form for shared normalized social life. Thus our attention should be directed more than ever toward viewing art from a mediacentric perspective, seeking first to discover as well as to clarify how television logic operates, and then to describe the television culture which follows.

Finally, some practical limitations in making any television program should be kept in mind. A television program treatment tends to emerge in one of two ways: (1) The producers start with the interpretation of a main idea for television, and modify it to fit the budget and facilities available, or (2) The producers decide upon a practical layout with workable mechanics and then, with these particular facilities in mind, build up an interpretation of the art program.

Chapter 6

The Way Toward Visual Culture: Comics And Comic Films

The Way Toward Visual Culture: Comics And Comic Films

Alphons Silbermann

The Technological Constituent

The relationship between the arts and technology has been discussed endlessly since the beginning of the industrial age. Mostly, those have been rather arbitrary arguments, predicting —with the help of unspecified slogans—the end of Occidental culture. Far from reality, these arguments prejudice the creative power of the artist by piling up in front of him all those overall social, economic and cultural misgivings, misgivings that are supposed to have their origin in a "technological age" of whatever kind it might be. The arguments deny the integrational process between art producer, work, and art consumer, a process that —besides being based on an historical and organizational/economical constituent—is also based on a *technological* one. But those arguments consider the arts only as a cultural element but not technology—however technology might manifest herself.

Under the technological constituent, we understand all those technological means through which groups of artists bring about and put in order both work and art experience and, above all, adjust this work and experience by way of integration to their time and their society. In other words, it is the means that binds artists and consumers to time and society. A pressing considera-tion of this constituent, as an element capable of integration, leads—from an historical point of view—over supposed divergencies between art and technology to a direct connection with the realities of the present. It also leads—and this seems essential to me, in light of the general theme of our conference, "Absolute Values and the New Cultural Revolution"—to a relation with the future.

In view of the obvious relevance of the technological constituent to the artistic creative process, it is hardly necessary to deal with the stereotyped and unrealistic charges that are often brought forward concerning that relation about "the corroding and degrading influence of machine technology," "mechanization of the arts" or the dictum full of prejudice that speaks of the "technical possibility of the reproduction of the work of art." And all this only because art producers and art consumers use technological arrangements which, like the print and electronic media, have become integral parts of our society. Let's say it candidly, that the frequent attacks on the technological constituent are no more and no less than a strategy of obscuration. That means that the blame for the incapabilities of producing artistic contents —contents that lead to artistic experiences and do not just consist of skill, diligence, precision and sweat—is thrown upon the technologies adapted and institutionalized by society.

Effect-Relations Between Art and Technological-Media-Induced Mass Communications

It seems to be necessary to talk about these technologies, called "mass media," within the framework of the subject of our committee, "Art and Technology." The reason for this is that the arts (understood as part of the culture referred to in the general subject of the conference as "cultural revolution") and mass communications (enabled by technological means) are put into a relationship with each other. Yet, a search for philosophically abstract "absolute values" may easily make us forget to realize that behind art (culture) as well as behind technology (mass communications) *human beings* are placed. Hence, the subjects we have to deal with can be understood and analyzed—in their relation to each other — neither in the abstract nor floating free in the scientific sphere. Here, a noological approach must fail in order to allow room for a psychological and sociological approach, i.e., for an investigation of the question of how the human being succeeds in grasping and appropriating intellectual contents. This procedure, if it is to be seized empirically, leads inevitably and directly toward a total social phenomenon (as conceived by

Marcel Mauss)[1], based on the triangle relation: person—society —culture. If we are asked which of those in this context could be the said total phenomenon that can be recorded in an observable way, we find ourselves confronted with that actuality generally designated as *effect*.

Indeed, a penetrating analysis of the effects will provide us with information about the relationship between art and mass communications. Most of the rather endless series of publications studying the effects of mass communications on the individual, on groups or on masses—whether they refer to attitudes, motives, habitualization, manipulation, consistency, predisposition, propaganda, conviction, exposure to stimuli or like in our general subject of "cultural revolution"—are based on psychological and/or sociopsychological cognitions with the purpose of understanding the mental treatment of communications after reception of a message. But this is by far not all. It is a fact that impacts and retroactions that come from mass communications proceed into so many directions that a one-sided view can lead to confusion and wrong analyses.

Therefore, the individual relations of effects are to be laid open and to be separated from each other. Such a procedure may seem very dogmatic and artificial. But if one wants to become aware of the different levels in order to clarify the relationship between art and those mass communications induced through technological media, or the other way around, it is inevitable to go this way. It just does not suffice to talk in this context simply about intellectual, moral, social or psychological effects or to summarize eye-catching impacts under those headlines. While circumscribing on the following pages the main circles of investigation referring to the relationship between art and the mass communications that are brought about by technological media, we expose at the same time the relationships within the communication system as those relations affect "absolute values and a new cultural revolution."

First, we refer to *the function of certain artistic contents in relation to the media*. This is a question of serviceableness, in which artistic contents of all kinds serve the media for the purpose of mass dissemination. The contents, whether on a high or a medium level, whether ephemeral or trivial, fulfill the function of a commodity that is demanded and bought, a commodity that the media produce, or have produced, and which is used and processed according to the present day conditions demanded by film, press, television or comics. On the one hand, one does not like to refer to this relation because it is thought that such an interpretation would pollute the "holy" field of art. On the other hand, it is taken as a pretext for cultural criticism on the basis of materialistic ideologies.

Next, we see *the function of the media in relation to certain artistic contents*. Here the problem is the artistically correct utilization of contents for the purposes of the media. If in this context cultural criticism comes to the fore, then, it is in the first place in order to preserve the unfalsified sociopsychological unity evoked by art experience. The point of departure of this criticism is the immanent meaning of art works and consequently—in view of their popularity and popularization—the fear that the media and the people behind them deliberately produce specific Kitsch-experiences. To be sure, the media personnel themselves don't like Kitsch, and they promote, in passing—if they are in the mood to do so—what they think is art; but they actually produce Kitsch. Whoever makes such a statement could, if necessary, present the evidence of a changing or decaying culture, but could produce neither the postulate of an intended general validity nor of an innovation.

If we connect both of the functions already mentioned with each other, then we are among those who want above all to save the mass media through the arts or the arts through the media. In this context the rather embarrassing inability of the television screen to replace communicative symbols by non-communicative ones is too easily overlooked, i.e., a sociopsychological process which, in the first place, is to be attributed to the traditional constituent of the collective memory.[2] Here, two groups confront each other like two hostile brothers. While the one says: "Hands off the art, you degrade it through the technology of your media to a penny gaff," the other insists on seeing the media as the art of today and tomorrow. Both views are untenable, in so far as their argumentation overlooks the reality of the relations between media producer and media consumer, that is to say, the *levels* on which these relations take place. It has to be pointed out over and over again that what is true for the sociopsychological functions of the arts in general is also true for the art disseminated by technical means, namely that the members of the societies look upon the arts on an aesthetic or a respectively social sociopsychological level.

These elements of thinking and feeling are inherent in the individual as well as in whole aggregates. They are impediments to all those people who are responsible for artistic programs. For once those people realize (consciously or unconsciously) that they disseminate cultural ingredients or aspects, they like to be regarded as cultural experts. This produces an attitude that elevates the technology they deal with (under whatever circumstances and in all its configurations) to culture. If a member of one of the various media organizations is a person who is deliberately concerned with the progress of political, moral, economic, social, religious or artistic culture, then he has to try to preserve his expert status. Now, if we add to the "genuine" expert the "bogus" expert as well as all those who are only interested in the material side of the media, then we are faced with a collective consciousness that makes all of them behave like artists. As it would be degrading for a "genuine" artist to admit that he does not serve real art (whatever this might be), he nominates the media as an art. And, as the "bogus" artist has to give proof of his right to exist more than anyone else, he proclaims the same even more vehemently.

This sociopsychological reaction brings us close to another moment in the relationship between art and mass communications, that is the accomplishment of a *process of identification.* Born out of the close connection between the function of artistic contents in relation to the media based on technology and the function of the media in relation to artistic contents — having always in mind that behind every content, for its production, its interpretation or its explanation, are human beings—the media themselves together with their technologies become producers of art. If, for instance, radio and television act in respect of the content "music" as entertainer, Maecenas, educator and impresario, it would be wrong to speak of an usurpation of this form of art. Rather, we should speak of *shifting of functions,* a consequence of the social change which animates the so-called "cultural revolution." This leads to the fact that the persons responsible for the artistic content "music" within socio-cultural institutions like radio or television identify themselves on the one hand with the institutions and on the other hand with the music they handle. This process of identification and, combined with it, the institutionalization of artistic contents by the responsible persons and by the socio-cultural institution itself creates the basis from which it is possible to analyze the *effects of the means of mass communication in relation to the arts and to culture in general.* The contents can now be regarded from various points of view: from the point of view of the communication process, the message,the feedback, the diffusion, the innovation and many other things.

Not all of these cultural complexes can be dealt with in the framework of this paper. We confine ourselves to two very important frames of references: *mass and popular culture* and the *system of perception.*

Mass and Popular Culture

Mass and popular culture, two related and often combined terms, represent measures for worthiness and unworthiness of respectively cultural artistic contents. The negative arguments uttered in this connection are directed in the first place against the manufacturers of artistic values, and especially against those who are in a position to diffuse those values with the support of the mass media. In other words, a *process of distribution* has come between art and communication and interferes in the process of communication. It is fundamentally different from the process of distribution of that time when society entered the age of industrialization. This new procedure, covering the entire production (i.e., also the cultural goods), was *primarily* not profit oriented; it was *primarily* need oriented. Needs, desires and aspirations of individuals and aggregates exerted a pressure on production, which led at the same time to the development of a mass society. One speaks reluctantly about mass culture. This is not only because it is assumed that mass culture was something despisable, but moreover, because it is said that it would serve to suppress, to ruin, even to make individuality, individual production and individual culture disappear—and together with this taste and the formation of taste.

The term "cultural industry" came up as a way of representing seeing in this way and summing up the entire set of problems in one expression. The social philosopher Th. W. Adorno claimed[3] that this expression was used the first time by him and Max Horkheimer in their book *Dialektik der Aufklarung.*[4] But this is not true. For, already in the middle of the 19th century, at the time of the great world exhibition, writers on aesthetics under the leadership of John Ruskin (1819-1900) and his follower William Morris (1834-1896) climbed the barricades and called for a clear separation between culture, on the one hand—with which they evoked the ideal of an artist (of an exceptional being, a dreamer, an artistic artisan)—and, on the other hand, the industry with all its ingredients like collectivism, standardization, mechanization and oppression of personal liberties.[5] What was then refused for aesthetic reasons is now discriminated on a socio-cultural and sociopsychological level with slogans like "cultural industry" or "industrialized cutlure." By these means, the attack is directed against mass and popular culture and their producers, as well as against the "uninformed" members of a mass society. Especially in this context, different phases of influence of the mass media are underlined, phases which reach from influence on taste via formation of taste, to organization of taste or even to dictatorship of taste.

Yet, already at the moment the attempt is made to define the effects according to term and content of taste, we are confronted with views that differ depending on their ideological approach to the matter. Those who are preoccupied with the relations between technology and society, as well as with the relations between technology and art, cannot, and must not, make inferences from the assumption that taste is a mystic or absolute phenomenon. It is an error to cling to an opinon according to which the arts exist in a world of the absolute, and to assume on that basis that the tastes of producers as well as of consumers are independent of psychological and social forces. On the contrary, we find that the forces leading to the formation of interconnected norms in respect to taste are socially and psychologically determinated and consist of social heritage, of social, psychological, biological and technological elements as well as the creative contribution of individual personalities. Hence, the fact that taste can be influenced and controlled by means of mass media technology is to be seen with respect to all those forces just mentioned above: they exert their effects in interaction.

If the relations between art and mass communications are to be kept in a cultural balance that serves society, the arts and the mass media, then means have to be found to guard against an undesired taste dictatorship. For this purpose, those pseudo-sociopsychological arguments are to be eliminated that are not only of the opinion that taste *eo ipso* could be measured statistically, but what is more, insist on being able to recognize, with the aid of philosophical analysis of a cultural item, the state of a society and its art consumers. Those alluring thoughts are to be eliminated in order to see clearly that the whole relationship between art and mass communications—as well as the one between art and mass media, i.e., the distributors—rests on the problem of the *connecting* of the control with regard to the *access* to the art by production and consumption and the control with regard to preferential treatment in questions of *distribution and mediation*. The popularization of the arts brought about through technological means is, above all, a question of the *separation* of the control over preferential treatment in questions of distribution and mediation. For the media of mass communications and the people behind them do not *create* the taste of their consumers—they only reflect it interactionally. Let us not overlook the fact that permanent, temporary and ephemeral art expressions are in the first place expressions of the activities of culture; in the second place mobile cultural changes; third, up to date circles of cultural effects that, fourth, lead in their totality to the worthwhile objective of making the individual consumer become a member of smaller or bigger groups of society through the binding power of the arts. Thus, we realize that in the relationship between art and technologically-determined mass communications, the logical end, i.e., the end of control over access and distribution, is confronted with a psychologico/emotional end that must not be overlooked and for which there is no rule of control apart from synchronization and social agreement.

The System of Perception

Up to here, we have looked at the set of problems from the point of view of production. By turning now to the side of the consumers, we intend to observe and analyze their behavior in view of the relations between art and technology and their effects. The frame of reference which will serve us for this purpose will briefly be called "the system of perception."

The passage of an artistic message to the recipient comes about via a process of communication which, temporally speaking, occurs *after* the creation, but *before* the effect of a cultural item embedded in technological means. Hence, this process reaches the recipient at a point where *perception and meaning* of a cultural item occur.

For a long time, scholars have been interested in perception and meaning in occurrences where these two stand closely connected, a situation in which the comprehension of the material and the immaterial environment played a most important role. According to our subject, we shall only deal with those technologies and their handling where the *visual* determines the artistic reality.

Already Leibniz, following the doctrine of Descartes about clear apprehension and thinking, was concerned with the phenomenon of perception and with realization of sensations and stimuli within the consciousness of men. From there onwards, a number of theories with reference to perception have been developed, theories which—guided in the first place by psychological knowledge—analyze the relation between associations of sensation and substantiality. Proceeding from the postulate that perception is due to the sensibility and the intelligence of men, it is held (stating the view briefly) that external objects and realities put into motion the totality of those faculties of the nervous system of men. By this, sensations and imaginations are evoked, which man, thanks to his intelligence, is in a position to interpret, to adjoin to an object, to place within his environment, to understand—in brief: to perceive.

Perception is obviously a psychical fact and the question is, to what extent should sociologically-based contemplations and analyses take it into consideration. The answer to this question is that the sociologist will be interested in regarding the *process of visual transmission of culture* through perception in so far as this relation between the visual message and its consumers is established.

This problem—a problem that includes questions of distribution, pedagogics, listening, viewing, reading and images of the mass media—has been examined from many sides; historians, philosphers, psychologists and natural scientists have, each in their own way, taken exception to the fact that what is presented to perception and is connected with a centralized sensation never coincides with the object of perception, for instance, with the environment. Therefore it is frequently proclaimed that things that are to be seen visually in perception are never completely the things that are meant. This attitude, however plausible it may be in its simplicity, never convinced the sociologist. Sociology and social psychology have not, for nothing, been trying to grasp the "meaning of meaning"[6] and to conduct experiments to measure meaning.[7]

The *meaning* of a visual communication as a step following perception is based on the fact that contents and forms of an object or an expression can be understood in their peculiarities. These peculiarities can be pursued by different approaches. If, for instance, we consider that words and music develop their effects in time, and paintings their effects in space—in other words, if we not only try to trace out the essence of words, music or paintings, but also to seize their existences as an operating force—then this approach leads necessarily, via a *phenomenology*, to the facts themselves. Indeed, through a phenomenological approach—used as an aid in returning from the "transcendental" and its power to the substantial—the character of perception and meaning, as well as its direct presence in the process of visual cultural transmission, can be elucidated; it is the conversion of the perceived object and its meaning into a permanent property of consciousness. Existentialist thinking, brought about by phenomenology, also tries to grasp the realities of the living and active being and to assess this being in its relation to the absolute being; in other words, to lead it back to its essence—another modality for the return from perception and meaning to the substantial, i.e., to the process of transmission of a message referring to an environment, constituted by art.

Another way to pursue the understanding of contents and forms of the expression in their peculiarities is the analysis of symbolic forms and symbolic thinking, of which Ernst Cassirer was already aware.[8] Erwin Panofsky followed the same direction; he tried to prove the identification of the symbolic function of an image with the deeper layer of its meaning, about which, according to his opinion, the entire constitution of time gives ingenious proof.[9] While the followers of this school of thinking examine the *symbols* and the *symbolic meaning* of a visual message, the psychological school, headed by R. Arnheim comes to the conclusion that the symbolic meaning is only *indirectly* expressing what intelligence and experience tell us about the content of an image.[10] From an epistemological point of view, opinions of this kind refer us back to the impossibility of getting close to visual communication, its reception and its decoding on the basis of logical discussions and empirical findings. This led to an abundance of pamphlets, all to the debit of the so-called "lecture culture," of artistic realities using technological means.

The sociologist is less interested in deciphering contents and in the demystification of the relations between art and technologies, but more interested in the welfare of mankind in its visual and cultural being. He wants to comprehend perception, meaning and their consequences as *topical human behavior.* With this (with or without inclusion of historical, anthropological, psychological, structural or linguistic knowledge) we turn to the consumer in so far as he is dependent on a cultural *process of diffusion* to obtain a gratification of some kind from the visual message.

To get close to this process and with it close to the behavior of men, we will have to turn to observable patterns of behavior. In this respect, there are, from the point of view of the recipient of the visual message, three central patterns of behavior at our disposal which are methodologically seizable. First, there is the procedure of *seeing,* which is the first contact between producer and consumer. It puts in motion certain neuralgic factors that bring about, apart from acceptance or refusal, structural elements of the personality. Many studies have already dealt with this physio-pscyhological problem in order to find an answer to the question of *what* the onlooker sees in an image and *how* he sees it. On a politico-cultural level, this leads to a very well-exploited field of education that tries to explain to the onlookers of objects what they have to see, i.e., what it means to see in the "right way." All those attempts are similar to that "science of the soul" which tries to lead the procedure of seeing into the field of action, known as consciousness.

After seeing follows a second pattern of behavior, namely the *understanding* of a visual performance. Now, a procedure is set in motion by the observer with which he tries to enter into the eye and the brain, even into the souls of those who produced the representation. The message, as it is said, is to be grasped in this way, i.e., the observer participates directly, but mostly indirectly, unconsciously, quasi-automatically in the genesis of the visual performance. This is a behavioral pattern which stands in the center of all those considerations that try to assist in the interpretation of reality.

The third behavioral phase is that of appreciation, of *estimation.* Starting from the optic material—elevated by a systematic work of thinking and intuition to the level of an analysis of meaning—individual value judgments are established which are finally meant to be axiomatic. Never mind whether it is about reading, listening or seeing; in relation with understanding and judgment, aesthetic and social elements are never to be confused. An exact separation of the two levels on which perception can take place allows us to realize that the three phases of visual communication mentioned above are in a dialectic relation of tension with each other. On the other hand, this makes us recognize that *possibility* of communication and *receptiveness* of communication are two opposite poles which have to be brought together for the aims of perception—they level out the embarrassing inability of visual media based on technologies to transmit a completely different world from their own.

With this I want to point out finally that the mechanisms of visual communication—from their reception until their effects—are the results of a *socio-cultural phenomenon* that is to be ascertained in its identity. It is precisely this socio-cultural phenomenon we deal with here if, according to the subject of this committee, the role and the consequences of the relationships between art and technology are to be discussed within the framework of interdisciplinary research. "Interdisciplinary" means to me not only that the different disciplines of art can make use of the existence of mass media technology for their research work, but—for the scientist interested in the behavior of men, and especially the society—above all that the existing socio-cultural phenomenon can be approached and explained theoretically, practically and methodologically from various points of view.

Without establishing a hierarchy, we present as a first and often-used way of analysis the one showing that visual representations do not come into existence *ex nihilo*. This approach, however, is often wrongly applied for all when extended into ideologies. In this case, mechanistic causality becomes a method against which the sociology of knowledge has protested since the days of Karl Mannheim. The empirical sociological analysis goes exactly the opposite methodological way and tries to grasp the *situationary*, i.e., it examines the influence of society on the role of mass media technology along with its socio-cultural phenomena.

Another orientation proceeds from the creation of visual performances in their various forms of expression and thereby derives questions of general character that refer to the social, moral,

Fig. 6-1:

GERMAN LABOR MOVEMENT (1904)

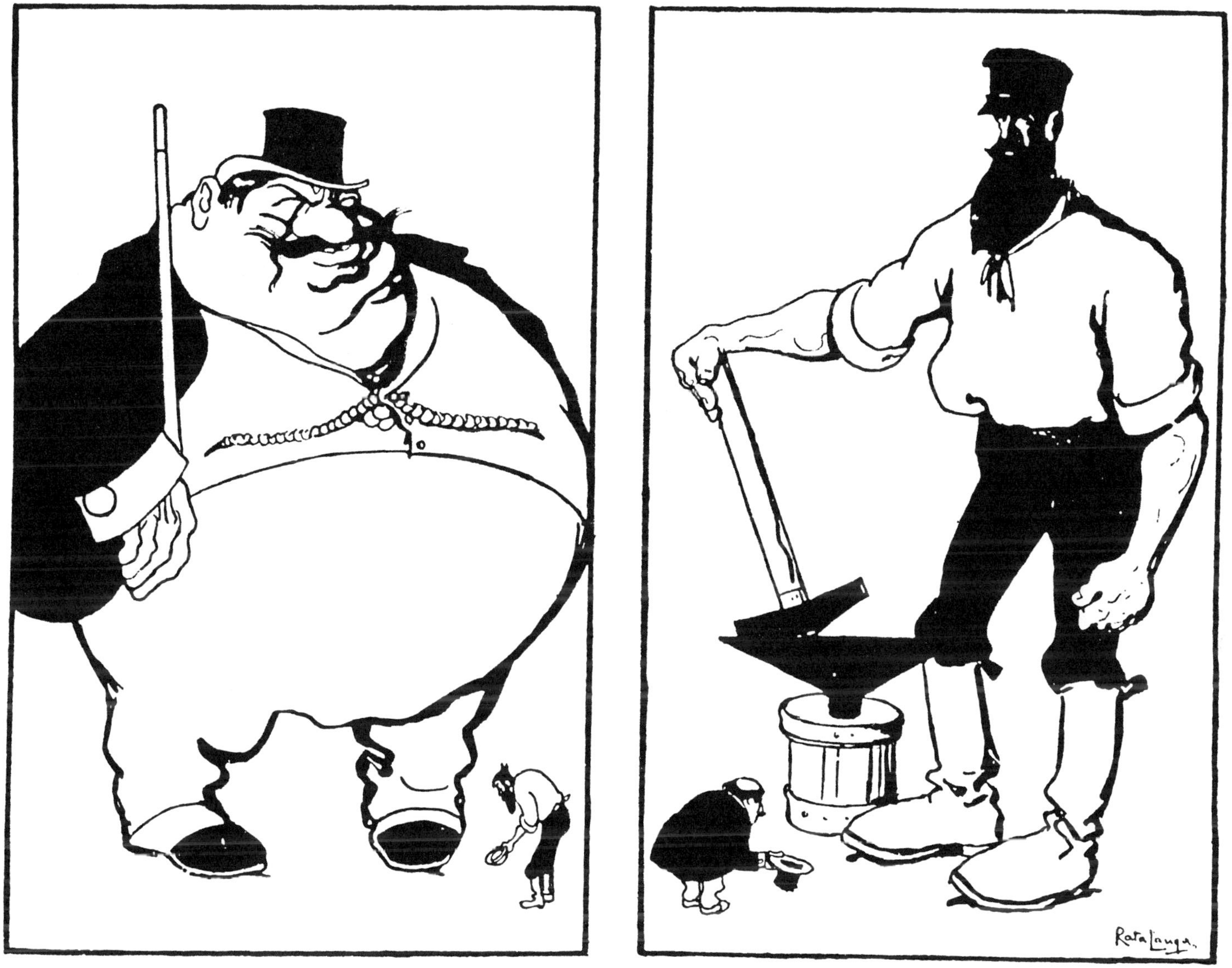

(CAPATALIST AND UNORGANIZED WORKER) (ORGANIZED WORKER AND CAPATALIST)

Source: Friedrich Knilli, "Der wahre Kakob"-ein linker Superman? In: *Comic Strips*. Akakemie der Kunste. Berlin, 1969, pp. 166, 17.

economic or cultural ideas of a certain class or stratum of society.

–Or there is an attempt to prove on the basis of socio-cultural phenomena that man is to be seen in the environment of a spirito-social existence.

–Or the socio-cultural phenomena are put into relating to social change, to politico-cultural developments or the web of social organization.

–Completely different from this is the approach that pursues the social fundamentals of the creation of a visual presentation as well as the aesthetic enjoyment obtained by it.

–Another approach to the socio-cultural phenomena interests itself only in questions concerning the general validity and the constancy of the concretized visual phenomenon.

–A broader approach seeks to find out the possibilities of expression within the individual and collective discussion between the specific realities: e.g., environment, visual presentation and society. With this, we are in the midst of the theoretical and practico/methodical approach called *environmental studies.* The origin of this approach of registrating environment, and from there the uncovering of circumstances, is generally attributed to Hippolyte Taine.[11] Yet, other scholars also, devoted to sociological thinking, had already recognized, before Taine, the "milieu" as a factor of influence on creation and behavior. We only have to think of Alexis de Tocqueville who, in his well known work *De la Democratie en Amerique,* found, like Taine, that every situation, every environment, creates a mental state that takes effect on perception and reception.

Certainly, with this reference to the environment (also referred to as "socio-cultural" or "apperceptive background"), a lot of analytical possibilities of the phenomenon discussed here are revealed. But they can easily lead to an apriorism when they are used to prove that it is, or was, the aim of some visual representation to disclose, *according to the opinion of the researcher or the producer,* more clear and complete than would be possible by factual reality, an idea that is important for society. We must prevent our becoming the victim of that false reasoning which attempts to understand society as an essentially total unit. After all, the many aspects of society are not just manifestations of a primary spirit whose origins they all are tending toward.

To solve the set of problems raised, it seems to me that the means nearest at hand is to pursue a line of inquiry which will show the role of visual technology in respect to perception and meaning of the presented matter, i.e., the relation between art and identity within the *environment of symbols*. However, we do not want to investigate symbols as such, but the dependence of the perceiving beings on their environment of symbols and the creation of this environment by the perceiving person as producer and consumer. The study of the milieu presents itself as a sociological theory of action; a (objective and subjective) human approach, implicating the actions of individuals, groups or collectives, allows us not only to centralize human actualities like love, hatred, humor, fertility, aggression, etc., but also culture and cultures as "total social phenomena." When well understood, the total phenomena of which I speak are not just simple juxtapositions of various cultural aspects in society, but are to be considered as their incorporation into individual or collective experiences. Indeed the study of the environment of symbols provides us with the possibility of comprehending those mechanisms that easily can lead to a *wrong* socio-cultural perception of human relations to art and technology.

Herewith we join the knowledge that every agreement in human communication — also the one between art producer and art consumer—makes demands for cognition and for meaning of symbols. They have to be learned and diffused for the production of technologically transmitted socio-cultural interactions. Only in this way can they be used effectively and contribute to the creation and maintenance of a viable environment of symbols. It would mean closing our eyes toward reality if we do not realize that nowadays it is, in the first place, the task of the technologies of diffusion to bring about the common possession of symbols. It has been said with good reason, "Symbols are the stuff out of which cultures and societies are made."[12]

The Comics in the System of Perception

The epistemological design pronounced up to this point, as it is no more than a rough outline, has now to be completed through a *thinking at the object*. And if we turn for this purpose—especially with regard to the general theme of his conference ("Absolute Values and the New Cultural Revolution")—to comics and comic films (we shall refer to them together on the following pages as "comics") it is for the following reasons. Over the years, when book printing developed, the so-called picture-sheets appeared on the market. They were the ancestors of the comics and served elevating, entertaining and informative purposes. Without dealing here with the development and contents of comics, comic strips and comic film,[13] we can say that comics, whether as printed products or as films, are to be considered nowadays as mass media. In view of their production in masses for masses, as well as in view of their effects, they are as a socio-cultural institution in no way inferior to the other mass media. It is all the more astonishing that scientific research hardly dealt with this medium. While press, radio and television have been scientifically examined in all their aspects, with regard to comics (apart from historical and aesthetic reflections) we find only a few empirical studies and no specialized research institutes. This lamentable state of affairs caused us to make comics the subject of our reflections. But also the fact of enormous economic importance of this medium is clearly shown through a study which we carried out for the International Social Science Council of UNESCO in ten countries. But finally, we made this the subject of our study because the comics became, as an artistic creation incorporating the technological constituent, an *integrating component of the system of perception* affecting people of all ages and social strata.

However, before we deal with this central question, we deem it necessary to submit, at least in *key words*, some characteristics of the material we are dealing with, in order to avoid possible misunderstandings in respect to the nature, the importance and the influence of the matter called "comics."

The following points are to be made:

–Comics get their topics out of the socio-cultural environment from which they originate and for which they are produced. Accordingly, they are also of importance for the creative sector, even if they are, on account of their figurativeness, classified as "fantasy killers" or, on account of their value as entertainer, as "rubbish," "Kitsch" or, in brief, as "trivial."

–The mutual dependence of picture and text is a characteristic mark of comics.

–Comics tell a story both on different levels of the pictures and different levels of time. Hence, something also happens between the pictures, and the consumer is called upon to discover the development of actions within the individual picture.

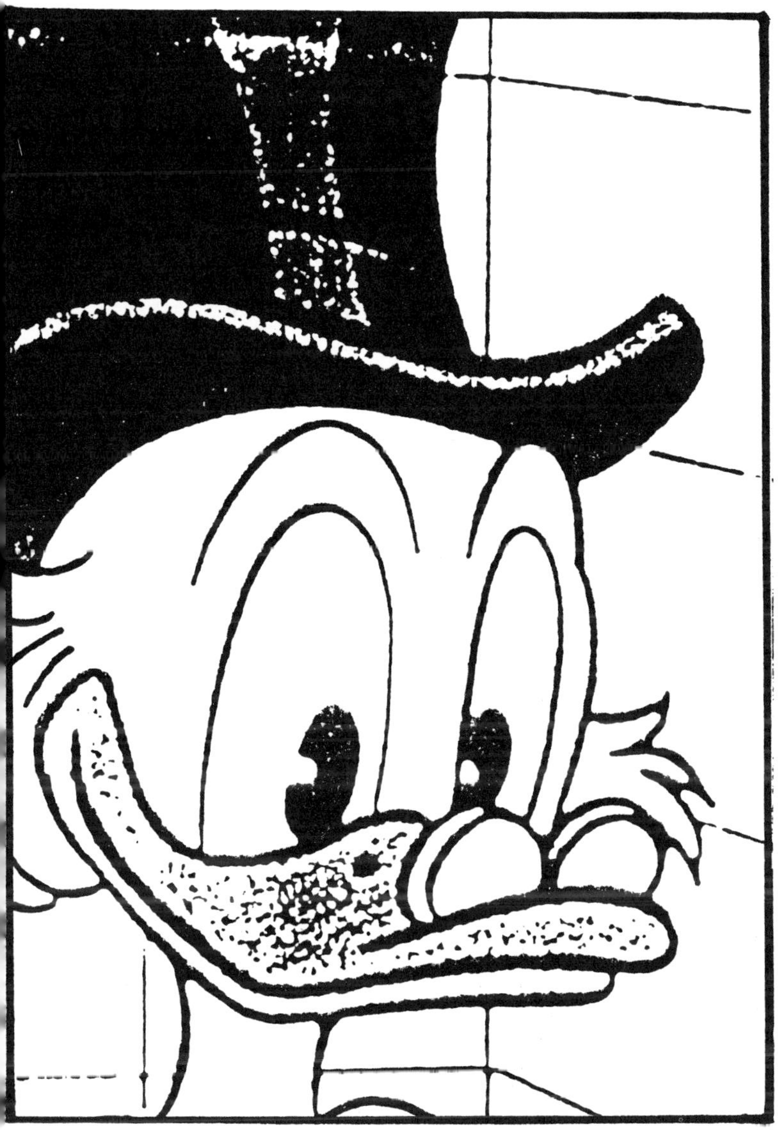

Fig. 6-2: Perception Unit: Close Up.

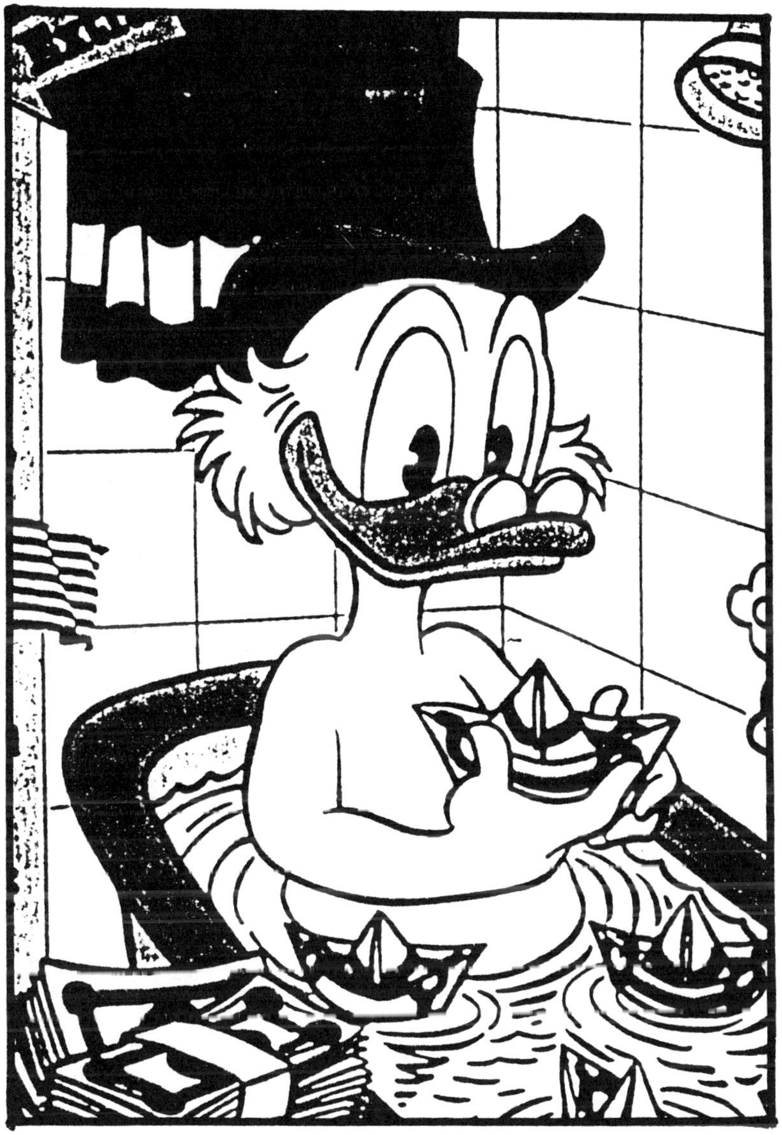

Fig. 6-3: Perception Unit: Medium Shot.

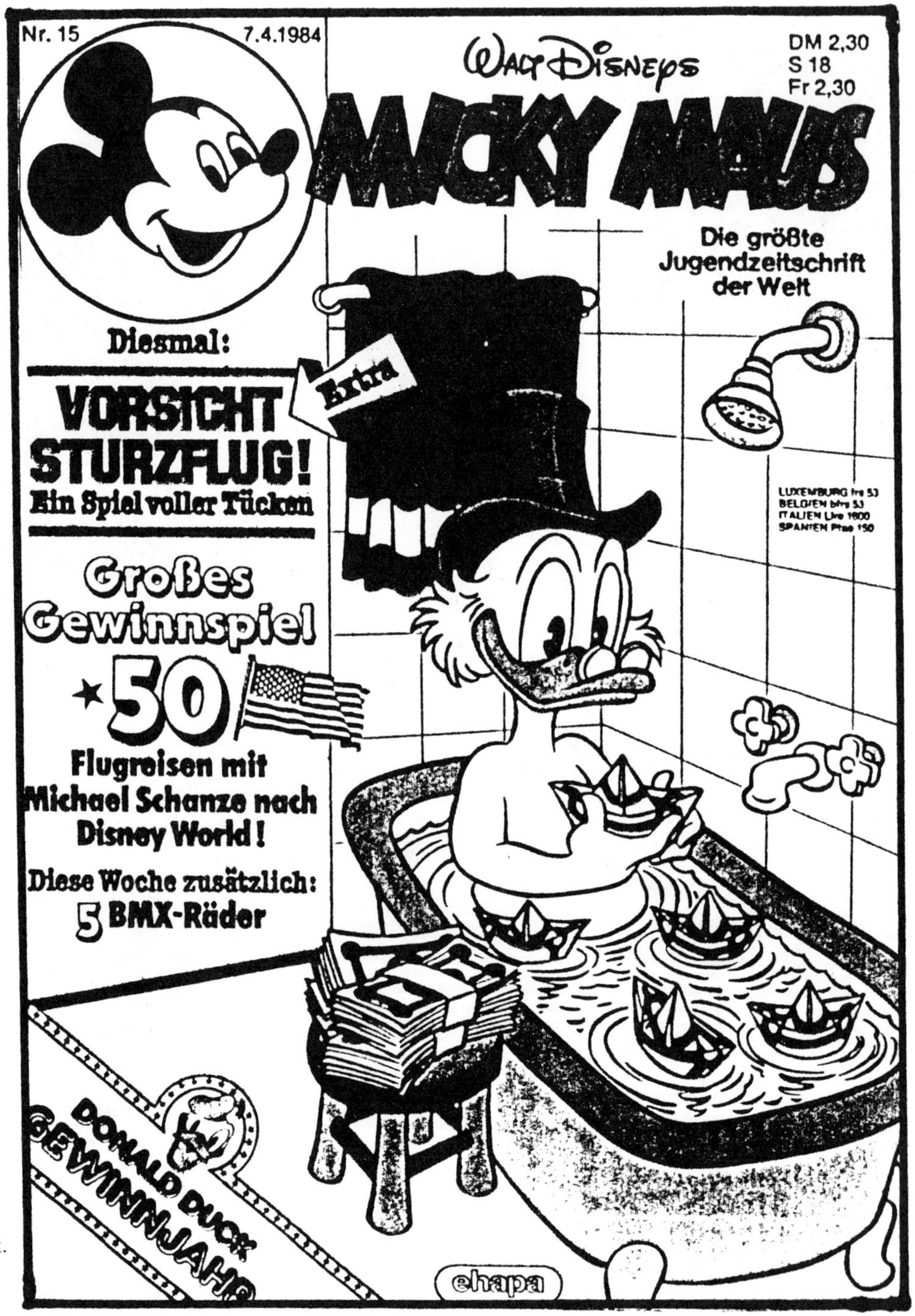

Fig. 6-4: Perception Unit: Long Shot. (© Walt Disney Productions, used by permission.)

–To render action as an essential ingredient of comics, intelligible short forms of expressions are used to describe atmosphere and character. They enable the interpretation of the pictorial information.

–Comics mediate, even as pure entertainment, certain mental values. Since the fact is that entertainment and information do not exclude each other, comic are latently or overtly open to any ideology.

–By means of speaking bubbles, comic pictures talk. For that reason, the reception implies two processes of reading that go parallel, intersect, complete or support each other.

–As far as entertaining action comics are concerned, they are based in their graphic presentation on often-changing perspectives, frequent cuts, bright colors, pictorial symbols, noise imitations and onomatopoeia. Among the main possibilities of picture adjustment are (as in feature films) the total or semi-total, whereby the figures are, in almost all cases, shown from the front. Very rarely, the contents of the picture are shown from a bird's-eye or a worm's-eye view.

After this short survey we can now take up our issue, in pursuit of our previous explanations, by discussing the example of the comics as an integrating component of the system of perception, the present *trend toward visual culture* through the function of art and technology.

The Trend Toward Visual Culture

Nowadays comics are used mainly in the following sections of cultural life: in entertainment, information, learning, education and publicity. Specific types of contents and forms of presentation have been developed for each of these over the years. These different forms show that the same performance of perception can be taken over by completely different cultural processes. It is in this sense that the assertion is to be understood that, at the moment, a lot of people—whether little ones or grown-ups—learn more relevant things about culture via comics and comic films than by pure seeing with their eyes. This shows that whatever is said for or against the cultural contribution brought about through the symbiosis of art and technology within the ramification of that part of the mass media called comics, refers to *function*—to respectively functional or dysfunctional effects, those being the consequences of functions.

Fig. 6-5:

MONTAGE

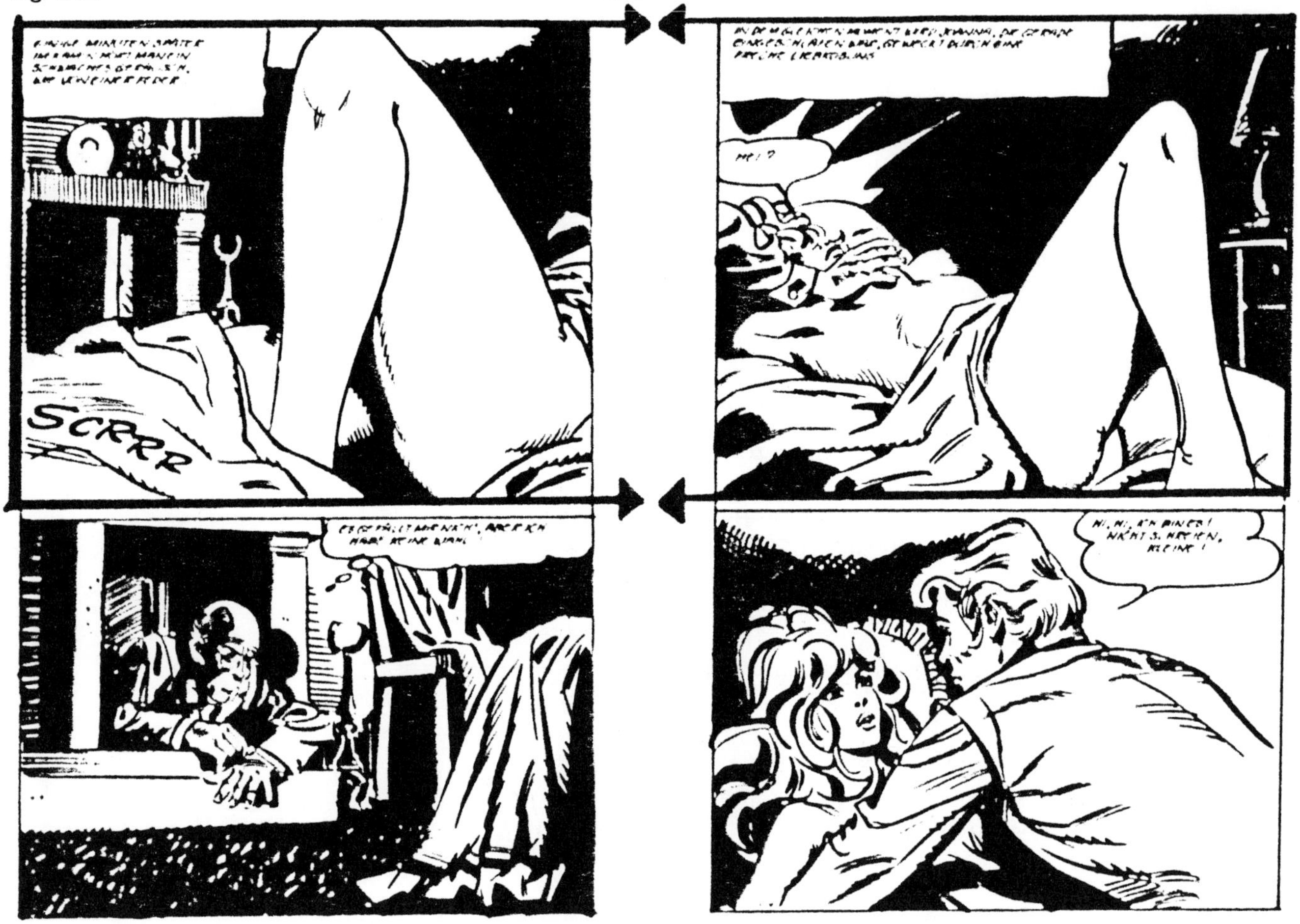

Source: Friedrich Knilli u.a.: Comic-Vorschule des Fernsehens. In: Dietger Pforte (Ed.): Comic o, asthetischen Unterricht. Frankfurt 1974, p. 274 (Reader on Comic Strips and school).

It should be understood well that, if we talk here about functions, then we refer to cultural elements that contribute to the continuance and/or development of socio-cultural configurations. In this respect, the function is *direct* if a specific need is satisfied by it; it is *indirect* if it contributes to the possibility that other cultural elements can become functional. Speaking of socio-cultural configurations, with this we refer on the one hand to the relationship between art and technology, and on the other hand

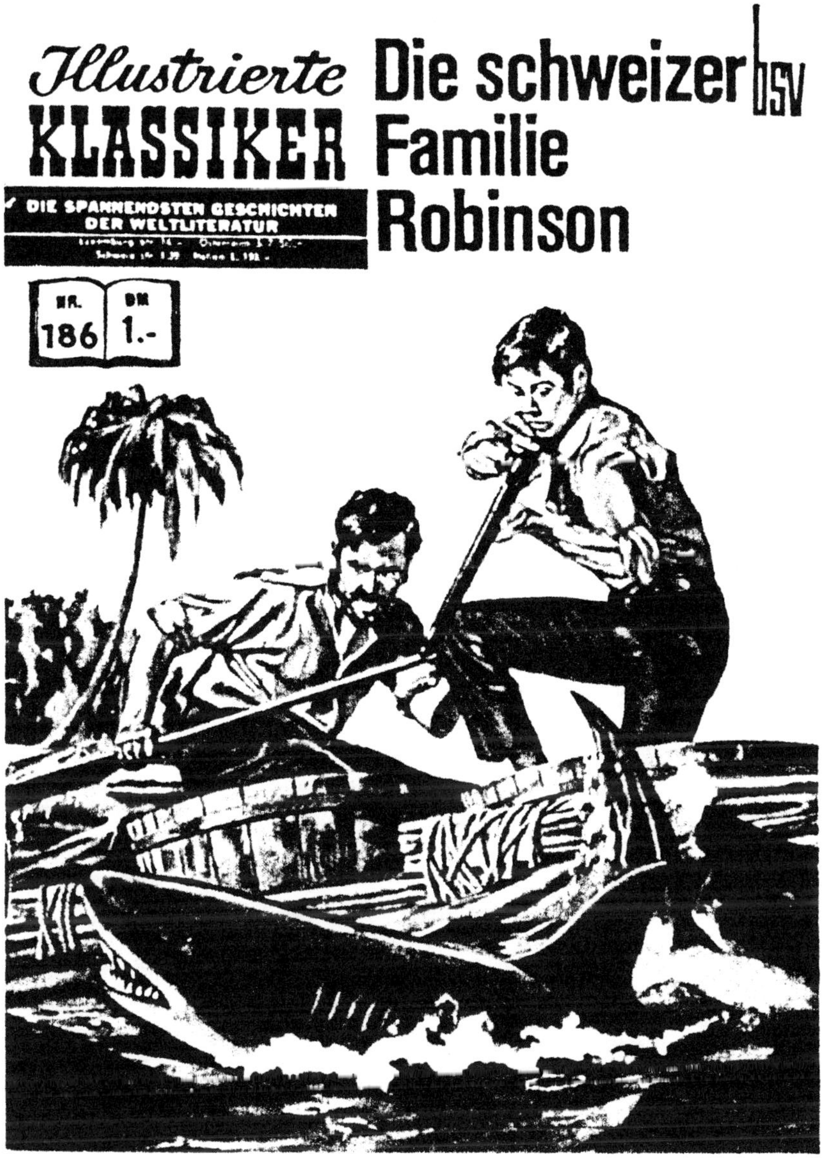

Fig. 6-6: The novel *Der Schweizerische Robinson* (1812) adapted for a comic strip called "Die schweizer Familie Robinson" (1960).
Source: Friedrich Knilli, ed., *Literatur in den Massenmedien-Demontage von Dichtung?* Munich, 1976, p. 83 (Literature and the Media).

to a system of perception bound to visual culture. Both are connected with each other in view of the needs that people try to satisfy. If this is to be made comprehensible, not only in an epistemological way but also close to the actual state of reality, then it will be necessary to show *for whom and for what* the above mentioned socio-cultural configurations, serving as a basis for "comics," are functional.

As far as the usage of comics within the socialization process of children is concerned, we know (speaking *grosso modo)* of three stages, corresponding to the various stages of childhood development.[14] At the first stage of early needs of presentation, there are the animal-comics as objects of perception. At the second stage of development the need germinates for the fulfillment of the search for reality, and is satisfied through the realistic appearance both of the physical scene and of heroes who are invulnerable or almost invulnerable. Only at a later stage, about the 12th year, does the need to become confronted visually with the real world grow and demand satisfaction, and real stories and facts that are close to reality are then on the agenda.

Social norms and relations, age, sex, and mental abilities come to the fore—as integrating components of the cultural system of perception—when the confrontation with the visual, evoked through the relations between art and technology, takes place; the relations between comics, child and visual culture are polarized. Whether the visual offers possibilities for experiences of fantasy or reality, the contribution of the child to the system of perception that is to be satisfied consists of internalized social norms as well as of the positive or negative socio-cultural relations, which, respectively, either demand a confirmation of reality or an escape from reality, or a substitution of it.

As far as adults are concerned, the analyses show that they form a big, if not the biggest part, of comic consumers; their functional satisfaction of needs is mostly done away with by arguments like semi-education, distraction or evasion. Going deeper into the matter, it is stated that a great number of the adult comic readers identify themselves with the hero of one or the other comic type created by artist: an energetic personality who, led by a vision of reality, dares to think and to act independently. The heroic qualities are transformed into visually recognizable actions, and by this the vision of an ideal is created which indeed is always and everywhere present, but becomes by visualization an ingredient of the *everyday moral.*

The foundation of this everyday moral, as far as it is relevant for society as a whole, has always been laid by situationary elements or phases that manifest themselves by social trends that also include, of course, cultural elements. They determine the creative side of the comics, i.e., that of producers and their consumers, and they are as little static as a society and its cultural phenomena. Producers and consumers, just as anyone of us, are subjected to social change which finds expression in *typical trends* of a society. These typical trends form the basis of any social judgment and, by this, also the position of comics in societies of today. While some time ago there was talk about the trend of "marketing-orientation," then of "other-direction," then of "identiy crisis" and aphoristically of the "permissive society," we can talk today about a trend toward visual culture. We have to reckon with socio-cultural trends of this sort above all because the stipulation of trends is not personal and the term has no inevitable moral implications. "Inevitability" means in this context that neither an individual nor an institution can be successful in changing the direction of a socio-cultural trend by personal or institutional efforts. No individual and no group or institution can detain the current trend toward a visual culture.

Let us look at some examples taken from different cultural sectors: schoolbooks for various subjects are enriched by comic drawings; Third World Comics are used for educational purposes: Bible workshops use comic booklets for the preaching of the gospel; the life of Pope John Paul II is distributed as comic biography; UNESCO is preparing the history of Africa in the form of comics; political comics with titles like "Marx for Beginners" appear on the market; television programs regularly use comic drawings for the explanation of economic, social and scientific information; youths are informed about professional chances with the aid of brochures with comic drawings, e.g., about the building trade, with indications concerning training, risks, difficulties, places of information and many others. Wherever we look and in whatever stage of the socialization process, we find ourselves constantly confronted with "comic art" as an expression of visual culture—regardless of whether we consider it to be trivial or to be a subculture.

Fig. 6-7: **PORNOGRAPHY**

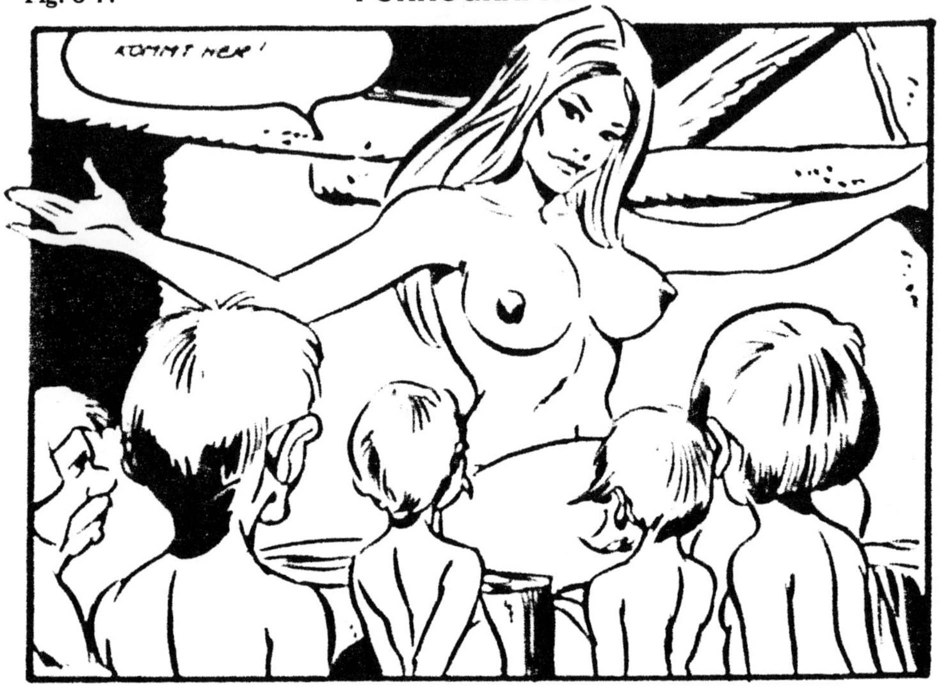

LUCIFERA TO THE DWARS: COME ON!

Source: Friedrich Knilli, ed., Medienmagazin 1. Im deutschen Reich der Zwerge, Riesen und Schutzengel, Munich, 1974.

Moreover, we also have to mention the use of comics in the *publicity* that is always present in everyday life—whether that publicity is intended to be informative, competitive or persuading. And it goes without saying that certain comic figures have become objects of worship and are used for advertising series of products. Indeed, if publicity uses the comic-form and the comic-style, this publicity follows and sustains the trend toward visual culture entering into the system of perception.

At this point, we have to stop, in order not to produce the impression that comics and comic films, in the framework of our multi-media society, are the only thing that is responsible for the way toward visual culture. Also, we should pause lest we lose ourselves in details, and neglect consideration of the relation between visual cultural phenomena and the people who are creating and supporting these phenomena.

This central, even commanding relation is immersed in invariable *dialectic tensions* which determine finally the nature of visual culture. And these dialectic tensions (because of their variety only a few prototypes can be mentioned here) lead—consciously or unconsciously, directly or indirectly—the way toward visual culture embedded in actions and behavior of people.

Dialectic Tension Between the Individual and the Mass

The interhuman nature of the communication of visual cultural items demands, in the first place, that there be an individual face to face approach, i.e., a communication process where any communicator meets with any recipient under his own specific conditions. That means, although mass media using visual culture are meant for an anonymous mass of people, every member of this mass is to be reached by the visual message. Hereby a tension arises that is at the basis of the questions: Is the message destined for anyone or nobody? Are certain persons addressed or just the impersonal abstraction of the so-called average citizen?

Dialectic Tension Between Two-way and One-way Communication

The tension arises from the question whether a two-way communication is evoked and corroborated by means of visual culture or does it remain a one-way communication process, whereby an alternating effect between communicator and recipient is not possible and desired feedback is missing.

Dialectic Tension Between Instantaneity and Absorption

This tension refers to the demand for immediate instantaneous communication by means of visual culture in contrast to the demand for absorption of the message. After all, the constantly growing complexity of our society, the technological development as well as the enormous value of information in all fields of human activities make it more and more difficult to understand the real meaning of an event or an invention without the necessary references to the background.

Dialectic Tension Between Subjectivity and Objectivity

This tension is the most discussed one in reference to cultural utterances. For although we know that both these criteria are unattainable in absolute perfection, subjectivity and objectivity attract each other like magnetic poles as far as visual culture is concerned. Thus, communications, based on the visual, vacillate between the demand for absolute truth and the justified want for personal interpretation, whereby the question arises where the right emphasis would be.

Dialect Tension Between Private and Public Interest

This tension (whether it is based on political, economic, cultural, social, moral or religious interests) is equally noticeable on the side of the producer as well as on the side of the consumer of visual messages.

Dialectic Tension Between Independence and Compulsion

With regard to this tension it may be sufficient to point to the dependence of producers and consumers of visual culture on tradition, cultural heritage or pressure groups, whether they be called church, state or political parties. Personal integrity and vested interests are inevitable contrasts.

Inherent dangers for our entire cultural life are attributed to any of those relations of tensions with regard to a progressive development of visual culture. Whoever holds that these tensions are inherently dangerous insists on the duration of absolute values, refuses to acknowledge a dynamic and evolutionary change of attitudes, and overlooks or wants to ignore the basic attitude that governs any change of attitudes, that is to say, the integration of a dynamic view on the rapidly proceeding socio-cultural change. The artist as well as the technologist has to behold the practical actions of everyday work in the light of the dynamic tendencies of the socio-cultural change.

If the general subject of our conference refers to "the new cultural revolution," then it is to be understood that any new state of affairs demands a solution, and that an adjustment has to be found between the dialectic powers of opposite poles of tension, an adjustment that represents a positive contribution to improved conditions of living. We hope to have clearly expounded that—in relation to the socio-cultural change—only the acknowledgement of a mutation of cultural communication together with a dynamic and progressive attitude, can lead on a leveled way toward visual culture and, in this way, reduce to a minimum the formation of a most undesirable cultural lag.

Fig. 6-8: **COMIC STRIPS USED AS INSTRUCTIONS FOR THE USE OF A PRODUCT**

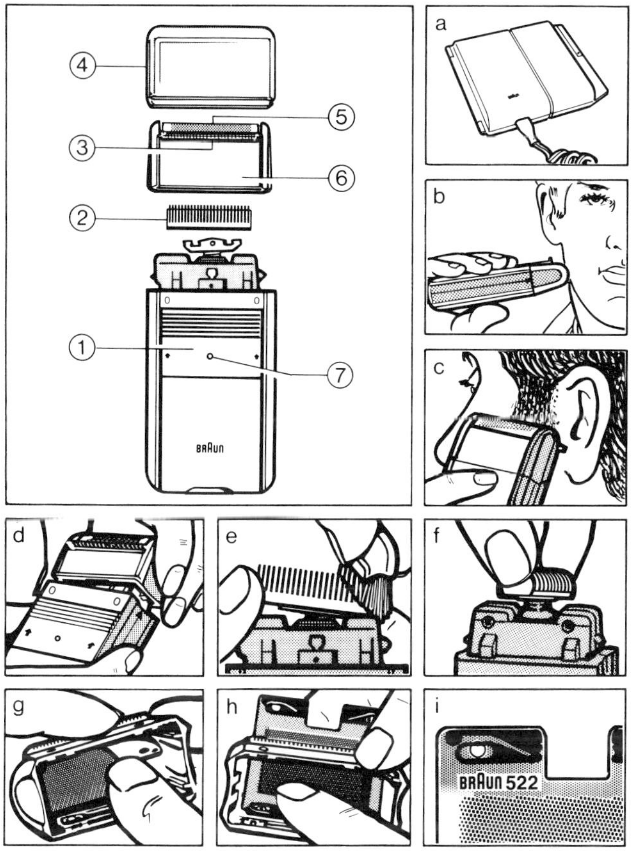

Source: Friedrich Knilli. Aloys Struck: Problematisierungs-analyse der Bedienunsangleitung fur den Braun Sixtant Compact Two-Way. Berlin, 1984. A study on the use-instructions for this shaver.

The Braun sixtant compact two-way
This shaver can be operated without converting on all international direct or alternating current outlets.
Its built-in, rechargeable cells permit 2-way shaving: either without the cord via the rechargeable cells or with the cord connected directly to an outlet.

Cordless shaving
The quick charging used by Braun in this shaver allows the shaver to become almost charged within 1 hour. The unit then automatically turns to continuous or trickle charge. This method is very preserving on the cells. The best environmental temperature range for the shaver is between +15 °C – +35 °C. The shaver should never be exposed to temperatures of less than 0 °C and more than +40 °C for longer periods of time.

Recharging the shaver is independent of local voltages and frequencies due to the built-in electronic control.
When recharging, set the switch to «Off» = [0 0]. The red pilot light ❼ must light up (permanently or intermittently).
The shaver may be recharged while stored in its case (a).

One hour of charging is sufficient for approx. 30 minutes of cordless shaving. The exact shaving time is, however, dependent on the type of beard. The shaver can also be recharged at voltages below international standards. In cars and boats, etc., however, a standard 12 V outlet is required. For your safety when recharging the shaver in a car while driving, please ensure that the shaver is lodged firmly and that the charging cord will not hinder the driver. Avoid direct sunlight on the shaver (see temperature range). The lowest admissible recharging voltage is 12 V direct or alternating current. Recharging this way (not quick charging) takes a maximum of 16 hours. For safety reasons use only the Braun charging cord No. 5-001-687 (available at your dealer or from a Braun Service Centre).
Continuous charging – regardless of which method you use – will not damage the shaver.

Shaving with the cord
When shaving with the cord (i. e. when the shaver is connected directly to an outlet), the shaver is ready for use immediately.

If the shaver should, however, not run immediately due to empty cells, recharge the shaver for approx. 1 minute (switch set to «Off» = [0 0]), which is sufficient to permit shaving with the cord again (after turning the switch on). Shaving directly via the charging cord while connected to outlets below international standard voltages (e. g. cars, boats, etc.) is not possible.

Notes

1. Marcel Mauss. "Essai sur le don." in *Annee Sociologique*. Vol. I. (1923 24).
2. See M. Halbwachs. *La memoire collective* (Paris. 1950).
3. See Th. W. Adorno. "Resume uber Kulturindustrie." in *Ohne Leitbild* (Frankfurtam M.. 1967).
4. Th. W. Adorno and Max Horkheimer. *Dialektik der Aufklarung* (Amsterdam. 1947).
5. See John Ruskin. *Selections and Essays* (New York. Chicago. Boston. 1918). chapter IV.
6. See. for instance. C.K. Ogden and I.A. Richards. *The Meaning of Meaning* (New York. 1945).
7. For instance. Charles E. Osgood. G.J. Suci and P.H. Tannenbaum. *The Measurement of Meaning* (Urbana. Ill.. 1957).
8. Ernst Cassirer. *Philosophie der symbolischen Formen*. 3 Vols. (Berlin. 1923–1929). See W. Nutz. in A. Silbermann. Ed.. *Klassiker der Kunstsoziologie* (Munchen. 1979)
9. Erwin Panowsky. *Meaning in the Visual Arts* (New York. 1955). See A. Hanseroth. in A. Silbermann. Ed.. *Klassiker der Kunstsoziologie* (Munchen. 1979).
10. See R. Arnheim. *Art and Visual Perception* (Berkeley. 1954).
11. Hippolyte Taine. *Philosophie de l'art*. 2 Vols.. (Paris. 1882). See L. Kofler. in A. Silbermann. Ed.. *Klassiker der Kunstsoziologie* (Munchen. 1979).
12. H.R. Zetterberg. *On Theory and Verification in Sociology* (Totowa. N.J.. 1965). p.1.
13. See A. Silbermann. *Handworterbuch der Massenkommunikation und Medienforschung* (Berlin. 1982). p. 49ff.
14. K.M. Wolf and M. Fiske. "The Children Talk About Comics." in P.F. Lazarsfeld and F.N. Stanton. Eds.. *Communications Research 1948–1949* (New York. 1949) p. 3ff.

Discussion of Alphons Silbermann's "The Way Toward Visual Culture: Comics and Comic Films"

Friedrich Knilli

I should like to discuss and critically appraise Alphons Silbermann's excellent study in three steps. In the first two I shall introduce historical and systematic constructions, as well as facts, which will serve as my point of departure for discussing the theme of the conference (Absolute Values and the New Cultural Revolution) and the theme of Committee III (Art and Technology). After this presentation of my criteria, I shall proceed with a critical appraisal of Silbermann's paper.

Absolute Values and the New Cultural Revolution

Two facts are, in my opinion, of revolutionary importance for the discussion of the theme of the conference: The genesis and development of a permanent program and, at the same time, the ahistoricity of this media culture.

The Permanent Program

When radio was introduced at the beginning of the 1920's, a new audio culture began in the history of mankind. Its new quality was exclusive acoustic communication. Also new was the fact that millions of widely dispersed listeners were able to receive information, education, and be entertained, for communication could take place not only with the velocity of sound, but also with the speed of light. A new feature about this form of communication was the permanent program which was planned and produced by authors, directors, and editors for politics, economics, science, culture, music, education, entertainment, sports, family, religion, region, school, and children. For over sixty years now they have been presenting programs for listeners around the clock. If all radio stations of an area are taken together, the total amounts to more than several hundreds of hours of programs which accompany the listeners from morning to evening and way into the night, and which take into account both their working and leisure time. The radio awakens the listeners, entertains them at breakfast, accompanies them on the way to work, keeps them in a good mood there, accompanies them home again, entertains them during their leisure time, and helps them to go to sleep. But the permanent program is not only geared to the daily routine, but also to the progression of the week: "blue" Mondays, "short" and "long" weekends; and finally to the seasons of the year and all the religious, cultural, and official bank holidays; uprisings, coup d'etats, revolutions, and world wars.

After 1945, a number of permanent television programs came into existence alongside of this nonstop radio programming. And in the eighties these were supplemented by the permanent programs of cable television and direct satellites.

Ahistoricity

The history of the permanent and nonpermanent programs cannot be written because of two insurmountable obstacles. On the one hand, the majority of these programs have not been conserved in archives, but have been systematically destroyed year after year for various reasons. This applies to both the printed and the audio-visual media, for the latter even more so. Secondly, had they been stored in archives, it would not be possible to either analyze or interpret them for sheer reasons of quantity. The length of time of the programs exceeds human life spans.

There have been attempts to take action against this lack of history: I am thinking here of the annual random samples of programs which have been collected in the Museum of Broadcasting in New York. I am thinking also of the Cultural Indicators Approach which attempts to register certain program trends on the basis of random samples. However, this does not constitute a history.

Does this not mean that with the genesis and development of the "industry of consciousness" we have, of necessity, fallen into a state of ahistoricity which is without parallel in the history of human thought?

Art and Technology

That this subject can be approached from many sides is obvious. I have chosen an approach which I should like to call semiological in the broadest sense of the term. According to this, art and technology are defined as follows:

Technology

In communications engineering, information theory and cybernetics (Shannon, Weaver, Wiener, Meyer-Eppler, etc.) communication is understood as "the recording and processing of physically, chemically, or biologically demonstrable signals by a life form. The living and inanimate organs which are involved in an act of communication form a chain of communication. Typical forms are the observation chain, the diagnostic communication chain, and the linguistic communication chain."—"The simplest linguistic communication chain demonstrates a *double connection* between the communicating parties as opposed to the observation chain and diagnostic chain. Apart from the real *signal* connection, demonstrable by physical methods, an agreement exists with regard to the *sign* function of the signal on the level of presentation."[1]

These signal connections, demonstrable by physical methods, are, in the case of face-to-face communication, sunlight and air, and in the case of indirect communication, technical appliances for the generation, transmission, and recognition of the signals.

Art

In the communication of art, there also exists between artists and recipients a double connection. Apart from the real connection by signals, which is demonstrable by physical methods, there is agreement about the use and representational function of the signal. The description of these agreements between the artist and the observer is given by the semiotics of art. A work of art is distinguished from entertainment according to the theory of signs, e.g., by the fact that it is a sign of itself. The sign-vehicle 'work of art' becomes both a sign and a signification in one. It is a new symbol. It is a specific code which constitutes the individuality of a 'work of art.' Roland Posner on linguistic art works: "This 'second' code operates partly on linguistic elements and thus has as a prerequisite the knowledge of the language in which the text is written (of the 'first code'); but it differs in two points from the linguistic sign system: The text by which it is realized is not preexistent but constitutes itself primarily in this and is not known to all members of a linguistic community but is established only by reading."[2]

'Popular' literature also operates with the 'second code,' which is based partly on linguistic elements and thus the knowledge of the language in which the text is written in is a prerequisite (that is, knowledge of the 'first code'); but it differs not only from the 'second code' of the work of art but also from the 'first code' of the linguistic sign system in that the text by which it is realized is preexistent and does not constitute itself in this. In addition, it is known to the majority of the members of a linguistic community (one of the most important prerequisites for mass entertainment). The 'second code' of 'popular' literature is thus a code between the 'first code' of the linguistic system and the 'second code' of the work of art. And this 'intermediate code' consists of cliches, stereotypes, set phrases, and standard expressions which were known to the majority of the members of a linguistic community long before they were used in the context of a specific novel. They constitute a code which, although it operates with elements of the linguistic sign system, differs from this in that it is a coding of the 'first code.'

The same applies to the 'language of images,' to the visual and kinetic codes which are utilized by the entertainers in film and television. Film art and television art are also signs of themselves. The sign-vehicles 'art film' and 'television art' are signs and signification in one. They not only refer beyond themselves to people in time and space, but also back to themselves. They are special visual and kinetic codes which constitute the singularity of a special 'work of art' in film and television. This 'second code' operates with visual and kinetic primary codes and has as a prerequisite general knowledge of the same. It is not preexistent in the film or the television program which realizes it, but is first constituted in the 'work of art' and can therefore be first identified only when seen.

The Way Toward Visual Culture: Comics and Comic Films

The Technological Constituent

I should now like to consider individually the chapters of Silbermann's paper.

The hostility of artists and art critics to technology is criticized here by Alphons Silbermann. It is also to be criticized from the standpoint of semiology. For although technology constitutes the sign connection between artist and art consumer, the work of art constitutes itself as such on the level of signification: that is, as a second code whose characteristic is that it is not preexistent.[3] Certain constituent interdependencies may exist between this second code and technology in particular with respect to analogical representation. This kind of interdependency of time and space in any technical means of representation was already recognized by Lessing in "Laokoon" as being the essential difference between painting and poetry: "If it is true that painting requires completely different means or signs for its imitations than poetry does, that is, the former requires forms and colors in space whereas the latter articulates sounds in time; it is indisputable that signs arranged side by side can only express objects which exist side by side or whose parts exist side by side, and signs which follow one another can only express objects which follow one another or whose parts follow one another. Objects which exist side by side or whose parts exist thus are called bodies. It follows that bodies with their visible properties are the proper subjects of painting. Objects, which follow one another or whose parts follow one another are in fact called actions. It follows that actions are the proper subject of poetry."[4]

It is self-evident that painting can also portray action, but only symbolically and without similarity or, as Lessing intimates in outline, because bodies exist in time and may also be the center of an action.

Effect-Relations Between Art and Technological-Media-Induced Mass Communications

Silbermann puts forward the argument here that the two most important effects are, on the one hand, "the function of certain artistic contents in relation to the media,"[5] and, on the other hand, "the function of the media in relation to certain artistic contents."[6] In the first case it is a question of the marketing of art through the media; in the second, of the correct usage of media for art. This twofold relationship—according to Silbermann—also comprises other, important problems: the problem of identification, the problem of functions-shifting, etc.

The significance of these aspects, also from a semiological angle, is so great and so topical that I should like to draw attention here to the branch of research which, in this connection, is increasingly intervening between certain artistic contents and the media. I refer to *formative research*, which today is being utilized for very different types of programs. Here is an example of the work of the *Children's Television Workshop* (CTW), New York: "An important part of the formative research which went into the year-and-a-half planning for *Sesame Street* involved measuring the audience appeal of possible programs before they went on the air, since appeal was a vital ingredient if *Sesame Street* was to reach and keep its audience. Small groups of children of the appropriate age and apparent cognitive skills were recruited to come to the Children's Workshop studios to view proposed programs and segments of programs. The program was projected on a monitor set while on an adjoining wall slides were projected at an angle to the child. Observers rated the proportion of each 7.5 second schedule that the children viewed the program rather than the projected slide. Producers could then relate the content of the program (or segment of a program) at any one point to the degree of attention that program or segment attained from the test audience of children. From these findings, in addition to accepting or rejecting versions of programs, generalizations were drawn about program features that did or did not attract the attention of economically disadvantaged children. The formative research played a crucial role, producers believed, in reducing the risk that *Sesame Street* would not hold its audience once it reached them."[8]

Mass and Popular Culture

What Silbermann elaborates in this chapter is, in my opinion, undisputable. And, in my opinion, there is nothing to be added to his remarks on the taste of the audience.[9] With regard to the historical aspect I should like to underline this more forcefully. For that which we term today Popular Culture can only be understood if we take into consideration that its genesis and development is inextricably bound up on the one hand with the emergence and the development of the media into the mass media, and on the other hand that it stands in competition to the movement of lay culture of certain social classes of the 19th century, e.g., peasants, artisans, and workers.

Here, the *workers' cultural movement*, which was organized politically and in trade unions, exerted a very great influence. In the Germany of the 19th century, trade unionists, social democrats, socialists, and communists realized very quickly that they needed leaflets, newspapers, magazines, brochures, books, libraries, and many other forms of information material for the political and cultural 'Kampf'; not only the printed media, but also social gatherings and festivals were very important communicators. The drama groups of the numerous workers' educational associations put on plays about strikes and didactic plays on the theory of surplus value, equal rights for women, the Basel Resolution of the First Internationale on the question of land reform, etc. They aimed to educate in an entertaining way. At the end of the 19th century, began the rise of the German Social Democrats and their trades unions to a mass organization, and the expansion of educational and cultural organizations of their press and publishing houses. First-rate publishing achievements like the Arbeiter-Illustrierte-Zeitung will not be forgotten, alongside the great productions of the Prometheus-Film company. With the introduction of radio and the refusal of licenses to the workers radio stations began the decline of the workers' cultural movement. The Nazis brought it to an end in 1933. After 1945 there was no place for the laymen's culture. The culture industry catered to the workers in their leisure time.

In the USA the New Deal inspired a short blossoming of the workers' culture.

The System of Perception

This is the longest chapter and Silbermann deals in great detail with perception. He describes an inventory of perceptive situations,[10] offers a possible division into three of the process of "seeing," and deals with numerous other aspects. Nevertheless, I should like to suggest an addition which I consider to be important for the perception of comics and comic films. I am referring to the so-called *Gestalt principles of perception:* proximity, equality, unity, form and background, thing constant, size constant. They describe the process of seeing under the assumption that we do not perceive a set of stimuli atomistically, but as a Gestalt, whereby Gestalt is understood as an entity that cannot be destroyed by transformation. An example: A melody which can be recognized as the same although it is played in different keys. This basic law of Gestalt theory was founded by *Christian Ehrenfels.*

The Comics in the System of Perception

In this chapter Silbermann criticizes the extremely inadequate state of research on comics and then proceeds to provide a catalog of the most significant components of comic strips and cartoons, to which there is nothing to be added. I would like to comment on the perception aspect with regard to comics for, in this connection, a comparison is often made with film and television. This is at least partly due to the fact that these also narrate stories in words accompanied by sequences of pictures. Looked at from this angle, the history of comics would appear to be no more than the prehistory of film and television. However, this would ignore essential differences of these media and differences in the codes. Thus comic strip authors present their stories as 'dead' information; 'live' information is that provided by film and television. It is 'live' because the information is supplied together with the energy necessary for its reception; by contrast the 'dead' information requires external energy to 'activate' it. The information stored in the 'dead' comic strip has first to be reawakened to 'life' by the actions of seeing and reading, which require a special type of reception. There is a further difference: the camera shot in film is a continuous 'take' which is delimited by camera angle and cutting. The shot is only interrupted when the change of camera angle remains visible and the transition of one perspective to another is omitted. The film maker narrates his story in such *perception-units*.

The comic strip authors also operate in this way; they also construct their stories from such perception-blocs. Each new picture with its 'bubble' for speech and space for commentary signals a change in perception; with each fresh picture of a comic strip a new perception-unit is reproduced—naturally in a different way from film and television where, on account of the different media structure, very many more analogical representations of that which is to be perceived are possible. With regard to the comic strip, similarity only exists with respect to visible spatial conditions; time and auditory conditions are communicated, for the most part, by typographical and graphic symbols (speech bubbles, symbols of sounds, change of type face for change in volume) or within the dialogue context or commentary. For this reason, the comic strip reader must use his/her imagination much more than the cinema goer or television viewer. That which is imagined is not just somewhere in the mind of the comic strip reader, it is firmly connected with the perceived picture by the referring text; sphere of perception and sphere of imagination are identical. But fantasy is not only demanded of the comic strip reader; the film or television viewer also has to supplement the perceived image with his/her imagination, e.g., in order to 'see' the image three-dimensionally or to localize sounds. For sounds are only provided with spatial coordinates, and the film picture only with two spatial coordinates, the film or television viewer only 'sees' these images three-dimensionally in his/her imagination. Such units of perception and imagination are represented both by pictures with speech bubbles in the comic strip and by camera shots in film; it is in these units that the invisible narrator tells his story. The perception elements thus used in the interplay of words and pictures—the indicating words and pictures—constitute the inner montage of the individual image.

The Trend Toward Visual Culture

In the final section of his paper, Alphons Silbermann develops a dialectical system of six pairs of opposites (individual and the mass, two-way and one-way communication, instantaneity and absorption, subjectivity and objectivity, private and public interest, independence and compulsion) within which the trend toward visual culture operates. I should like to add to this an aspect which seems to me of importance: the symbolization, or to be more exact, the *emblematization of this visual culture*. Emblems are images, which are endowed with additional meaning by means of inscriptions (*inscriptio*) and signatures (*subscriptio*) thus becoming symbols. This art of symbols is ancient. In late Greco-Roman times, hieroglyphics originated which were used as a model for a new pictography in the Renaissance. During the Baroque age, emblematics underwent a consciously artistic transformation, which corresponded to the Baroque penchant for antithesis with its opposition of image and meaning. But this art of symbols survived the Baroque era and exercises an influence up to the present day. We find evidence of it in advertising, in the feature film, in documentaries, and in the comic strip.

Research on the Effects of Comics and Disinformation, A Discussion of Alfons Silbermann's "The Way Toward Visual Culture: Comics and Comic Films"

Walter Nutz

In his paper, Professor Silbermann makes three chief statements, statements which I consider to be most important. They are:

> —technology appears as an obvious and integrated part of the mentioned artistic creations;
>
> —comics in the system of perception illustrate the trend toward visual culture through the functions of art and technology, and
>
> —the dialectic tensions represent the dramaturgic structure of everyday life where man simply uses comics for various needs. The simplicity and self-evidence with which the comics are used already constitute a certain cultural behavior which leads to "the new cultural revolution."

I do not want to take critical issue with these points because I largely agree with them. Instead, I wish to complete and extend Professor Silbermann's arguments. My completions refer primarily to our experience with these media on the basis of scientific and public occupation in the Federal Republic of Germany, while my extensions are of a more general nature and deal in particular with the informative or disinformative use of the media.

Professor Silbermann told us that integration of technology in a work of art is often rejected, and that many arguments are given for this rejection, or at least the relationship between art and technology is dealt with in an ambivalent way. As a completion to Professor Silbermann's remarks, I want to discuss the relationship between art and technology.

It is not just the thinking of elitists which leads to the interpretation that an artistic creation is *only* an expression of free ingenuity, and an expression which is by no means based on social factors because technology forms part of the "low" substructure. The problem is that there is a widespread fear of establishing contacts with technology in general. Not only scholars but many other people, in particular young people, often view the term *technology* in connection with the destruction of the environment, fear of the future, missiles, nuclear annihilation, etc. All these terms have a negative meaning. This attitude is very evident in the Federal Republic of Germany, where, moreover, a cultural-critical, cultural-philosophical and socio-psychological approach prevails over the empirical-scientific approach. *So-called research on the effects of mass media has been given priority.*

As a result, the emphasis of research is not on the question of what a medium *is* but on the question of what a medium *should be*. The possibilities of the media are assessed in an absolute way. These possibilities are then classified according to moral criteria, and brought into a categorical imperative of specific cultural criticism. In other words, *cultural criticism is given priority over empirical research.*

Before these researchers are even willing to meet Emil Durkheim's demand that social phenomena be dealt with as "facts" (French: *choses*),[1] and before they even know the interdependent structures between the media and society—in other, more precise words: before the social relationship of the producers, contents and consumers of comics is definitely defined—they not only talk about the ethical confusion created by this medium, for instance, or harmful kitsch or leveling trash, but they classify this medium in its totality in a specific stereotyping philosophy.

Such a stereotyping philosophy naturally has an ideological basis and is aimed at assessing the manifestations of a popular culture as being of inferior value; in addition, it aims at disqualifying popular culture on moral grounds as dangerous, harmful to youth and morally corrupting.

I have observed this phenomenon in particular in countries (such as the Federal Republic of Germany) where a dichotomy exists between "higher" and "lower" literature. Literature sciences, proceeding from an aesthetic valuation, always differentiate between "good," or artistic literature, and trivial literature. It is impossible in this way, of course, to establish a relationship with reality and meaning, by means of such literatures. In other words: trivial literature is *absolutely* disqualified and is not regarded as a *different type* of literature which must be related to different sections of readers as well as to a different cultural level. Like comics and comic films, this so-called trivial literature uses stereotypes and a specific symbolism.

Although, in my country, the discussion of these media at schools and by the public apparently has been influenced by the ideologically, rather than the empirically-oriented disciplines, the empirical approach has not yet been exhausted. Besides the already-mentioned activities regarding research on comics in their social reality, I want to point to the fact that, to date, no detailed analysis has been made on the symbolism of comics. The basic insights of Ernst Cassirer[2] and C.G. Jung[3] (whose studies have been continued) are not yet exhausted in this respect. At present such an approach is possibly more productive than research on the effects of mass media.

Many publications, at least in the Federal Republic of Germany, show that under the generic terminology of *research on the effects of mass media*—according to which comics and comic-films create and foster anxieties, primarily in children—both cultural criticism and certain sciences can be misused for certain purposes.

"Anxiety" is the big magic formula which, with a socio-philosophical and socio-psychological palliative, constitutes the basis and starting point of many approaches, an anxiety which allegedly is related to technology in general and—considering the subject we are discussing—to comics in particular.

Of course we know that this represented anxiety from the past. This fear of technology was formulated as early as the twenties in the movies of that time. For example, Fritz Lang's movie *Metropolis* of 1927 was interpreted in Germany primarily as an antitechnological assault on machines. The French, in contrast, "were worried," because this movie "was to them a crossbreeding of Krupp and Wagner."[4] Thus anxieties can have many forms, depending on the interpretation and individual arbitrariness that are brought to the situation.

Our conclusion is that overvaluing the effects of technology and (in our specific case) comics has caused empirical research on facts to atrophy in the shadow of cultural-critical slogans.

Regarding the extensions I mentioned earlier, I first want to prevent a possible misunderstanding. As media, comics convey information, *inter alia*, which may be disinformative. This true statement, however, could be taken to apply to the sphere of research on the effects of mass media, which is unjustified and would be extremely misleading.

What I am addressing here is the capacity, inherent in the media, of passing on information only with the possibilities that apply to that particular medium or instance. No matter whether our world is represented or described by movies, television, radio, novels, the theater, etc., we always experience it as a section, as the representation of a specific dramaturgy. And since it is impossible to totally convey the whole, this world always remains in the mirror of *structural* disinformation.

These few words already indicate that the problem of specific information with regard to the comics is a problem of immanence of the relationship I described above. In other words, research must approach the medium and its interactional and interdependent relationships, and not the question of whether the consumer is disinformed.

What I mean takes place against the following background: social existence is always a communicative existence, with communication never being static but always dynamic and in a process of development. Every instance of information given through a medium is a communication and gives us the facts through an "intermediary." This "intermediary" and this communication are limited by their own possibilities and functions. Therefore, the medium changes the information in its form and content. *It is this change that I mean when I am talking about disinformation.* Moreover, if we know that more information does not mean better information, we have defined the background of what I illustrated as the concept of disinformation.

In this, we see one specific problem with which I wanted to extend Professor Silbermann's arguments concerning the approach and directions in which scientific activities should be pursued: *we should give priority to empirical research on the medium and its interdependencies*, meaning research not only on men as partners in the communication process, but also on technology.

Possibly I have created the impression that I underestimate the problem of the effects, or that I even reject research on the effects of mass media. The contrary is correct. I also believe that the poor results of the research which has been conducted on the effects of mass media are attributable to the fact that this research is not concerned with the empirical social relationship, but is allowed to speculate in isolation from any relationship.

One thing is certain: while cultural criticism still formulates its slogans in the field of absolute assessment, that is to say, in the field of a negative assessment, consumers throughout the world have long adopted and accepted the media, and are "revolutionizing" their own culture in a permanent social process. Whereas the consumers of the contents of mass media, including those of the comics, have long embarked on the way toward visual culture, cultural criticism still reacts with the above-mentioned verdicts.

Thus the comics and comic films do not constitute a problem for those who use them—to quote a German scientist—but they constitute a problem for those who deal with them critically and are unable to cope with them.[5]

Notes

1. Emil Durkheim, *Les Regles de la methode soziologique*,20th edition (Paris: Presse Universitaires des France, 1981), p. 15.
2. Ernst Cassirer, *Philosophie der Symbolischen Formen* (Darmstadt: Wissenschaftliche Buchgesellschaft, 1977; first published, 1923), 3rd Vol.
3. C. G. Jung, *Der Mensch und seine Symbole* (Olten: Walter Verlag, 1968).
4. Sigfried Kracauer, *Von Caligari bis Hitler. Ein Beitrag zur Geschichte des deutschen Films* (Reinbeck bei Hamburg: Rowohlt-Verlag, 1958), p. 98.
5. Hans Dieter Zimmermann, "Das Vorurteil uber die Trivialliteratur," *Akzente* (5/1972), p. 386.

Chapter 7

Social Aspects of Computer Art

Social Aspects of Computer Art

Abraham Moles

Art is not an object such as the Venus of Milo, or the Empire State Building; it is a relationship which man enters into with things, an active relationship for the creative person who changes them and orders them at his own will, a passive relationship for the consumer simply enjoying the shapes and the patternings of sensorial elements. For the consumer it means a surplus of life, a programming of sensuality or an experience in the sensualization of patterns. Whatever else it may be, it is playing the game of mastering one's environment or of being mastered by it.

Art remains, for the most important part, a programmed sensualization of the environment. The conventional arts have for long been only that. We now have to consider, in this new society which is now taking shape with the impact of technological means—in particular: the computer, the data bank, the programming techniques, the instantaneous flow of information the situation that art in general or the activities connected with aesthetic pleasure, will occupy in the immediate future. "Computer Art" is an expression covering a wide range of aesthetic products which have been made in connection with, or directly by, a computing machine equipped with terminals which have made it possible to produce, from an assembly of elements, a directly acceptable product such as sounds (talking and singing computers), music (computer composition) shapes (sketching, designing, architectural projects), and more recently, colors and tactile patterns. It has become quite fashionable with the public, prepared already by the schools of optical and visual art in the sixties, and has been dealt with in a large number of exhibitions, competitions and art reviews. What will become of it now, when computer sciences and technology reach the social scale?

It can be legitimately contended that computer art is doomed to disappear, the victim of its own success. The very expression "computer art" is becoming increasingly obsolete, in so far as all forms of artistic creation are becoming more and more perfused and relevant to the field of "programming." In this, the world of art is simply reflecting the evolution of society as a whole, which is more and more dependent on information and communication processes—this new science which has come to be known as "telematics."

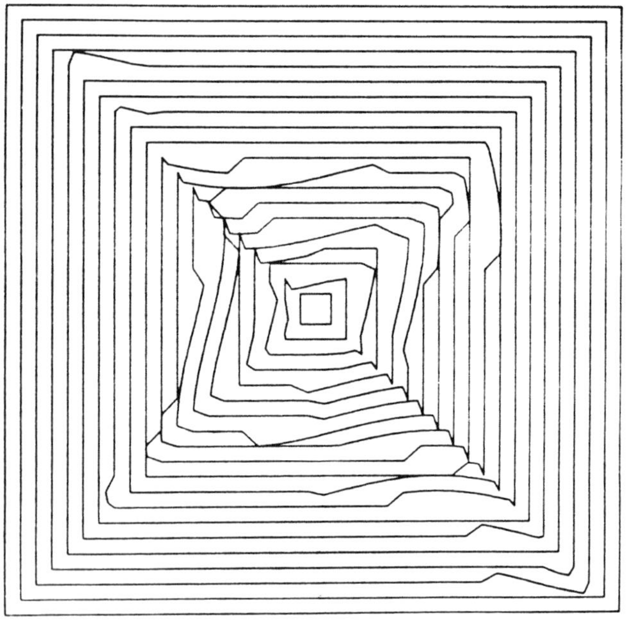

Fig. 7-1 Vera Molnar—74.388/ 14.29.00

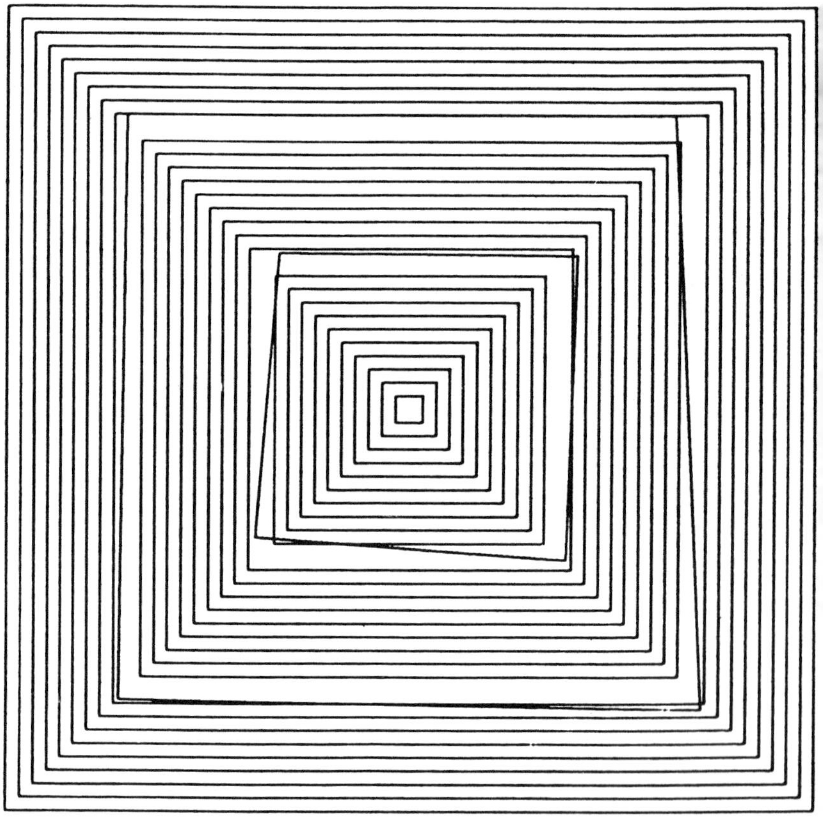

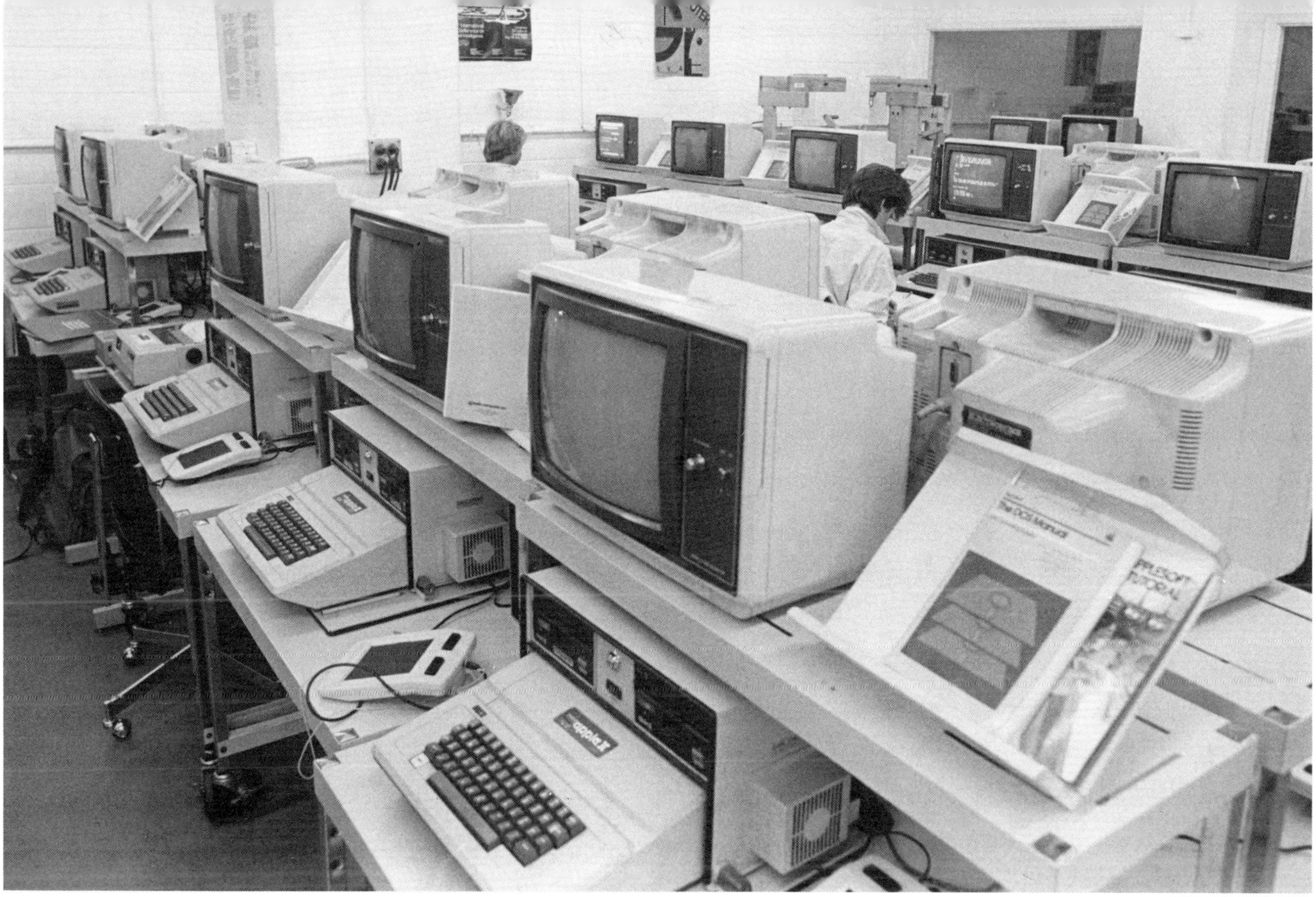

Fig. 7-2: A typical computer installation, used for computer art and other purposes. This is at the School of Visual Arts, New York City. (Photograph by Lloyd Eby)

Form generation, volume patterning in architecture or sculpture, combinations of colors, of shades of light, elementary morphemes picked out from anthropometrics of the human body, sounds and chords, of lights and moments every imaginable type of sensory activity or receptivity is coming, faster and faster, within the range of mechanical, artistic devices which are now available to be controlled, organized and programmed by a computer-issued message, the programming of which is the truly creative side of art.

Thus, computer art no longer exists as such because all forms of art are now within the reach of computerized thinking and automatic designing. It is now just a question of time, of technological capabilities and of the will or interest of the artist whether or not any particular art is a computer product.

Computer art is fundamentally a structuralistic process; it is one of the most perfect examples of structural thought applied on

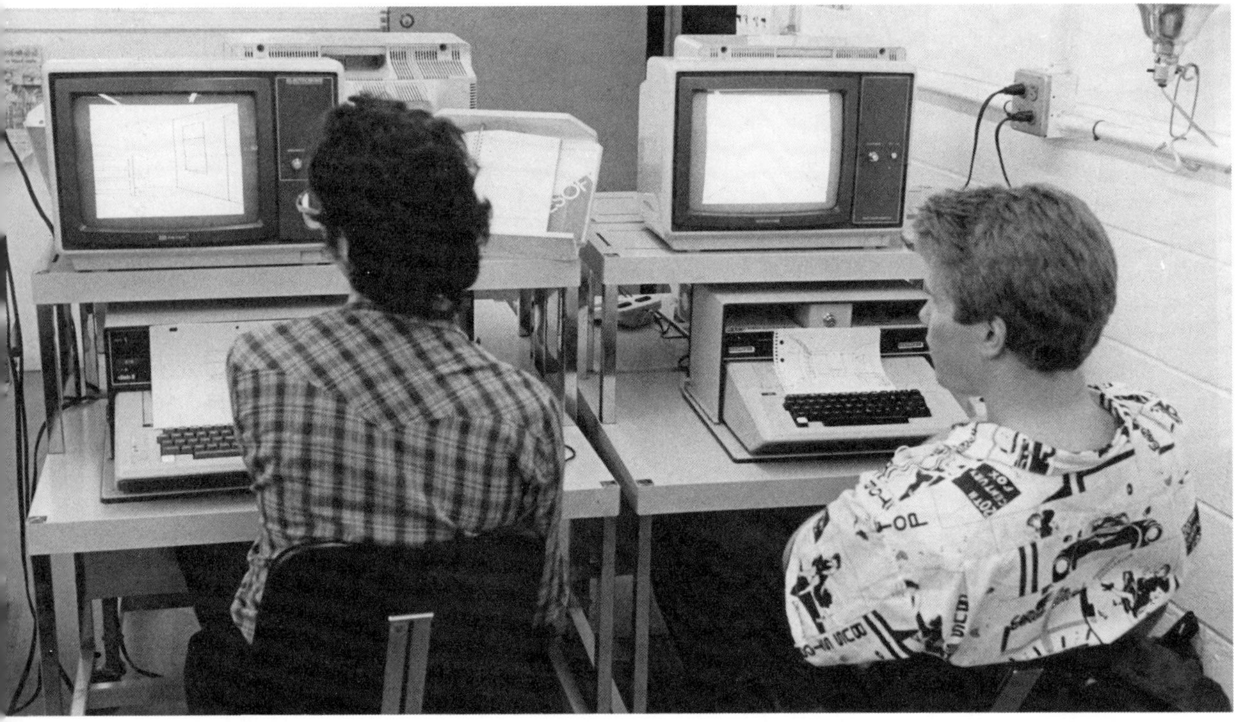

Fig. 7-3: Two students working computer art projects, at the compu er center of the School of Visual Ar New York. (Photograph by Lloyd Eb

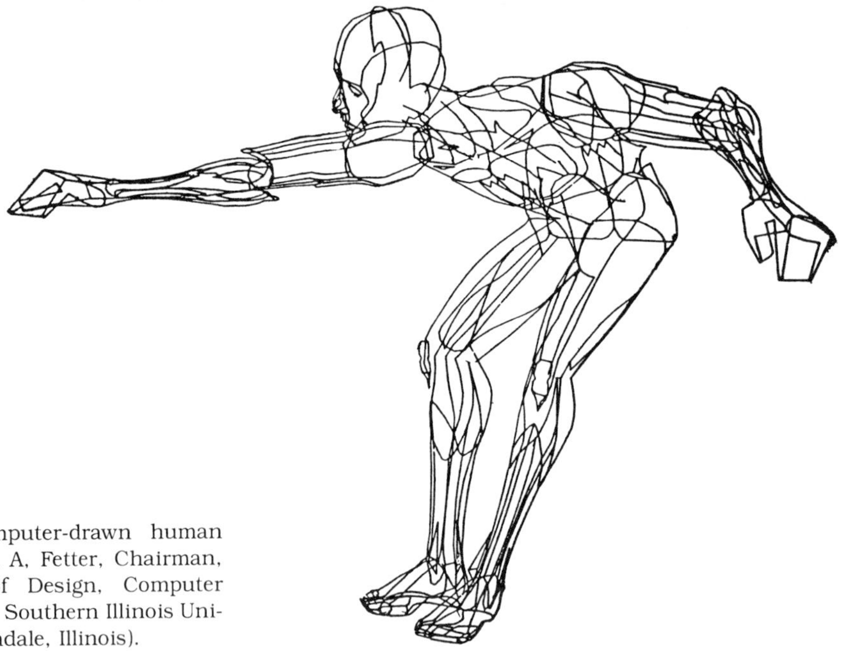

Fig. 7-4 Computer-drawn human figure (William A, Fetter, Chairman, Department of Design, Computer Graphics Lab., Southern Illinois University, Carbondale, Illinois).

An Analysis of the Aesthetic and Psychological Profiles of Some Recent Tendencies in Art

Fig. 7-5

++ + • - —	Classical art	Surrealist art	Computer art	Conceptual art	Minimal art	Geometric art
Rational/ Irrational	++	+	++	•	-	++
Integrate/ Segregate	-	—	-	•	+	•
Manner/ Work	•	—	+	+	—	-
Object/ Project	++	++	+	—	+	++
Hot/Cold	+	•	—	+	++	
Novelty/Finish	-	++	-	++	++	—
Fine/Thick	++	++	+	-	—	+
Aleatory/ Determined	—	-	•	•	-	—
Natural figurative/ Non-figurative	++	++	•	-	-	—
Technological/ Non-technological	—	—	++	•	—	

LEGEND:
- ++ Very positive aesthetic/psychological profile
- \+ Mildly positive aesthetic/psychological profile
- • Neutral aesthetic/psychological profile
- \- Mildly negative aesthetic/psychological profile
- — Very negative aesthetic/psychological profile
- Blank Does not apply or undetermined

a very large scale. This, according to the famous DENIS saying, means that any shape in space, time, color or shade, appears ultimately as a right combination, respecting certain rules (a code or structure), of simpler elements selected from a restricted number of types which can be properly analyzed, identified, stored and recalled at will, a sequence of orders in accordance with correctly designed programming.

The burden of creation thus moves away from the former characteristic hard struggle of the artist using his mind and hands to master matter, paste, color, musical instrument, etc, . . . to become the correct way of programming elements which are more or less available, ready-made beforehand. This is

Fig. 7-6

A possible image of the large art center

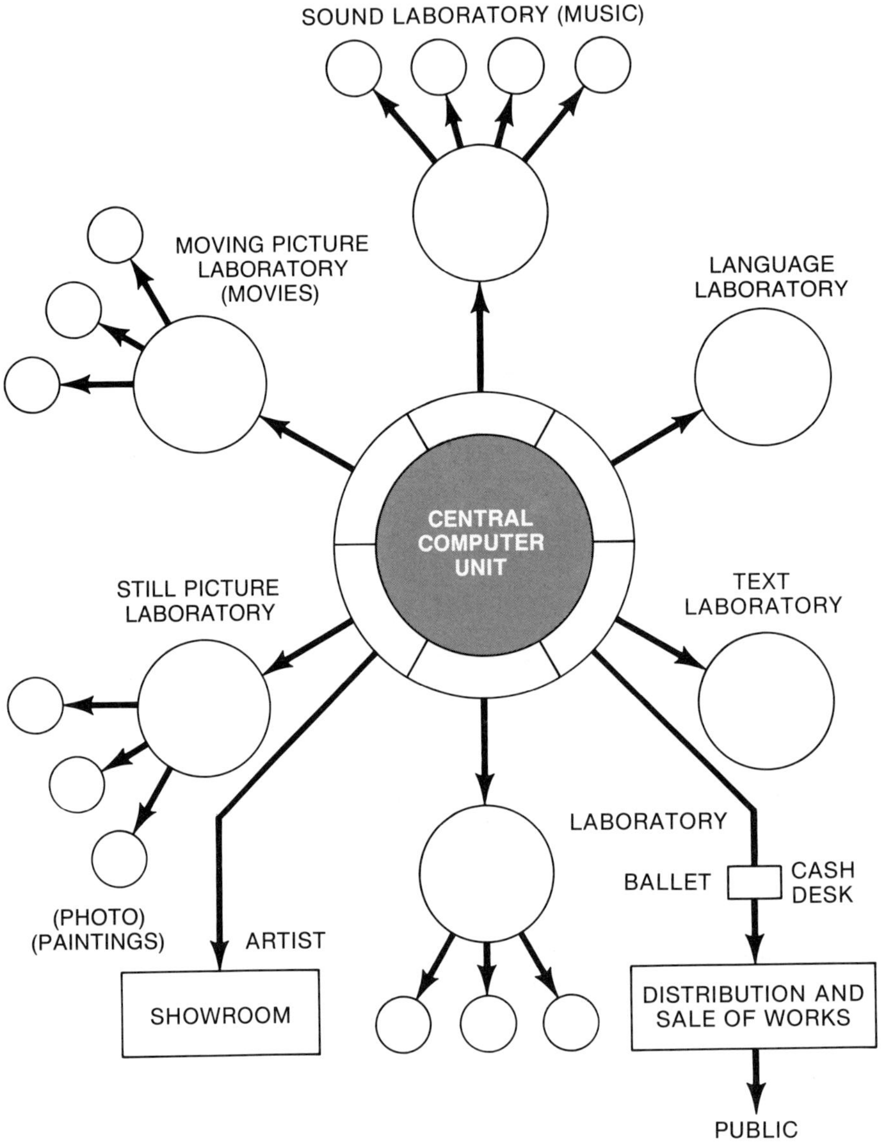

A possible image of the new computer art cultural center. This model, which is already valid in a number of cases, enhances the use of a big central computer unit for processing data and combining them according to more or less complicated pieces of programming. This central unit is surrounded by a variable number of applied aesthetic laboratories dealing with terminals and analysis + synthesis processes with various sensorial "show stuffs" to be processed and elaborated. The showroom, the studio, the reproduction equipment take charge of the testing of the reaction of the public, and the final diffusion and consumption of the elaborated products.

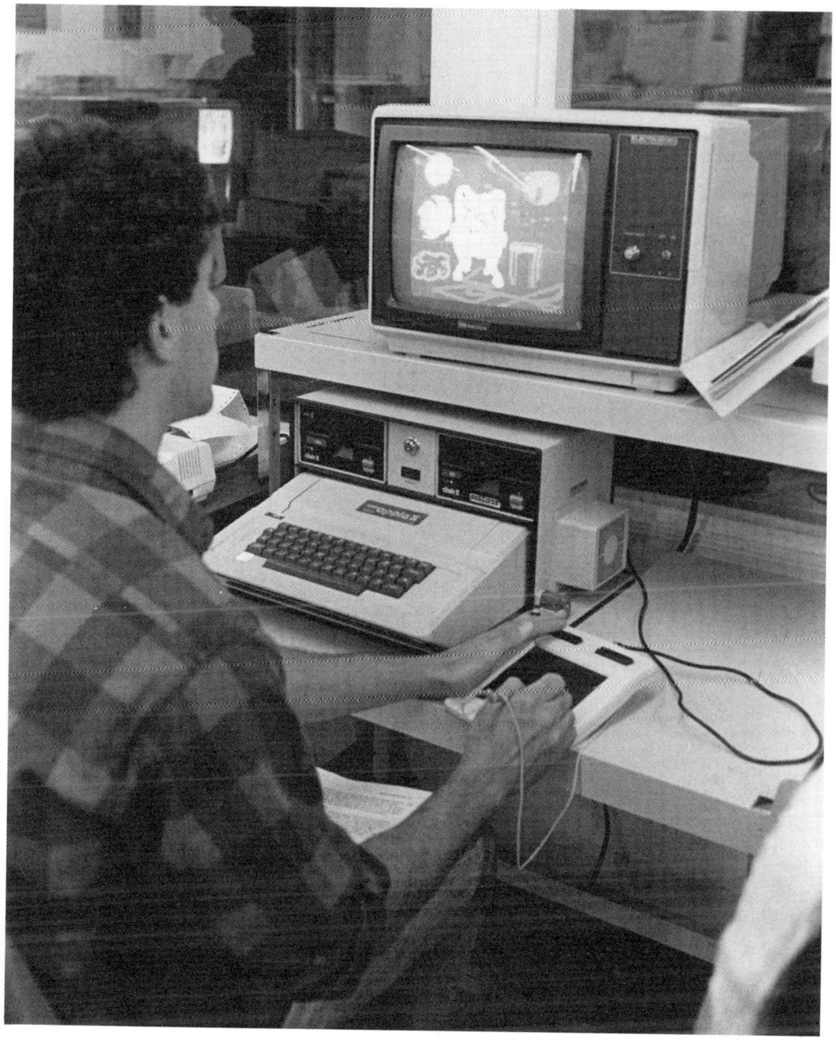

Fig. 7-7: A student at the computer center of the School of Visual Arts New York. (Photograph by Lloyd Eby)

a revolution in the process of the creation of art, and consequently in the very life and work of the artist. It is no longer a sensual contact with shapes in the environment but a conceptual effort to combine banal elements and thus induce intelligibility and originality which, in their turn, and in accordance with good dialectics, lead to novelty.

In fact, most contemporary computer art still tends to be rather passive: art to be "seen," art to be "heard," art to be "received," but all too rarely art to be practiced. The modern citizen has been too harassed by life and time budgets, to be really able, until now, to consider artistic creativity as an integral part of his free time. He grasps what comes his way, but does not generally take a hand in artistic creativity as such. This relative passivity is closely related to the pattern of contemporary life through the ratio between free time and constraint time, whether this latter is real "work" or simply material chores. The sociologist tells us that this ratio, if it changes, will only do so very slowly; thus we have not to expect any drastic change in the above-mentioned passivity in art consumption, except maybe among children who may devote more time to a kind of aesthetic creativity inherent in electronic and computer games.

But this makes the exceptions to this rule which set behavioral leaderships even more interesting: art as composition to be practiced, which is no longer the private pleasure garden of the full-time artist, but whose updating implies a creative will and power, possibly giving rise to series of products to be consumed by oneself or by others around one. This is a new phenomenon which, even if it is real, must be carefully studied. Art as play, playing as art, with the tremendous power computerized devices place to hand, suggests a new type of behavioral leadership. It should be the task of the art sociologist to describe this new "dynamic entrepreneurship" which Schumpeter had not envisaged.

For ten years now computer art has survived the strains of fashion and of the social value attached to something purely novel in its early stages; it is no longer something new, whether it is design works made by powerful firms (IBM, Bell, Boeing, etc.) and taken over by art galleries, or whether it is the product of academic laboratories (Colonna, Huitric, Risset in France for example). Thus, the simple fact that a work has come off an automatic drawing table, out of the memory of a computer, is no longer sufficient for it to be attributed the quality of "being artistic," simply because it is new.

It is no longer new; we get more and more used to it. It must be seen as it really is: a new channel for use by the artistic mind, oriented toward the injection into the world of forms which were not already there.

We know that machines are now able to absorb, with no loss, the entirety of sensorial stimuli, to grind them into grains of coded symbolic elements and to store them in cells of the computer's memories. We are faced with the basic question: how is it possible, out of this heap of grains, to assemble a new pattern which is not a conventional shape—more or less modified—and already familiar, but one which injects *novelty* into the world.

To a large extent, despite the efforts of conceptual artists such as Forest, art is a programmed sensualization of the environment: *visual* environment, immediate and semi-passive, scanned by the eye in a movement so spontaneous that we can use it freely; *time* environment, that of the sonic or acting process; *spatial* environment, which man explores at length and discovers to be a whole. Visual arts, auditory arts, arts to be practiced, whatever their form—the artist is the programmer. The artist, alone, semi-determines micromovements of the eye, or paths of urban rambling, games of mazes and gardens. He, alone, *conditions* these movements, with due respect for the major rule of the game: the provision of pleasure for those who will play it. The computer, with its databanks, its subroutines and its algorithms, provides the artist with the means to elaborate this programming of the sensorial game, on a scale which artists of the past were incapable of imagining. It is worth mentioning in this respect the sixteenth century mannerists, whose obstinate rigor grasped the vertigo of infinity in the field of permutation they were trying clumsily to explore.

We now see the actualization of their dream resulting in the emergence of a *neo-mannerism* of the computer, i.e., an operational style in which the *manner*, the procedure, is more important than the actualized form, since a form is only one particular solution from a large number of other possibilities which can be realized in the same way simply by making a different choice of parameters. The concept of *algorithm*, which is now widely accepted in the field of information sciences, is basic in this new way of thinking; an algorithm is a rule for conducting operations of the mind with a definite result.

Finally, computer art is always a product of a basic idea crystallized in the programming, and of a series of actualizations obtained by giving various values to the parameters of the program. It means that for any *one* idea, there are a number of realizations, all different, but all basically identical aesthetically. All of these "permutational works" are thus, at the same time, "equivalent" and "different." They provide an interesting solution to the idea of diversity in unity.

In this way, many people can own the same work of art from the point of view of the artist who determined the algorithms, but for them they are all *different*, since they are other instances of the unique program. Multiplicity is not a property of computer art

WORK OF ART AS AN ATTITUDE: AN AESTHETIC BEHAVIOR

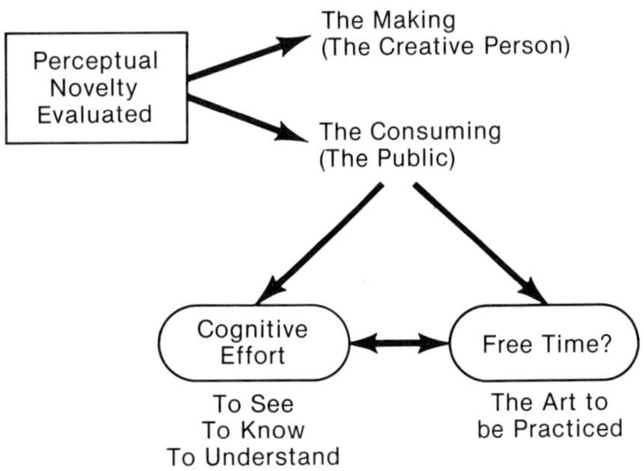

PURE ART	APPLIED ART
Created for Itself	Created to Accompany Other Productions
Appropriated by Attending to the Art Itself	Appropriated by Receiving it Along With Other Products

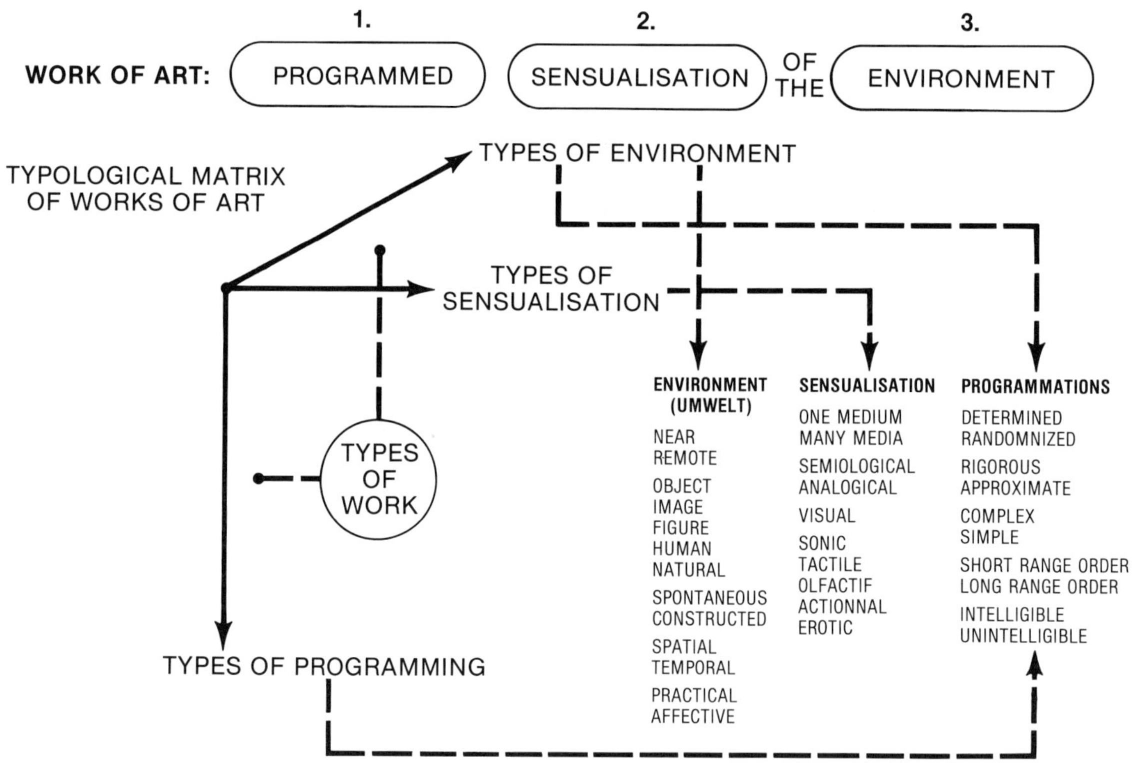

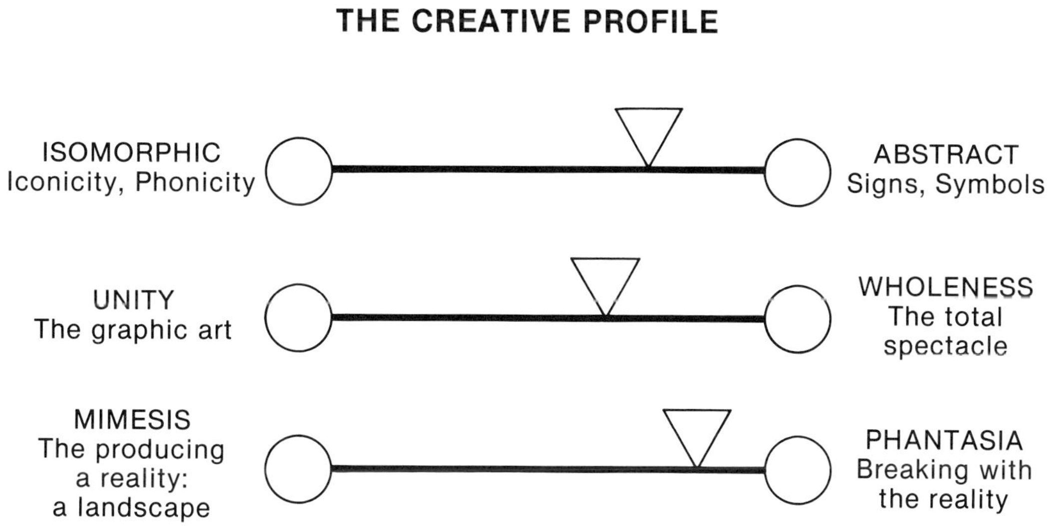

alone (see lithography, xerography, engravings of all kinds), but is a feature bound by essence to computer art; what has been done once can be done again and be exactly the same.

Fascination is the quality of the relationship between the spectator and the forms which leaves a lasting impression of the latter on the movement of his eyes. It operates in numerous ways; the single "strength of the pattern" (the pregnance of the Gestalt theorists) is only one of them. Intricacy, double meaning, symmetry, perspectivist vertigo, perceptive integration, color shock, imposed refinement of a given detail, mastering of an apparent complexity by means of distinctive clues—all these processes are the tools of fascination. They are the ones the artist exploits at his own will, those which the innovations of optical art have systematically inventoried.

Art with computers has new resources for fascination at its disposal: e.g., the easy tracking of a lengthy design resulting from accurate and numerous variations that the watching eye will integrate into the mind at a glance; the discovery of regularities

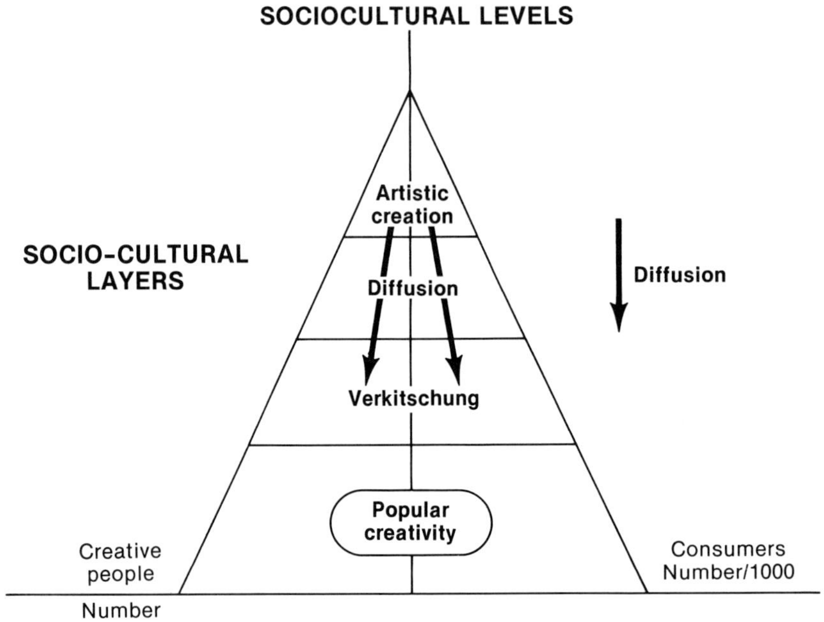
SOCIOCULTURAL LEVELS
SOCIO-CULTURAL LAYERS
Artistic creation
Diffusion
Verkitschung
Popular creativity
Diffusion
Creative people
Number
Consumers Number/1000

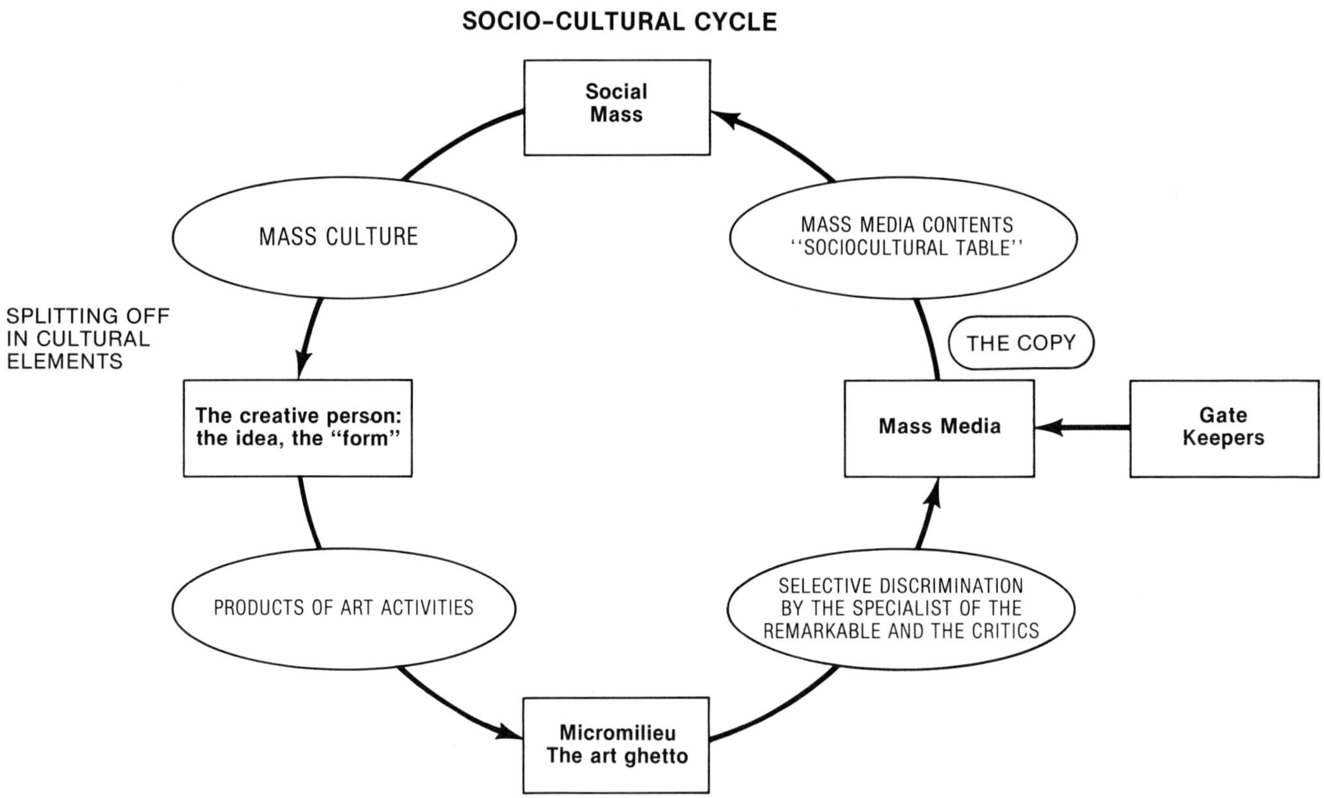
SOCIO-CULTURAL CYCLE
Social Mass
MASS CULTURE
MASS MEDIA CONTENTS
''SOCIOCULTURAL TABLE''
SPLITTING OFF IN CULTURAL ELEMENTS
THE COPY
The creative person: the idea, the "form"
Mass Media
Gate Keepers
PRODUCTS OF ART ACTIVITIES
SELECTIVE DISCRIMINATION BY THE SPECIALIST OF THE REMARKABLE AND THE CRITICS
Micromilieu The art ghetto

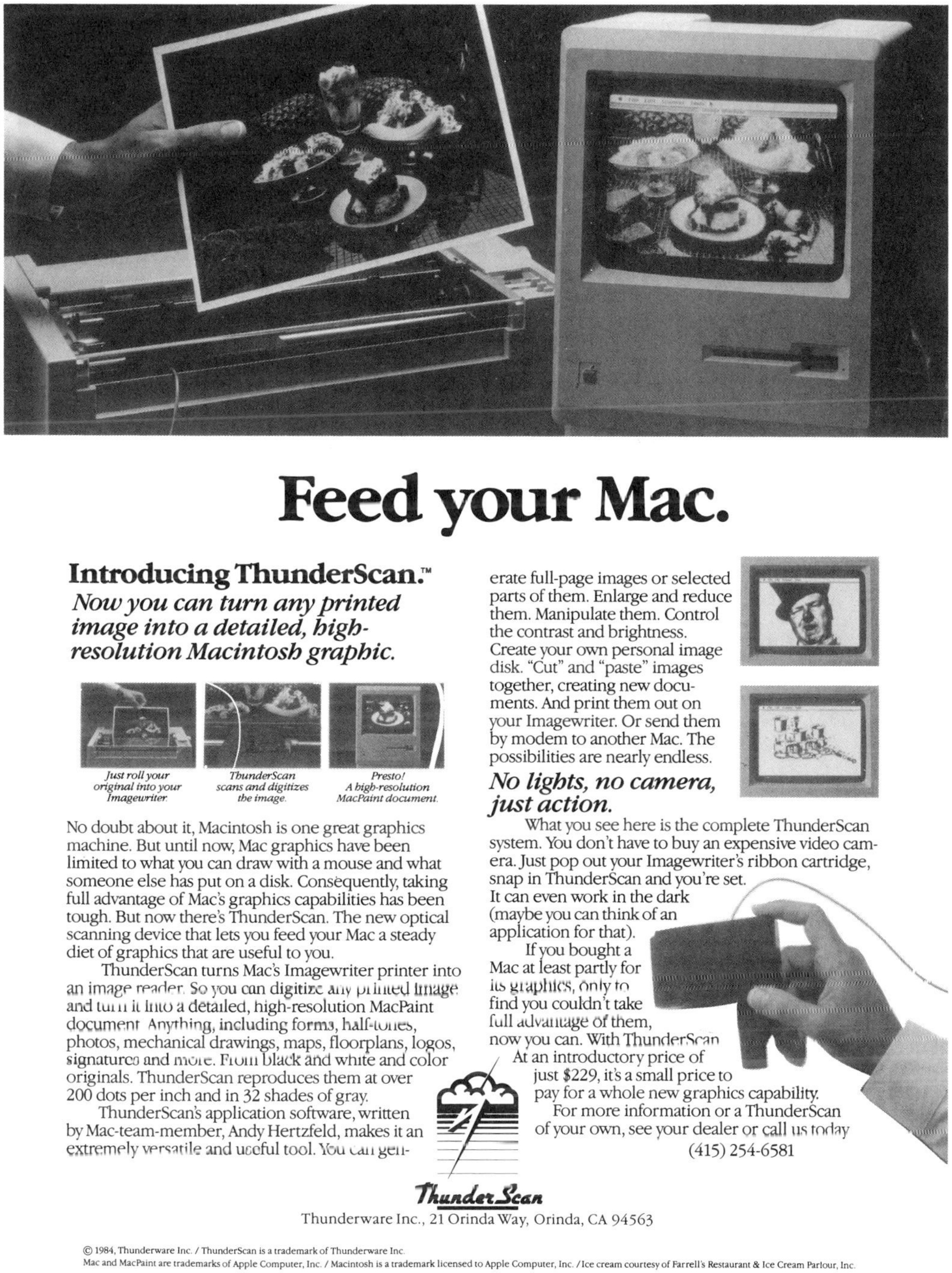

Figure 7-11. Computer software for converting art work to computer information. (Courtesy: Thunder Scan, Inc.)

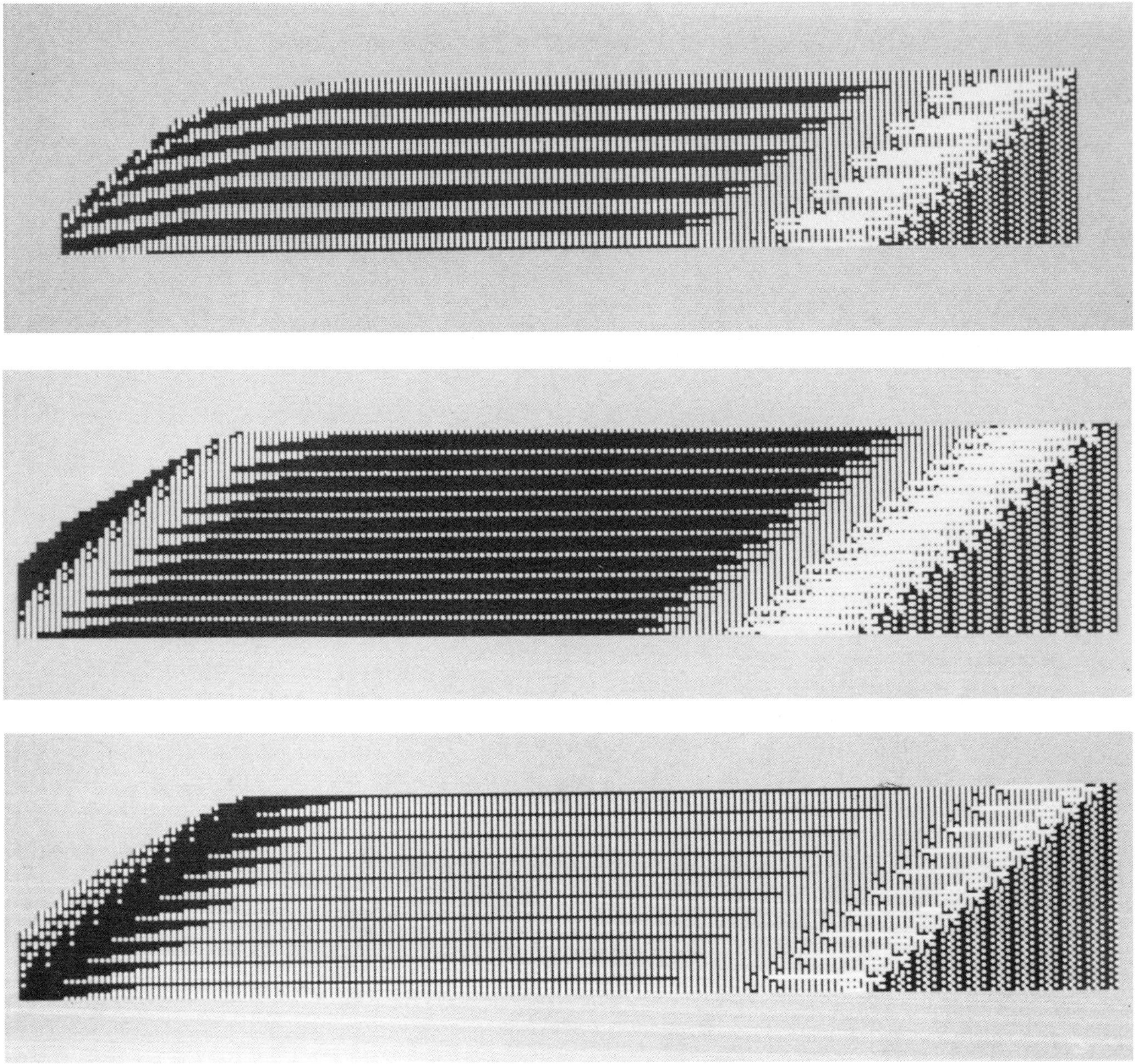

Fig. 7-12 From the series "Folded Structures" created in 1983 by Terry Blum of the Fashion Institute of Technology, New York. This series was created using *Cromemco* software to create and fold shapes using the data from one image to create the next. Spatial movement and dimension is contained within the boundaries of the structure, which itself exists as a unit in space. In this case, the results are enhanced by the low resolution. The work, itself, is a response to the medium.

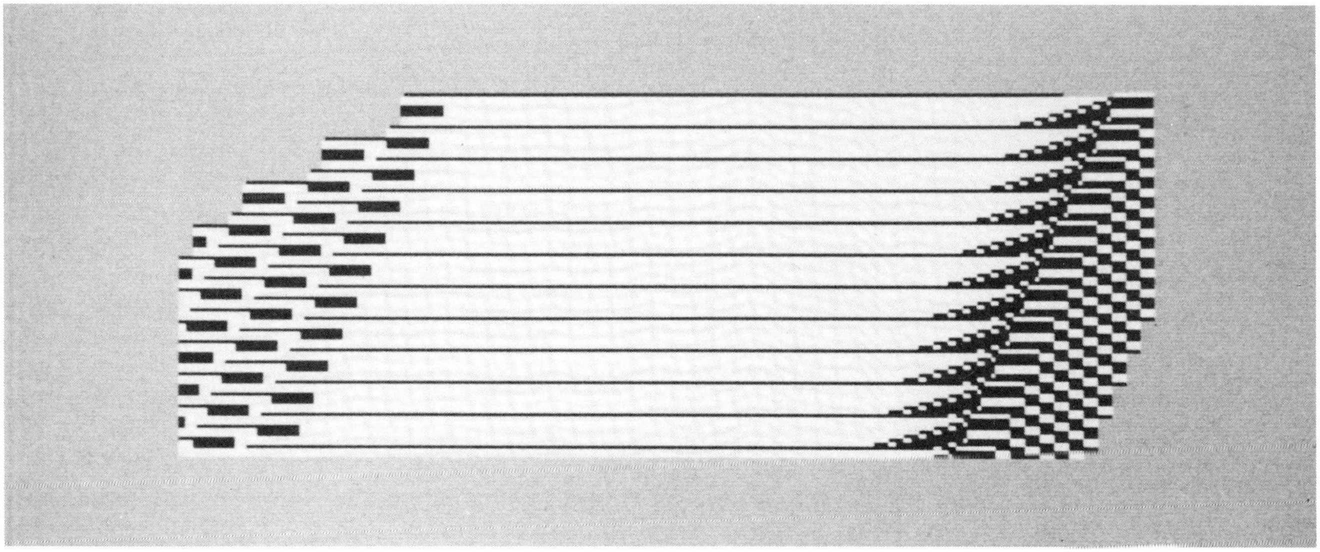

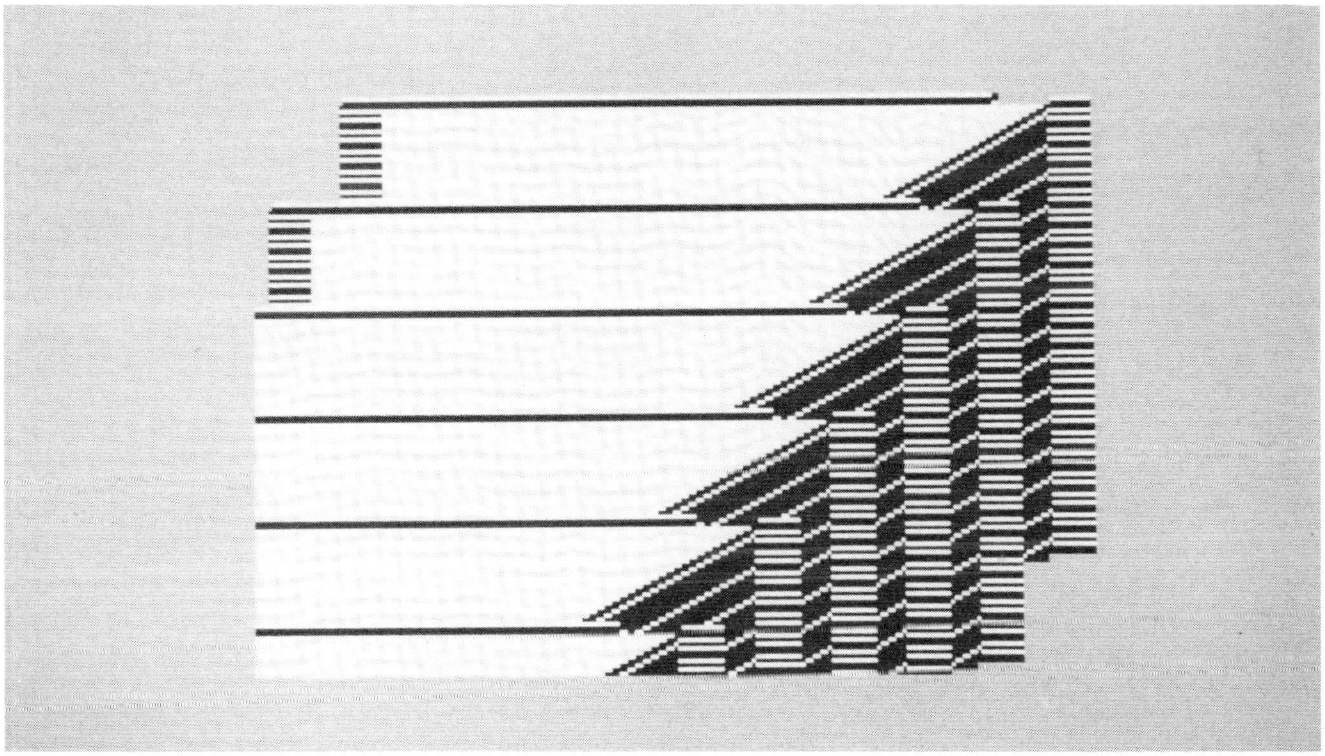

behind the local deviation; the rule behind the exception; distant order behind close disorder. These are only a few of these new possibilities which come directly from this "accuracy of detail in the enormity of the purpose" which the "sequence controlled processing machines" offer. Our artists, in computing, now have to list the algorithms arising out of this new possibility, and to assess the power of fascination they can extort from them.

Computer art goes directly into the imaginary museum, it goes only incidentally in the actual Museum of Fine Arts. For the time being, owing to the structure of our society, computer art is *de facto* a *passive* art: i.e., a thing to be looked at, but this is not necessarily so *de jure*. If the amount of free time changes substantially, and if informatics becomes increasingly popular, which is already happening, with microinformatics and drawing machines, then why not computer art in everyday life? The recent expansion of video games and informatic games of all kinds is a hint of art creativity in everyday life.

Now all the conditions are fulfilled in the United States and in Japan, for instance, for the development of a vast process of "Computer Kitsch." Out of the unlimited capacities of small computers already linked up with the TV screen and the automatic printer, a huge field emerges at the overlapping of:

— video games
— computer games
— optical designs

This new style of computer kitsch will find its place as a major application of the home computer for which the industry is at loss to find convincing arguments. The whole series of variations on Op art designs, more and more confusing art and crafts, will probably cause an inflation of visual geometrical arts, since they enable everybody to feel himself the equal of Vasarely, Morellet or Picelj, thanks to very easy-to-control devices promoted by the small computer industry. We have to consider the impact of this enormous factor upon art in itself. One might well envisage a descrimative factor of quality arising out of the very idea of fascination power, already intrinsic in Op art.

The true basic dichotomy of the modern technological world is the one lying between the invention of new forms, new patterns, and the copying process which obtains many results from one

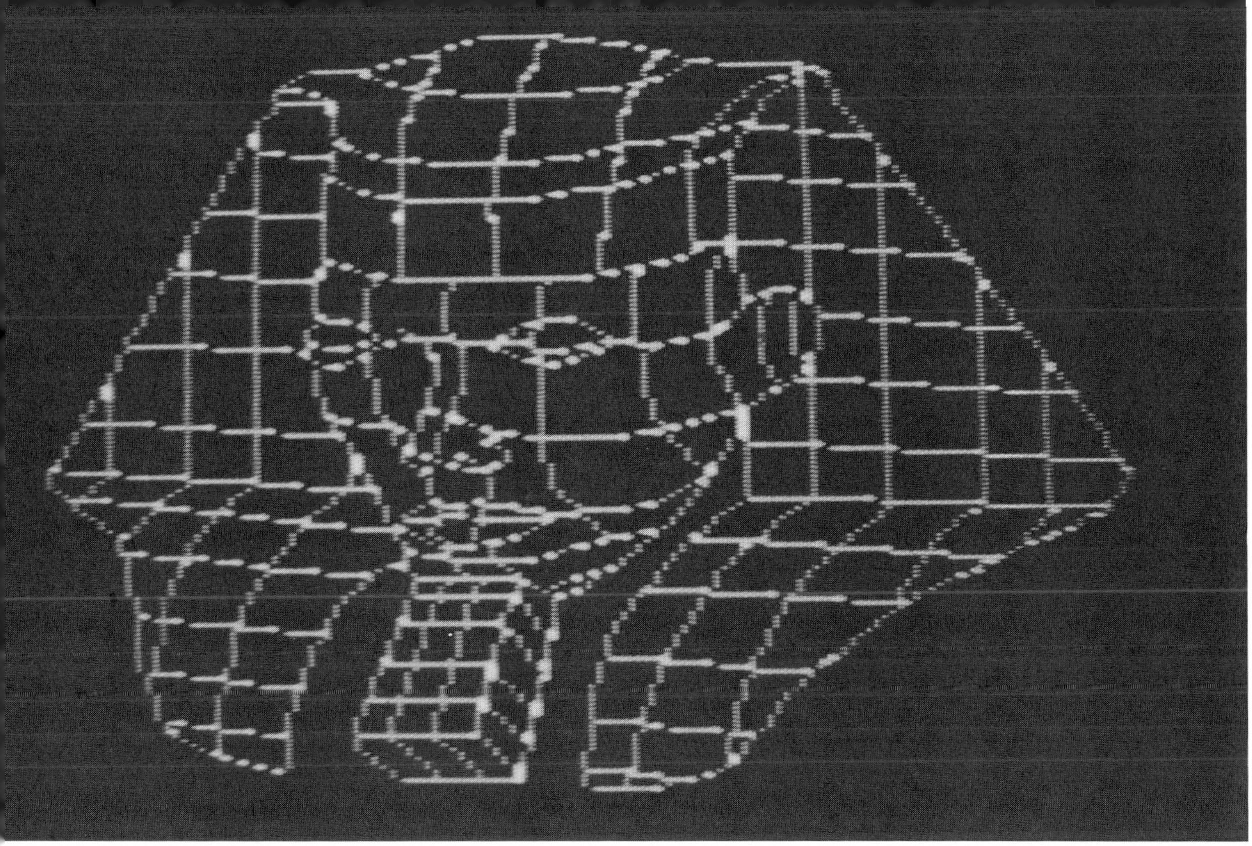

Fig. 7-13 Untitled; video art by Tyrone Depts, School of Visual Arts, New York.

basic shape. At present, more and more clearly, a new concept is emerging of the responsibility of the creative person, formerly the "artist," but now the one who discovered the algorithm, which is a basic idea embodied in a program from which all the possible realizations have stemmed.

To go further, this includes a concept which is relatively new in the field of artistic practice, the one of the art patent, i.e., some kind of moral right, or "ownership" of the basic algorithm behind the concretized idea, independent of the practical result it has led to. The surrealist attitude had hinted at this idea, but never fully grasped it. It is obvious in computer art, which places the artistic realization exactly at the same social level as the scientific one. Basically, this means there is no longer a true difference between the two. The creative activity of the mind is the same whatever the field of application; simply, the rules of constraints, the contact with sensorial reality change, the rules of acceptability are not the same, but the idea of novelty and compliance with rules is the same nature.

To the extent that the computer gives birth to a kind of neomannerism where the work in itself, as an actualization of an idea, is more and more *dissociated* from the idea, or from the more or less elaborate program which has given rise to this particular work, the position of the public as "art consumer" and the social frame itself undergo a drastic change. In fact, the artist finds himself more or less in the same situation as the scientist or the inventor. A need appears for what could be called an "artistic patent" system which covers explicitly by its "claims" the new process, combination, or the originality of the programming of

Fig. 7-14 Untitled; video art by Tyrone Depts, School of Visual Arts, New York.

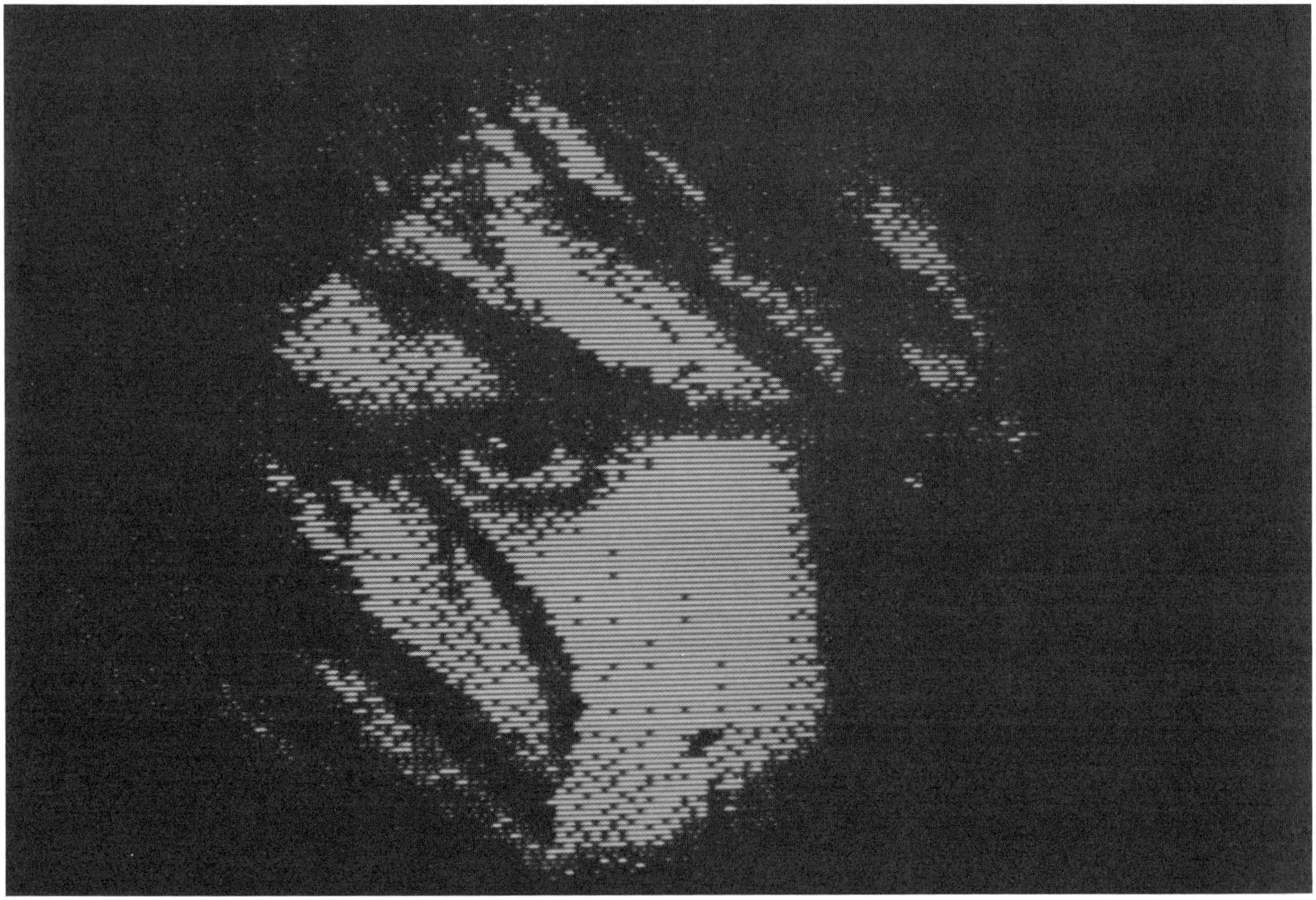

sequences, provided it can lead to one actualization (realizability principle) and makes subject to *copyright* the particular realizations which will be made in the future. Considering the fundamental role the artist plays in society as a source of new behavior, he should, at least in principle, have a warranted right of control over the various executions of his ideas. This might be a solution to the financial problems of the artist, although certainly not in the short term, but it is very different from the conventional subsidizing through gifts, grants and foundations, which have biased some branches of art evolution.

In the present situation the role of the art critic is to imagine, on a rational basis, what could become the word "computer," and aesthetic urge in a "computerized" society. In fact, the change is so great that what will be increasingly called "informatics" is obsolete, its field being covered by "telematics" with only the scantiest of references to the tools employed. This handling of logical operations on a large scale, widely trespassing the span of logical capacities of the individual, will result ultimately in some sort of logicization of the aspects of life, among aesthetic creations. "Logics" and all deductive processes linked to it, will simply be a basic component of the environment and time budget of every individual at all times (even when he sleeps) and in all places (whether public or private).

What might this aesthetic urge be? Firstly, art work properly. But, also, new types of visual surroundings in which young children will start out in life by playing, no longer with dolls or tin soldiers, but with liquid crystal cards which, upon pressing a key, display an animated cartoon, in color: a two-dollar gadget; acoustical surroundings where a Muzak bath of computer-arranged music, the melodic pattern of which is sufficiently faded or drowned in the musical paste for it *not to impose* its pregnance at any moment, but providing a sweet acoustical syrup; surroundings where the image screen on the wall of the private cell is fed at any time either with significant images supplied by the data bank, programs that the occupant may have forgotten to change or switch off, or with fascinating geometrical patterns issued by an automatically self-developing program generated by the ATARI sub-routine purchased at a give-away price by the owner of the system to rest his eyes when, in his inflated free time, he is tired of playing chess with the memory of the central computer which is more or less renowned in this field.

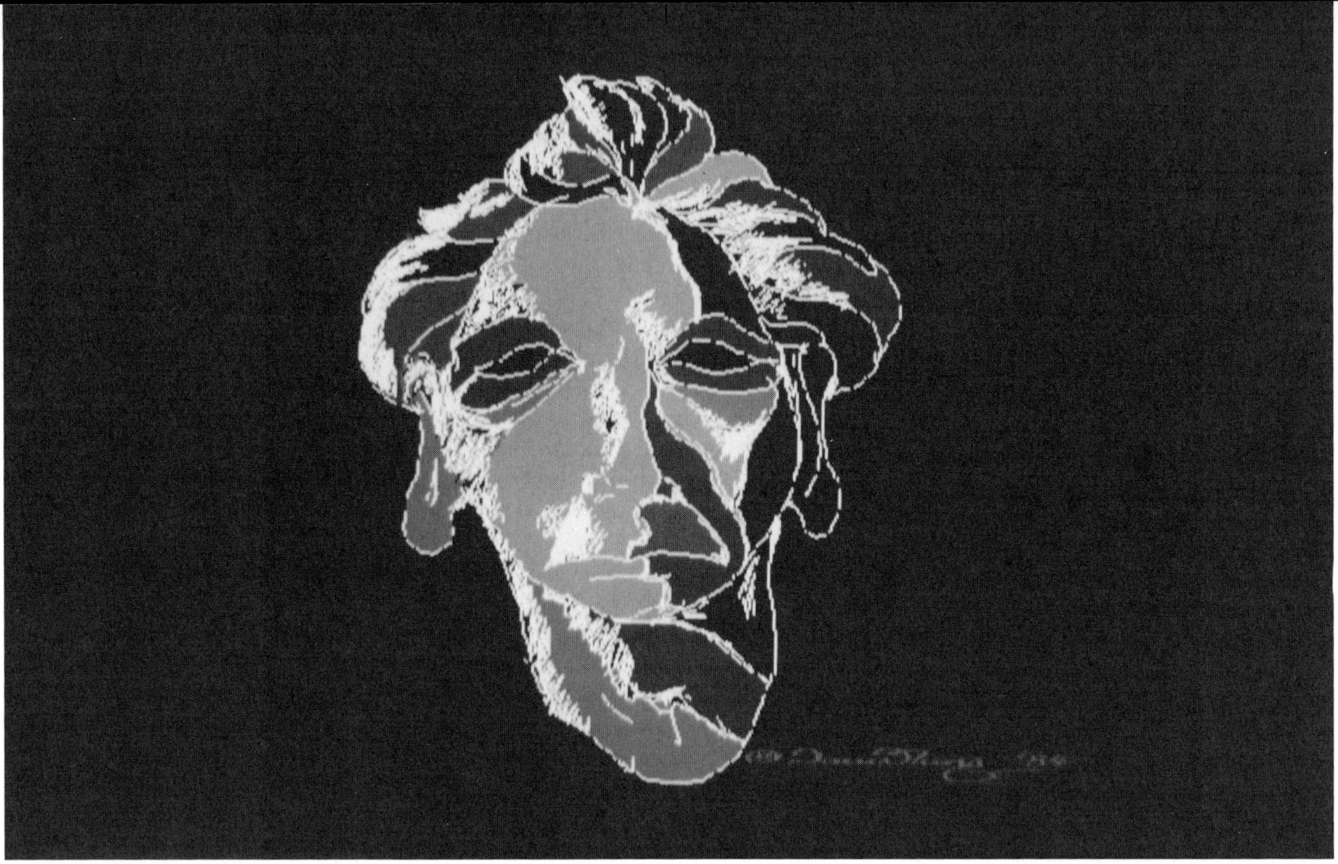

Fig. 7-15 African Mask; video art by David Tung, School of Visual Arts, New York.

What, then, would "Art" be like? Is there a definite meaning left for the word, corresponding to the condensation in space/time of aesthetic impregnation, whether passive ("just look and enjoy") or active ("I am now the master of the universe"—a small part of it—and I construct my surroundings for my own pleasure and that of others into the bargain). Art has been inescapably connected with *works* of art in the history of Western culture; this is the so-called "masterpieces" ideology. When art is indefinitely diluted, until the concept boils down to a sort of aesthetic itch of daily life, what becomes of the aesthetic experience, considered as a privileged moment in the lifetime of the individual?

Is there any longer such a thing as a privileged moment? When all moments are supposedly sensorially enriched, *no* moment is privileged and this is, in the long run, what computer art—what is now called by that name—is moving towards. This goes beyond the question of "computer kitsch": this step has already been overtaken.

The conventional art critic who has been educated basically to some concept of respect for the art work, and who has already found it so difficult to incorporate into his values the idea of art to be played, art as a game, art as a consumer product, has a very difficult time to do away completely with his "Kantian" concept of transcendental moments of existence, corresponding to Maslow's peak experiences. Such peak experiences generated at least a "meaning" for the other, second-rate experiences, which appeared "second-rate" simply because the atmosphere was not right, nor was the time, and the quality of the moment was declining, but bore, still, a vague remembrance to what it should have been if the receiver—the individual—had been in the right mood. There is no longer such experience in the computerized world at large, the cable society, the society of communicational affluence where everybody can have everything. If he is *entitled* to everything, which striving for value can he have left? In fact, this calls for the enforcement of a personal aesthetic discipline. It is a question of *art dietetics*, and this may be one of the major aspects of the future with regard to the role of the computer in social aesthetics.

Such a future may be a relatively distant one. It will emerge selectively as a way of life for those whom evolution has privileged(?) in the major focal points of culture such as Tokyo, Los Angeles, New York, Munich; but it will happen only gradually (if it happens) in the lives of the inhabitants of Truchtersheim or Peyton Place, leaving place for a sort of attraction/repulsion vector for the large metropolis, with a type of social artistical activity. We assume here that the West will not decline, despite the fact that our economic prophets of decay apparently consider this is a very real assumption indeed. Anyway, these processes will be progressive ones and, more accurately, they will happen only step by step.

This apocalyptic (for the "good thinkers" of art) view is only one side of the picture, the most remote one. For the coming years, the evolution may simply mean a tremendous power to shape new things, to introduce new forms, to multiply the faculties of the imagination by the power of logical deduction in so far as the "artists," as they are still called, having been born twenty, thirty or forty years ago, are willing to undergo the internal transformation of their minds and capabilities, i.e., their *new fields of craftsmanship*. In this respect, there will be a tremendous selection with ferocious struggles between, on one side, the successful

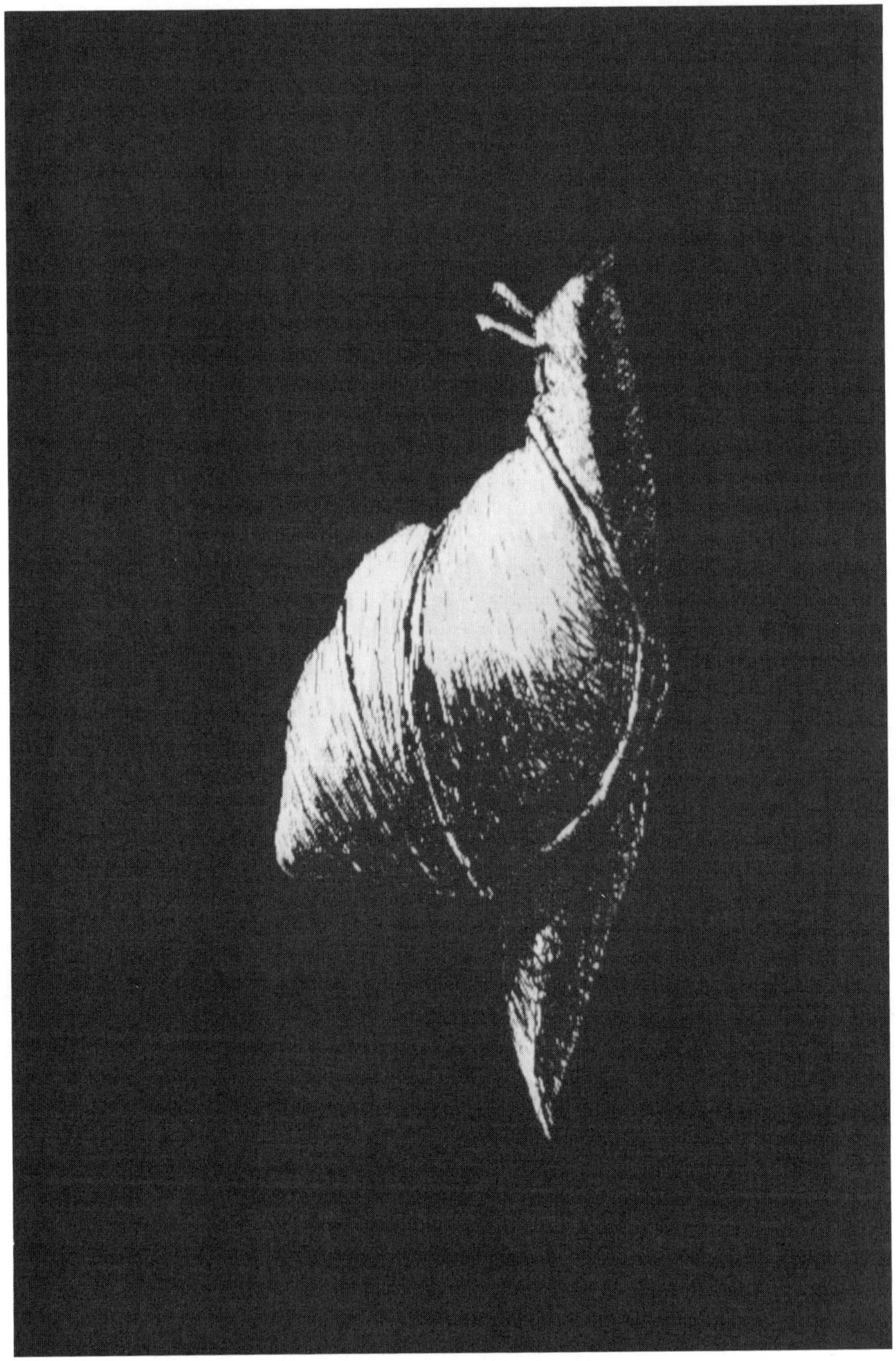

Fig. 7-16 Snail; video art by David Tung, School of Visual Arts, New York.

conventional Artist who nowadays satisfies the needs of people raised in a culture which acknowledges Art as a synonym of Works (of art) and, on the other side, the future "Umweltgestalter" who simply injects drops of aesthetic feeling into certain parts of the pattern of the surroundings without sufficiently mastering what he is doing, and consequently fails quite often.

A perfect example of the successful use of computer art is undoubtedly that of the evolution and destiny of animated cartoons. It is well known from developments which have already taken place in the most advanced centers scattered throughout the world that this technique of visual expression embodied in schematic images of partly-true, partly-invented "reality" is already promised to computerization. This will mean that the various features of an animated cartoon, the shapes of the objects, the movements, the colors and shades which can be superimposed on the shapes will, in some way, be stored separately in the computer in the form of sub-programs that the

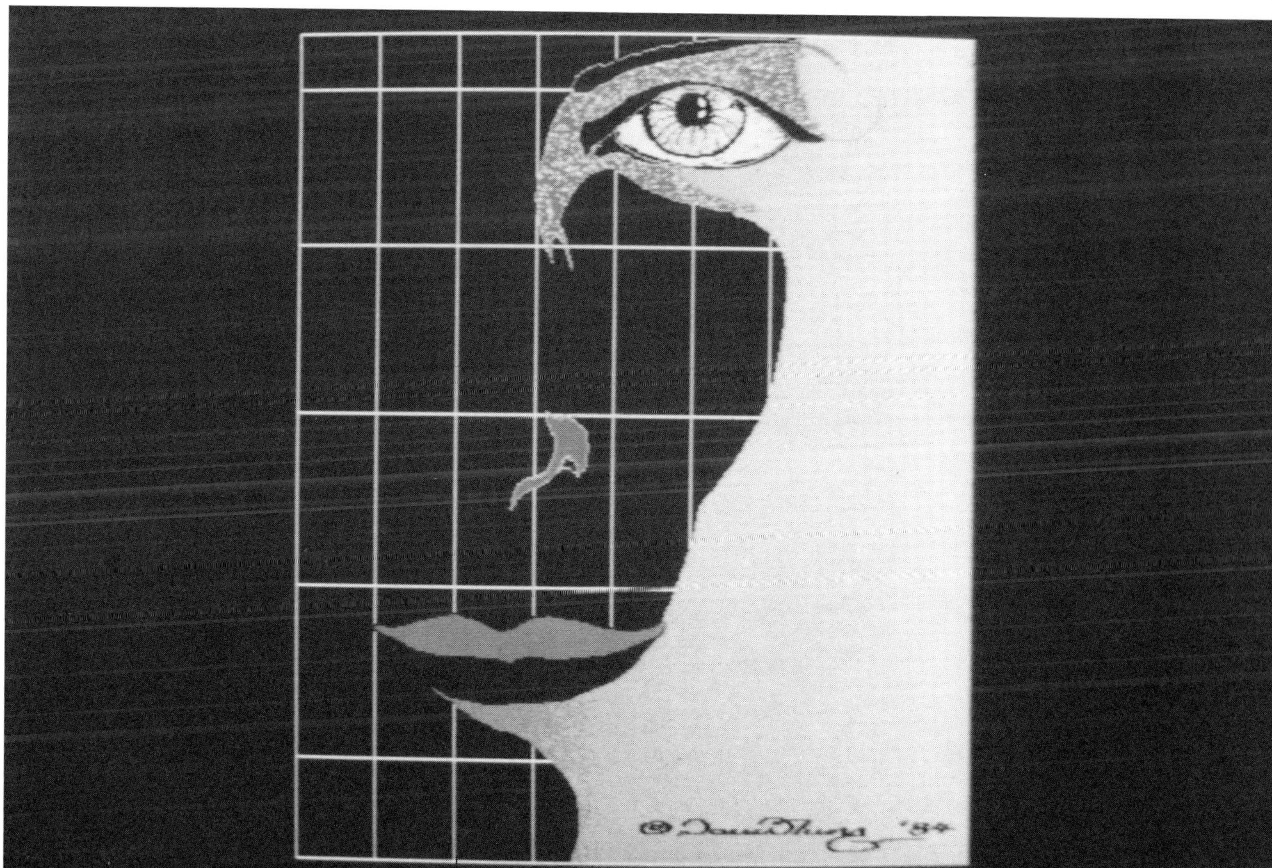

Fig. 7-17 Untitled; video art by David Tung, School of Visual Arts, New York.

composition master will combine according to his whims or to extrinsic rule (e.g., didactic, technical, scientific, etc.), while the computer takes charge of aspects such as consistency, non-contradiction, satisfactory redundancy, that is to say, the "figurativeness" of the image, as described by the author in 1972.

In past times the artist was an "intuitive programmer," nowadays he is a graduate from a programming school just like the research engineer; he replaces a vague-but-sure-intuition by the injection into his work of a percentage of randomness, duly measured and appropriately experimented. The artist—or the creator of new shapes who replaces him—may be trapped by his own game, if his activity seems to him sufficient justification for his existence, and he no longer cares about his public but shuts himself away in the ivory tower of the pure artist.

The product of the computer solves this dialectical game which is played by the artist as a pure creative being and the artist as a provider of patterns, which are sensually consumed by a spectator, unfamiliar with the act of innovation.

How can this dialectical game be introduced into a "well programmed society"? This is the question.

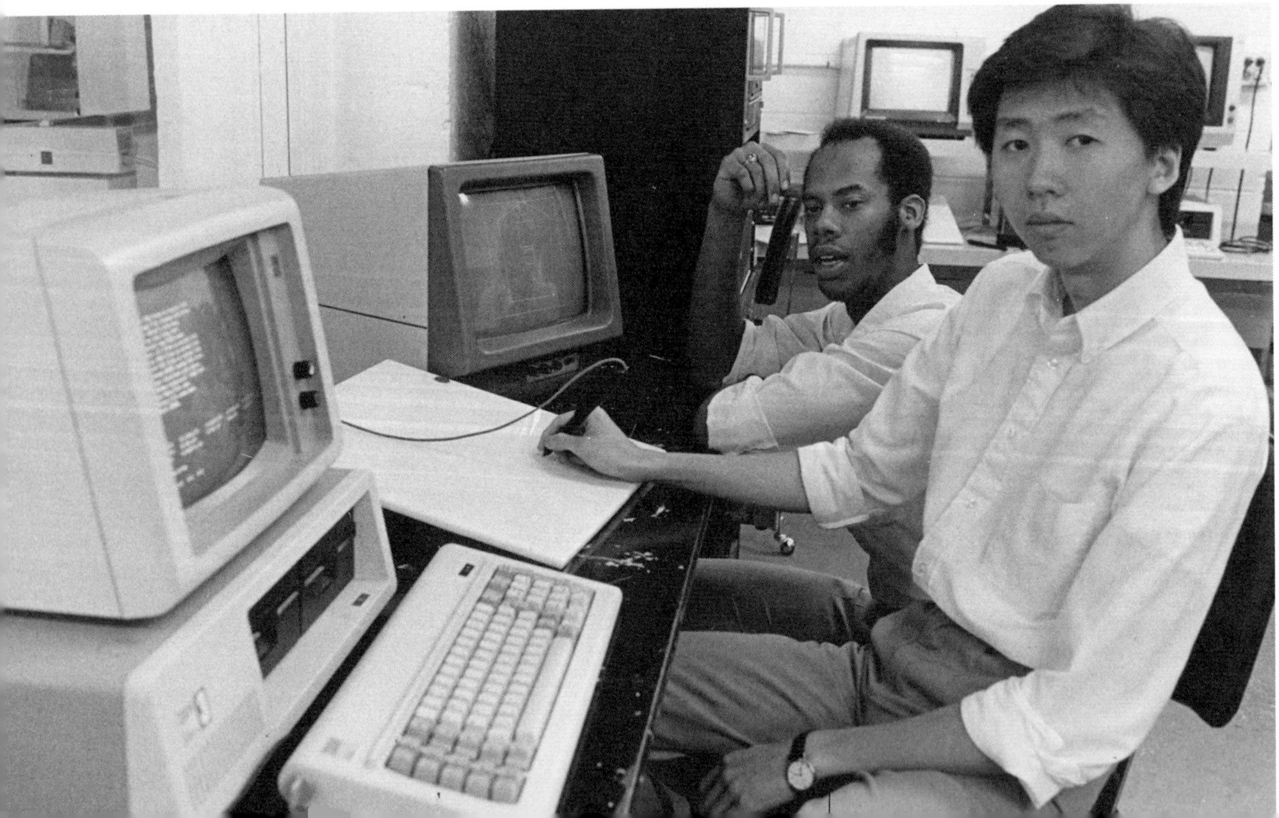

Fig. 7-18 Computer artists Tyrone Depts (left) and David Tung (right) at the computer center of the School of Visual Arts, New York, where they produced the computer art used in this volume. (Photograph by Lloyd Eby)

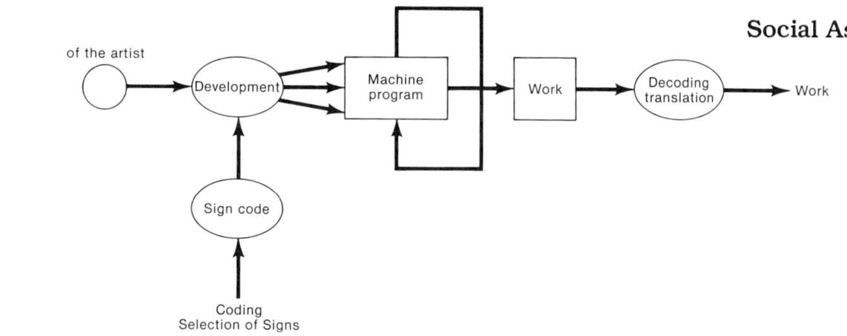
of the artist
Development
Machine program
Work
Decoding translation
Work
Sign code
Coding
Selection of Signs

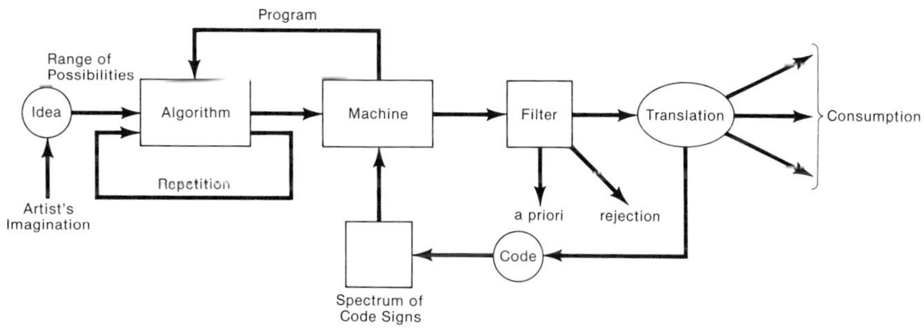
Program
Range of Possibilities
Idea
Algorithm
Machine
Filter
Translation
Consumption
Repetition
Artist's Imagination
a priori
rejection
Code
Spectrum of Code Signs

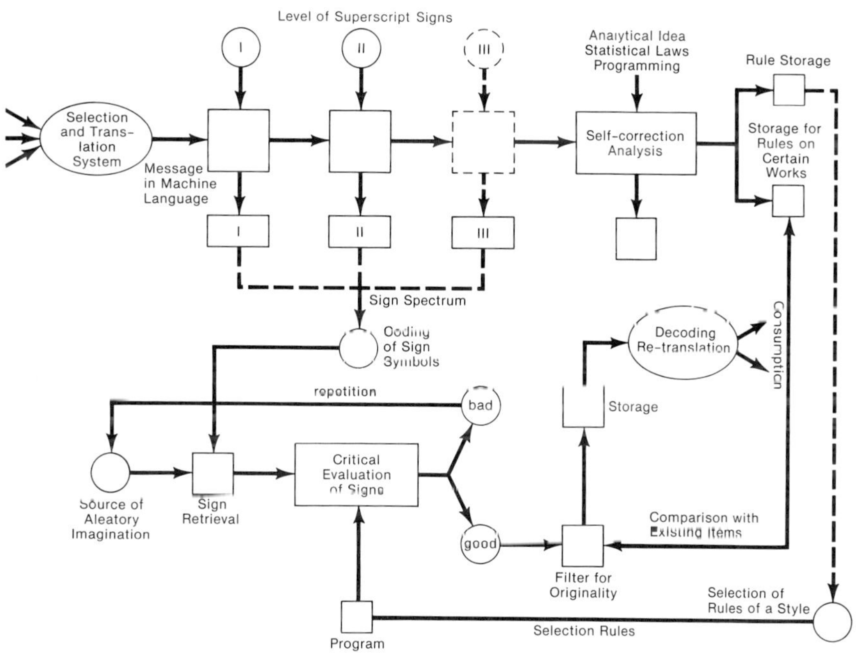
Level of Superscript Signs
I
II
III
Analytical Idea
Statistical Laws
Programming
Rule Storage
Selection and Translation System
Self-correction Analysis
Storage for Rules on Certain Works
Message in Machine Language
Sign Spectrum
Coding of Sign Symbols
Decoding Re-translation
Consumption
repetition
bad
Storage
Critical Evaluation of Signs
Source of Aleatory Imagination
Sign Retrieval
Comparison with Existing Items
good
Filter for Originality
Selection of Rules of a Style
Selection Rules
Program

Bibliography

Berger R., *La Telefission—Alerte a la Television*, ed. Casterman, Paris, 1976.

Coll. *ATARI Graphics*, published by Alfred Haney Guides, 1983.

Hiller, *Experimental Music*, McGraw Hill, New York, 1967.

Moles A., *Musiques experimentales*, Cercle d'Art Contemporain, Zurich, 1961.

Moles A., *Art et ordinateur*, Casterman, 1972 (exists in German, Rumanian, Bulgarian, Russian).

Moles A., *Of mazes and man*, Proceedings of the 8th symposium of EDRA, 1979, published in French as a chapter of *Psychologie de l'espace*, Casterman, 1977.

Morvan P., *L'ordinateur dans l'art*, Larousse, 1975.

Chapter Eight

The Challenge of Video Art

The Challenge of Video Art

René Berger

Introduction

Any civilization or society represents the deployment in time and space of a source of energy which combines the ability to *transmit*, *disseminate* and *control*.

In its broadest sense, the ability to transmit implies the *generation of something which did not previously exist*. It therefore involves the power of creation ("Let there be light and there was light") as well as that of *production*.

The ability to *disseminate* and/or distribute refers to the process whereby the "products" are directed towards target groups forming either micro-networks or macro-networks, such as those now established by the mass media.

The ability to *control*, in the American sense of the word, refers to the entire process whereby action may be directed towards the set objective. The deployment of social energy is thus structured in a series of clearly defined relationships which ensure, according to time and place, the predominance of a specific factor, whether religious, political, legal, economic or administrative.

If a society ceases to exercise its control, it lapses into anarchy; if it ceases to disseminate or distribute, it breaks down; if it ceases to transmit (to procreate or produce), it perishes. If a society abuses its control, it lapses into totalitarianism, a phenomenon which can also arise—as we are just beginning to see—as a result of excessive dissemination (the information overkill to which the modern world is subjected) or overproduction leading to economic imperialism.

René Berger

The Western World

The combination of factors which determine the structure, maintenance and security of any system is based, at the present time, on a *model of civilization* which has spread throughout the world, the blueprint for which is to be found in Western, and especially American, society. The concept which generates, underpins, activates and directs this model is that of *development* —witness the universally accepted distinction between the developed countries and the Third World, whatever subtle distinctions may be drawn between industrial and post-industrial societies, between underdeveloped and developing countries. Further evidence, if required, is provided by the fact that, despite these differences, all countries are evolving along similar lines, including China—the sole exception prior to the death of Mao—which is now following suit.

In structural terms the Western model is based on three key factors, whose combined action had transformed the modern world: science, technology and economics.

Broadly speaking, after Galileo, *science* first introduced, then imposed *experimental Truth*. It gradually evicted all other forms of truth—religious, philosophical, even ethical—or reduced them to the level of subcultures. Despite their influence, Mahomet, Buddha and Jesus are *regional* phenomena, specific to certain areas of the world. Science alone is *universal*. There are no more American electrons than there are Chinese protons. Scientocentrism is a growing reality.

Likewise *technology*, the total sum of the principles, methods and processes arising from the techniques applied by industry, has dispensed with all forms of traditional activity, from the religious ceremony to the skill of the local craftsman, or has achieved the same result by *romanticizing* them in order to adapt them to its own needs: *Europe 1* is heard in the remotest areas by the mere turn of a switch; the investiture of the Prince of Wales becomes a television extravaganza, like any other variety show; the Papal election is presented in the same way as the World Cup. Time is reduced to a technical factor. The transistor has replaced the tom-tom; Philips and Sony have colonized virgin forest and favelas. Aspirin has turned the medicine man into a museum piece (to the continuing fascination of ethnographers). *Technocentrism and scientocentrism go hand in hand.*

Economics as a discipline has always existed, but never as the exclusive activity it now represents in our Western model. Today there is nothing in East or West which is not convertible into an *exchange value*. Religious and moral values have been abandoned or set aside. At the very most, there is evidence or a certain resistance on the political front—or is this already a rear-guard action? There is no head of State who does not insist on the need to produce more and more, to export more and more, endorsing thereby the laws of the market place and heeding only the sole standard by which nations are assessed—the universal measure of Gross National Product.

Economicocentrism, scientocentrism, and technocentrism converge to promote the only value which is currently acceptable —that of efficiency, under the aegis of the most ambiguous concept of all, that of development.

The concept of *development* maintains the illusion, until recently a reality, of a linear improvement (as distinguished from evolution), just as it maintains the illusion, also a reality until recently, that it could be identified with the notion of *progress*. Over the last few decades, however, the concurrent and exponential growth of science, technology and economics has given rise to consequences, the shortcomings and dangers of which have been clearly established—pollution, disturbances, unemployment, genetic manipulation, inflation, fluctuating currencies; even the satellite, designed for worldwide communication, now looks like a planetary supervisor.

And yet—paradoxically—politicians, scientists and experts still maintain that development is at the service of mankind, that it can be directed, accelerated or checked, as if, for some, it depended on rational thought, for others, on the authorities, for all of us, on our freedom to make good or bad use of it. The question of making good or bad use of it, however, has become increasingly irrelevant. The last few decades have shown that science, technology and economics are *self-developing* (or self supporting) forces, each of which evolved in accordance with *its own specific logic*. Their approach is purely strategic, consisting of a series of goals to be reached; their sole aim is their own self-development, with no concern for other values.

This preliminary outline, however summary and controversial, may help to throw light on the question of the social energy represented by the mass media, especially the most powerful of them all: television, currently under pressure from, but also, if market trends are anything to go by, in collusion with the new media of video.

Video in the Modern World

Television: From One Form to Many

The market and technology in this field have changed so rapidly that a distinction must be made, if only as a precaution, between:

(1) *Macrotelevision* or mass-television, apt to be known as official or national television in Europe,[1] and free or commercial television in the United States[2] (terms which demonstrate the extent to which terminology is dependent upon its politico-

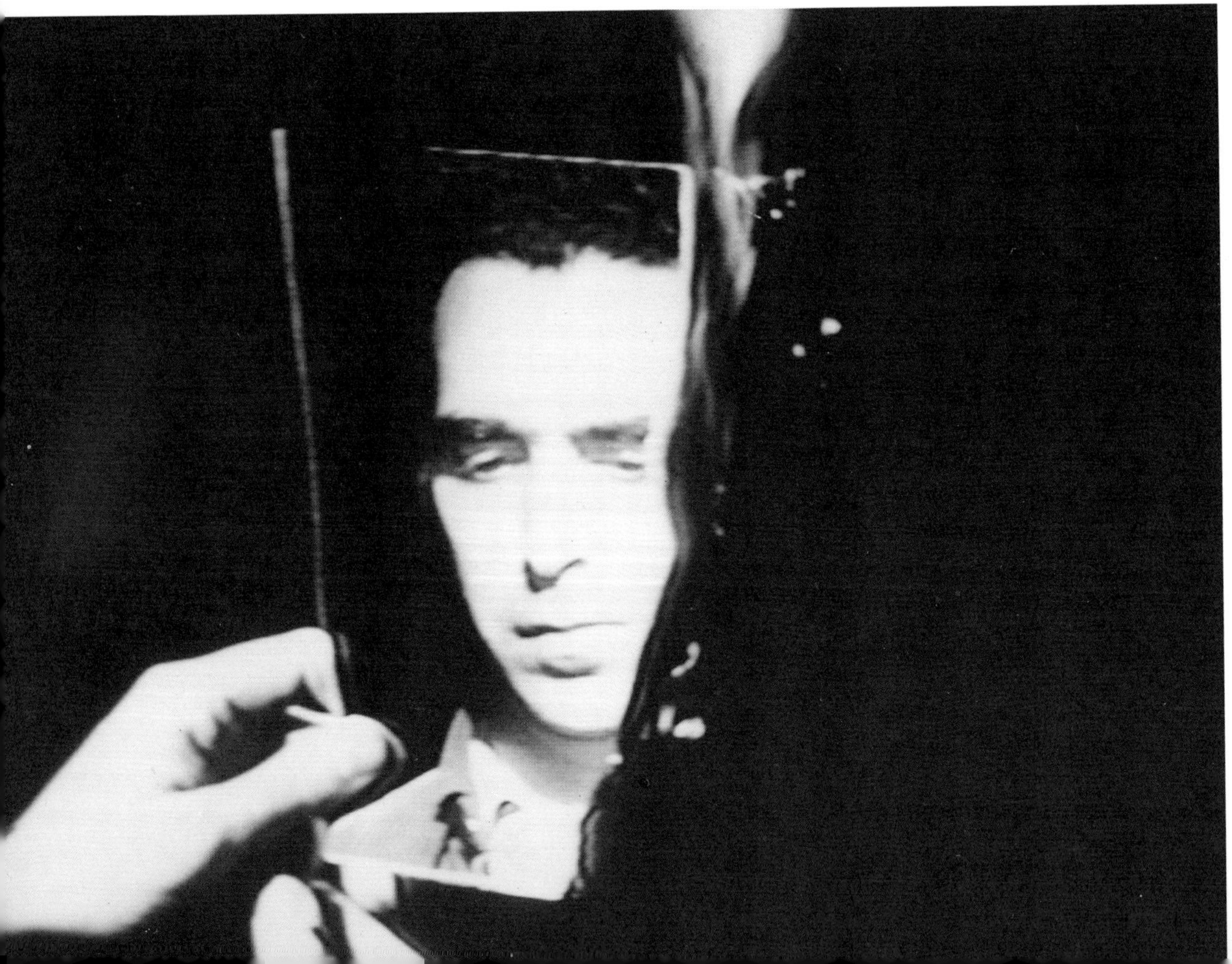

Fig. 8-1 Peter Campus, "Three Transitions" (Photograph by Nathan Rabin, from Castelli—Sonnabend Tapes and Films, New York).

economic context). This form of television uses radiowaves for its dissemination.[3]

(2) *Megatelevision*, or satellite television, which is already in use for certain specific purposes but which, because by definition it oversteps the boundaries of national sovereignty, has encountered political opposition. Here two options are possible: either direct public access to the satellite by means of special equipment, or indirect access via cable hook-up to a national retransmitter.[4]

(3) *Mesotelevision*, or *local/regional television*, also called *community* television, carried via cable (CATV—Community Antenna Television—in the United States).[5]

(4) *Microtelevision*, or *individual* or *group* television, which uses portable video equipment as its instrument.

This differentiation reveals not only quantitative or technical differences but more importantly, and less apparent, *the different structures used in communication and in social relations*. The rapid development of the media over the past decade must be taken into account as well. Therefore, the following remarks do not aim to delimit the scope of a field which is constantly changing; they are only intended to establish what we might call "trend models."

Macrotelevision

Regardless of the type of authority under which it falls, or the ends to which it is put—be they profit, culture or propaganda—it may be said that the whole macrotelevision is essentially characterized by its *desire to reach at all costs the largest possible number of viewers and by its use of all possible means to attain this objective*. Multitudes of individual receivers are linked to a powerful transmitter, and the program emanating from a central point and unidirectional, is ruled by a strict hierarchy. The power of the politico-economic authorities increases in proportion to audience size.

Owing to its macrostructure and its dependency on the authorities, macrotelevision tends to operate according to formulae designed to ensure efficient regulation in the form of as homogenous a product as possible. The classic aims of information, education and entertainment are pursued within a set of limits defined both within the program schedule and even the "genre" (news programs, dramas, variety shows, games, etc.). Thus news programs, which might be thought to be governed by the unpredictability of news, in fact conform to a ritual which balances news reporting in an imperceptible yet masterly way: sensational news stories first, followed by the more soothing items, such as exhibitions, cultural events, sports and racing results. Hence, too, the liturgical style today of televised debates everywhere, and the TV hostesses with their dual role of comforting mother and older sister. *Roles* are assigned and are rarely deviated from. Thus, the system operates according to a homeostatic model; variations from within or without must not threaten the maintenance of this balance. Control is maintained through audience surveys which, all-powerful in the United States, are considered only relatively so in Europe, or at least according to official statements. "New program" research is usually left to a research department whose suggestions are systematically ignored; program content, in whatever category (information, education, community or cultural affairs), is only accepted if it fits into the preestablished framework. Similarly, the education category, whether termed higher or continuing education, tends to become synonymous with teaching, on the lines of institutionalized courses; continuing education is confused with instruction, encyclopedic knowledge being preferred to the training of the critical faculties. At best, a few privileged directors are allowed —within limits—to play the role of *franc-tireurs* in order to assuage the conscience of the system, whose standards, however flexible, are not to be infringed.

Mesotelevision

This form of television is not widespread enough, especially in Europe, to be analyzed in a complete and relevant manner.[6] Nevertheless, based on Canadian, American and European experience, it can be said that a new "trend model" has appeared, creating an entirely *new type of social relations*.

One of the more decisive changes, as yet not fully understood, is the fact that the *roles* institutionalized in macrotelevision are *subject to permutations*. Instead of the strict division between transmitter on the one hand and receiver on the other, the citizen can be alternately in front of the screen, at the microphone, or behind the camera. Permutations of this kind have definite effects, as they can progressively destroy the concept of roles and the logic underlying them, and cause radical structural changes. These changes establish a relationship between new partners, whom I will call "interactors." Therein, the division of labor institutionalized by macrotelevision is invalidated, as is the distinction between professionals and amateurs, although in this case there is an obstacle in the form of the invcterate public bias in favor of the former. Mesotelevision is unique in that no one works for it in a professional status. Producers, directors, journalists and film editors are no longer considered specialists, and the rhetoric, drama and scenography characteristic of macrotelevision are challenged.

But things do not happen quite so easily. Experience has shown that the vast amount of creativity available to mesotelevision, instead of being apparent to those involved, seems to them at first a handicap. Thus, these interactors first base their behavior on the macrotelevision model, as if it were a matter of competing with professionals. For their part, viewers are just as wrong in demanding—accustomed as they are to macrotelevision—that local programming resemble what they consider to be "real" television. An appropriate parallel would be that of photography and film making, which both began by imitating painting. New techniques always take their inspiration from their predecessors, which serve as models. It is only with experience that the specific potential of the new technique affirms itself to become, after the fact, a new form which in its turn becomes a recognized genre.

But the public must also learn, however gropingly and hesitantly, especially as too often those persons reputed to be the most alert prove themselves to be strangely reactionary, downplaying innovative experiments or branding them as "awkward" without realizing that this awkwardness is itself a promise of fruitful experimentation. Therefore, if a certain amateurism, or at least what is generally perceived as such, is inherent in mesotelevision, we would probably gain by examining this particular facet of the experience, instead of criticizing it in the name of a concept of professionalism inherited from macrotelevision.

Therefore mesotelevision, still in its infancy, should not be expected to throw off its imitative tendencies and immediately define its own inventive modes, of which only a few have been explored thus far. However the most rewarding community experiments—especially those featuring live broadcasts—have already revealed that the classic triad of communication, education and entertainment[7] change, as do the concepts of "schedule," "program," and "genre" associated with them. These lose their value as categories at the same time as the division between professionalism and amateurism disappears. Mesotelevision reveals that variety shows, news, dramas, sports and cultural programs are only classified as such on the basis of a set of production conditions inherent in the macrotelevision structure, which itself derives from a particular form of socio-cultural and political organization. Televised debates, for instance, which are as privileged a "genre" as any, since they bring together the most eminent authorities on all subjects and from which we could expect truth to spring (is it not reputed to spring from the "meeting of minds"?), have, in fact, proved to be governed by rules as strict as those of any game: opinion must be evenly divided, with equal numbers "for" and "against." The host is not a referee but a program-keeper who must, at all costs, prevent any excessive sharpness or digressions which might abort the debate and upset its precisely scheduled duration. He must also ensure that the program is sufficiently "entertaining" to guarantee viewer satisfaction, whatever the topic examined and regardless of its outcome. This is not the case with mesotelevision. Ordinarily, citizens who generally know each other, sometimes very well, feel no need to make stars of themselves or to force an often scanty oratory talent. It is the weight of their conviction that counts, which leads them, instead of attempting to win the debate through reasoning or effects, to seek support by creating an image of personal commitment, the decisive factor for the community. Without seeming to, and by its very "awkwardness," the debate "genre" becomes a *discussion*. A new type of creativity appears, no longer obliged to compete with macrotelevision models, which tend to stand in its way. Rhetoric gives way to exchange.

Microtelevision

Basically, microtelevision, thanks to its lightweight and relatively inexpensive equipment, makes it possible for almost anyone to produce and edit his own program. Its fundamental interest, therefore, is not that it gives a new meaning to the ideas of education, democratization, lifelong training and even information, but that it *encourages creativity itself*, so that culture ceases to be something passively received and becomes a form of action in which one participates, which one creates. The relationship is no longer that of transmiter to receiver, or even that of "interactors," but what I would call "inter-operators." In so far as reality, even if regarded as sacred, does not exist independently of the communication process, but through collective techniques which involve specific social, economic and political conditions that are never sacred, to say the least, but social,[8] it becomes clear that video, by virtue of the fact that it is accessible to all, breaks with pre-established patterns and is capable of creating completely new structures. This possibility of dissociation and innovation is such a new phenomenon that it has been overlooked by Japanese, American and European manufacturers who continue mainly to base their advertising campaigns solely on the reproduction qualities of the medium when, in fact, they have marketed the most revolutionary instrument of all. Video in hand, the camera operator is in a position to decolonize established communications systems, challenge the *establishment* at its very core, and resolve the confusion existing between the natural and social orders. Nor is it just a question of insurrection; in the broadest sense, video allows one to challenge the most subconscious mental mechanisms in order to *liberate creativity*.

Unlike the homogenized space of macrotelevision and the multi-directional space of mesotelevision, it is difficult at the present time to speak of a space unique to microtelevision. At best we might speak of an ever-increasing number of cells which, unlike television channels or static cinemas, are most often mobile and even include the videobus; a cell on wheels. Thus, microtelevision differs from the homeostatic operation of macrotelevision and more closely resembles the operation of *open systems*, whose interactions cannot be dissociated from those of the environment.[9] Thus, video reveals itself to be one of the best tools for producing "neguentropy," as described by Schrodinger and Brillouin,[10] or even the inventiveness we attribute to children. Just as mesotelevision has brought back the communications biotope, video would seem to have rediscovered the road to creativity and, above all, to provide the medium for expressing it.

Unfortunately, things are far from having happened in this way. As we will see, video was very quickly put to serve more profitable interests than those of creativity. Nevertheless, there are artists ("Indians" I would call them) who have not lost their faith in the media. Before examining their situation, let us clear up a point which tends to add confusion to all issues.

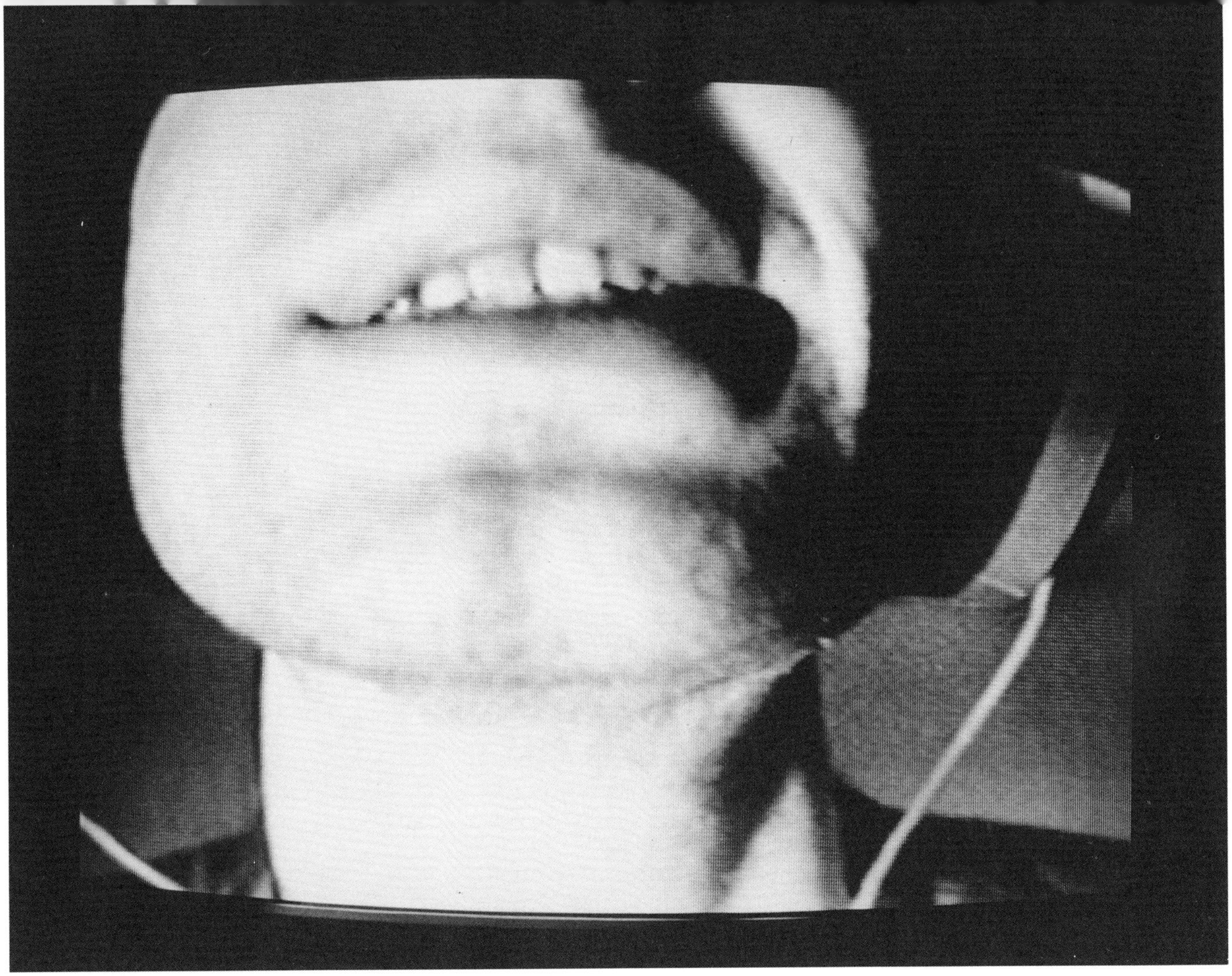

Fig. 8-2 Bruce Nauman, "Lip Synch" 1969 (from Castelli—Sonnabend Tapes and Films, New York).

Communication and Dissemination: A Matter of Scale

Mass media, as the term would indicate, are considered to be a means of communication. However, experience has shown that macrotelevision, which is broadcast unidirectionally from a powerful transmitter to a multitude of consumer-receivers, is less an agent of communication than an *instrument of dissemination*, not to say *dissemination apparatus*. Information sent out via the antenna is directed not so much to consumers as to target receivers who suffer the impact of the messages rather than choosing the content. It is thus a gross misconception to call television a means of mass communication. Ballistic information is more concerned with hitting its target than with establishing communication, which implies exchange.

This point takes on special significance as the mass media have made us discover a hitherto unnoticed limit and, through their very extension, have brought it into full daylight. Technological progress, however prodigious it may be, has in no way changed the fact that we are and remain, in the words of Leroi-Gourhan, "living fossils."[11] Nothing has changed since our accession to human status. Is the "global village" foreseen by McLuhan just an illusion? For electronics to take on a human quality *they must respond to the limited conditions of our state, for which meso- and micro- television could furnish the yardstick*. Not that macrotelevision is disproportionate but that, without our knowledge or with our complicity, it may have alienated us by extending our horizons beyond their normal limits. This is a hypothesis which has yet to be tested, but which may allow us to understand better why the progress of macrotelevision does not seem to have extended our consciousness as we might have expected and hoped but, on the contrary, has produced a glut of information whose cancerous effects are now quite clear. Although appearing late on the scene, meso– and microtelevision could provide the means for returning to that stage which macrotelevision went through prematurely and which megatelevision is unfortunately preparing to abolish. This is a challenge to the Nation State as it comes up against the new communication structures: hence, the resistance, the stonewalling, the prompt repression, in any case the active distrust.

While macrotelevision (and the approaching megatelevision to an even greater extent) is directed towards the "statistical being" that man has become in mass society, paradoxically associating the most advanced abstractions and the most intimate impulses, mesotelevision aims at restructuring human relations at the *community* level (region or city), within the boundaries of that limited area in which we move daily, surrounded by those we can still call "our own."

Microtelevision, for its part, aims at rediscovering in us the primary driving force that has existed ever since man first began to think and make tools, since creativity is nothing more than imagination at work with the aid of a controlled technique. It allows us, or can allow us, in the electronic age, to restructure human relations at the level of the *community* which consists of individuals united in a group, the primary social nucleus.

Meso- and microtelevision show up what our technological society seeks to hide or suppress: namely, that the more the media expand, the less real communication there actually is, replaced as it is by *dissemination*. Communication is, above all, a matter of scale, and certain limits cannot be overstepped with impunity.

This affords me the opportunity to pay tribute to those video artists I have, with good reason, referred to as "Indians," for it is they who blazed the trail before being placed in "reserve" (or on "reservations").

Video Art, The Unknown

It must be said that video art is not easy to get to know. Yet the video artist works with videotape which, by definition, enjoys the privilege of universality. Nothing can top the electronic image for speed—provided the reception *standards* are compatible, which is not always the case. These standards remain inconveniently ethnocentric, with the Japanese, Americans and Europeans jealously guarding their respective shares of the market. Rumors of an agreement between the major producers in these countries have recently been denied. Provided, too, that the tape can be *shown* (something so obvious that we tend to forget its importance), which implies the possession of equipment which, though much less expensive today than in the past, is still far from being within the reach of all. Provided also that the equipment is operating satisfactorily, which almost always requires—for programs of any importance—the presence of technicians or even engineers. Provided, finally, that video productions have access to (and this is far from being the case) premises and equipment specially designed for their presentation. How would painting and sculpture have evolved if they had not found in museums, palaces and salons the organized places where the public and private image of art could take shape?

Video art, outside of a few museums, cultural institutions, universities, occasional exhibitions and galleries, exists almost exclusively in the homes and studios of artists who make their own tapes and have the equipment to show them. In contradiction to the hasty and, hence, specious analogy with television, it is not easy to see video art; even though the videotape is shown on the television screen, it has no direct connection with television. The latter organizes the program, the broadcast and the reception. Video art, notwithstanding a few spectacular breakthroughs (Nam June Paik), is still in the troubadour era, which means in the Middle Ages.

Paradoxically, the latest electronic technology—at least in terms of the artistic usage to which it is put—requires the artist to move or, in parallel and complementary fashion, requires the amateur to travel.[12]

There is another, no less paradoxical, difficulty. How do we come to know modern art? Through exhibitions. But it must be admitted that much of our information comes not so much from direct contact with the originals themselves as from reproductions, often chanced upon, in reviews, press, articles, magazines, or even rumor or hearsay; second– or third-hand knowledge which we need not to go into here. On the other hand, it is important to recall that the *reproduction* of a painting gives one an idea of the original work, as does the reproduction of a sculpture, despite the lack of a third dimension. Thus, to a certain extent, an isomorphic relationship, or at least a credible resemblance, exists between the plastic work of art and its printed reproduction.

The characteristic of video, however, is its shift in emphasis from space to time as the privileged dimension. Hence, its incompatibility with topographical reproduction. In dealing with video, printed materials like books, magazines and newspapers are therefore obliged to select images and freeze them, in other words, to treat them in a manner which runs counter to their kinetic nature.[13]

My purpose in speaking of these problems, however summarily, is that unless we take them into account we get a wrong picture of video art, on the basis of which some people consider themselves qualified to be both critics and experts. As in all things, what is important is *experience*. I do not think it is too much to say that there are probably only a few hundred people in the world today who have sufficient experience to guarantee the necessary relevance of their views. Knowledge of the subject remains partial in all cases, including mine. This is, in part, due to the fact that the conditions of the art market make it rather difficult to obtain.

Art and The System

Unlike the plastic arts, which benefit from organized networks whose main agents are the art dealers, galleries, collectors, museums, critics, art festivals, auctions, etc., in other words all those who ensure the circulation of art works, there are few outlets for art videotapes. Although some galleries, such as Castelli and Sonnabend in New York, have taken an interest in this field (but only occasionally), as have Anna Kanapa, and notably, Howard Wise, who has attempted to set up a permanent rental service; and although some museums have already installed videocassette machines (the Everson Museum in Syracuse, New York, was the first), it is to be wondered whether in fact this preparatory phase will be followed by a second, which would require the creation of an organized distribution network.

Why does the market, which could be the driving force behind video art, seem so hesitant? Works of art—or at least some—may appreciate in value enormously. Originals, as well as limited edition prints, are goods with a market value based on their "rarity." Whereas a videotape which, by definition, may be reproduced indefinitely and may even be pirated, defies the concepts of both rarity and ownership.[14] With financial interest lacking, it is hardly surprising that the video art market is having trouble getting off the ground; in fact, the question is whether it ever will. Certainly, the major art festivals, following an initial infatuation with video art, quickly returned to "marketable products" such as paintings, sculpture and prints. Moreover, reticent as the art market may seem, the art press is no less so. Without going as far as to say that lucrative interests govern information, it is noticeable that information tends to concentrate on art that sells and to ignore that which does not. Information concerning video art, therefore, is still sporadic and only surfaces at international video festivals.

One unexpected consequence must be mentioned in passing: video unintentionally shows up the ambiguity of the plastic arts in their dual role of objects of aesthetic value and merchandise.[15] We thus find unsuspected convergences between video art, body art, ecological art, social art, etc., all of which operate, if not outside the market, then at least on the fringes.

There is yet another point which deserves thinking about. Whether we like it or not, and even when we buy solely for visual pleasure, most amateurs, including those who disclaim any idea of future appreciation, cannot help but think of paintings, sculptures, serigraphs and lithographs in terms of their place on the wall or in the living room. In short, at the risk of raising eyebrows, I would say that our typical attitude is to regard a work of art as, above all, an *ornament*. Witness the number of collector's homes open to the public.

Video art differs in this respect not in degree, but in nature. Walls, living rooms or gardens no longer enter in the question. The videotape is seen on a television screen and "exists" only for the duration of its projection. Despite the word "art," video *art* does not entirely correspond to what is normally understood by this term. It thus follows that we do not and cannot get to *know* it in the same way and through the same means. Contemporary art has not, over the years, entirely escaped economic considerations. Without going so far as to say that it is tied to the market, it seems clear that aesthetic value and commercial value often go hand in hand. This ambiguous relationship has become commonplace today.

It is difficult to study video art without "artifying" it by patterning it after our customary behavior and practices. To try to meet this challenge, let us briefly survey the first steps of video art: who first thought of using a videotape recorder as a creative instrument? In what circumstances? For what reasons? For whom?

On the Threshold of Video Art

From a purely provisional viewpoint,[16] we should first mention the role played from 1950 on by WBGH Television in Boston, with its resolutely experimental orientation.

In 1964, WGBH broadcast the series "Jazz Images," which was the first program to seek to break with the traditional portrayal of a band—whether in documentary or poetic form—in order to invent abstract imagery appropriate to the music. One year later, in 1965, Nam June Paik, the Korean-born artist, used the first portable VTR to record scenes from his taxicab, and showed them in a Greenwich Village coffeehouse: "Cafe a Go Go, 152 Bleecker, October 4 and 11, 1965 . . . 'Five Years Old Dream of Me.' The combination of electronic television and video tape recorder is realized."[17]

The tone of this note indicates that the event is the culmination, as well as the beginning, of a long search, which is illustrated by the now famous statement of the same artist: "As collage techniques replaced oil painting, the cathode ray tube will replace the canvas."

Nam June Paik's work was exhibited that same year at the Bonino Gallery in New York. In 1967, the Rockefeller Foundation granted WGBH-TV in Boston a subsidy of $275,000 to encourage artistic creation in television, which led pioneers such as Fred Barzyk to produce with artists like Paik, Tambellini, Piene, Padlock, Seawright, Kaprow, Stan Vanderbeek and Douglas Davis, the series entitled *The Medium is the Medium*, broadcast in 1969. That same year the first video art exhibition was organized in New York by the Howard Wise Gallery under the title *TV as a Creative Medium*,[18] with the participation of Serge Boutourline, Frank Gillette and Ira Schneider, Nam June Paik and Charlotte Moorman, Earl Reiback, Paul Ryan, John Seery, Eric Siegel, Thomas Tadlock, Aldo Tambellini, and Joe Weintraub.

Paul Ryan, introducing his video "Everyman's Moebius Strip," put forward a striking analogy: "A Moebius strip is a one-sided surface made by taking a long rectangle of paper, giving it a half-twist and joining its ends. Any two points on the strip can be connected by starting at one point and tracing a line to the other without crossing over the boundary or lifting a pencil. The outside is the inside. The inside is the outside. Here the power of Video Tape Recorder (VTR) is used to take in our own outside. When you see yourself on tape, you see the image you are presenting to the world. When you see yourself on tape, you are seeing your real self, your 'inside.'"

In 1968 Sony marketed a portable VTR deck called Portapak in the United States, which was to become the instrument of all pioneer video artists. Beginning with its first issue in 1970, the magazine *Radical Software*, edited by Beryl Korot, Ira Schneider, Frank Gillette, Juan Downey and others, featured critical, philosophical and technical articles by those artists who represented the most remarkable reflections on this subject. In 1970, too, Gene Youngblood published *Expanded Cinema*, and Nam June Paik invented the *video synthesizer*. Douglas Davis, for his part, devoted his attention and enthusiasm to the development of this phenomenon in a series of articles, while Michael Shamberg made known his thoughts on the subject and his experience in a work entitled *Guerilla Television*, published in 1971.

The West Coast was equally active and productive in this area.

In 1967, KQED-TV in San Francisco started the first video workshop and launched an experimental program under the joint direction of Brice Howard and Paul Kaufman. From this came "Music with Balls" by Terry Riley, "Sorcery" by Loren Sears, "West Pole" by Robert Zagone, as well as the breathtaking and unexpected "Homage to Descartes" by the poetess Joanne Kyger. James Newman, for his part, aware of the importance of television, reorganized his art gallery in December, 1968. In conjunction with KQED-TV, San Francisco, he launched "Open Gallery," a series of televised programs featuring artists such as Robert Frank, Ken Dewey, Walter de Maria, Yvonne Rainer, Anne Halprin, Julian Beck and The Living Theater, Robert Nelson, Frank Kappa, Ewin Schlossberg, Terry Riley and Philip Makanna. The success of the series when broadcast in 1968 owed much to the work of producer/director John Coney.[19]

Some of America's most active galleries began to feature video art: Bykert, Leo Castelli, Fishbach, Bonino, Gibson, Sonnabend, Ronald Feldman Fine Arts, and Howard Wise. The *Kitchen* was to play a decisive role in New York; a place for meetings and experimentation, it is one of the only places where one can watch videotapes almost everyday.[20] In 1971, the Finch College Museum of Art, in New York, at the instigation of its dynamic director, Mrs. John Varlan, held one of the first video festivals to take place in a museum. Thus, the new video art form continued to emerge with the participation of Les Levine, Vito Acconci, Dan Graham, Dennis Oppenheim, Steve Reich, Eric Siegel, Peter Campus, Robert Whitman, Michael Netter, Tsai Wen-Ying, Constantine Manos, Jackie Cassen, Russel Connor, James Seawright, Stephen Beck, Richard Felciano, David Howe, Keith Sonnier, Bruce Nauman, etc.[21]

Video art caught on very quickly in Canada. In fact, it has perhaps developed the most in this country, especially in connection with cable television, which is not my topic here. I will limit myself to the project which is generally considered to be the most original, and which, under the name *Videograph*, offers *videograms* "made by citizens for citizens." Founded in 1971 on the initiative of Roger Forget, the *Videograph*, which receives fairly heavy funding from the Canadian National Film Office, had produced some 140 projects by March 1973, having turned down 300 others. At 1604 rue St. Denis in Montreal, the *videotheatre* seats 100 and features a circle of suspended multi-screens. Each

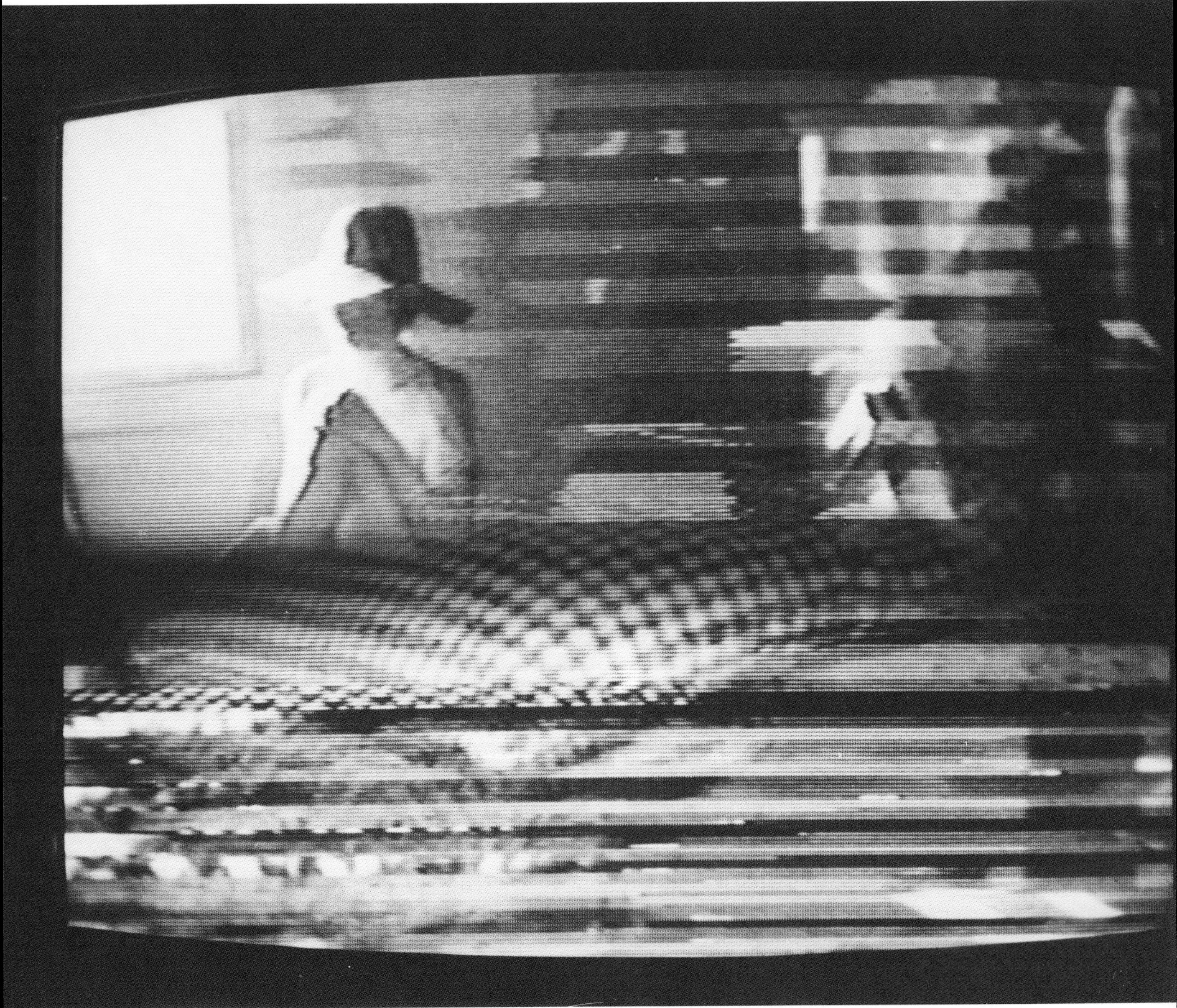

Fig. 8-3 Ralph Hocking and Sherry Miller, "Selected Works" 1975-1979 (from The Museum of Modern Art, New York).

new videotape is introduced by its author, and the discussion held each evening allows participants not only to exchange critical points of view, but to create a new type of relationship, as anyone can become a producer. The *Videograph* is designed neither to create "artists" nor to launch artistic careers; it is a community and political service. However, the distinction is not all that easy to make. It seems more and more difficult to classify videotapes in any of our existing categories. Videotapes produced in the context of the *Videograph*, for instance, cannot be classified as civic, social, or political purely because they pursue one of these objectives—just as video art tapes are not purely aesthetic simply because they are made by artists.

Despite all attempts, video art has a paradoxical and disconcerting tendency to escape any strict traditional definition. This is no doubt due to the fact that the electronic medium entails changes, not only in relation to dates and contents, but also in terms of the principles and methods to be followed, those which we have inherited from our literary culture having proved inadequate in this respect. Reasonable as it is to want as precise information as possible on the subject of video art, we find that it varies, often to a considerable degree, according to whether we look for it in the United States, Canada, Japan, or Europe; and this applies to an even greater extent if we include "pre-video" research, such as that presented in 1963 at the Parnasse Gallery in Cologne by Nam June Paik, Wolf Wostell and Joseph Beuys, or that of Luciano Giaccari at *Studio 971* in Varese in 1968.[22]

In the absence of a history spanning barely twenty years, whose elements are scattered throughout catalogues too numerous to mention, the main difficulty resides in the fact that video art only really surfaces at international or national events or exhibitions. Thus, unlike the plastic arts (as I have already pointed out) there is no permanent "display" of video art in an appropriate setting where the public could get to know it, as it gets to know painting and sculpture.

Video Art and Its Orientations

This problem is encountered once again when we try to establish a system of classification.[23] Based on the familiar categories of literature, we could attempt to establish a distinction between video performance, video theatre, video chronicle and video epic (why not? some videotapes, such as those made by the Italian artist Baruchello, run for over twelve hours!). We could also speak, more broadly, of realist video, of lyric video, of tragic and comic video (as illustrated by *Totologies* by Michael Jaffrenou). Nor are analyses based on film criticism any more relevant. It is difficult to align video art with the established practices it has continued to defy for the twenty-odd years of its existence. So, instead of differentiating between movements, trends and schools, I will confine myself to pointing out a few of the directions it has taken and which, we may be sure, are already changing.

"*My body, this unknown*" is the path taken by many artists, once it was discovered that the electronic camera allowed for immediate reproduction and practically costless and unlimited production (erase and start over!). Video is open to everyone and offers one the possibility to explore any subject, free of cost, for videotapes on end. The wonder of being able to examine at leisure one's own face, nose, hands, feet, arms, legs, forehead, neck, tongue, hair . . . a continent without a visa and without limits! And we can understand how grimacing, a ridiculous exercise in our eyes, but so important for children, inspires so many artists eager to bring us back into the game—and thus our childhood—in the twofold sense of playing and freedom of movement. Bruce Nauman, Keith Sonnier and Les Levine, after exploring parts of their bodies, move on to the elementary movements—eating, putting food into one's mouth, chewing, swallowing—and then to such basic activities as walking, climbing and jumping, which Marcel Mauss once studied from an ethnographic point of view under the title of body techniques. Painting ends at the portrait, video seizes on the movement both from within and without. Unlike film, which tends to focus on certain objects and is thus selective, video has the power and the luxury not to have to choose, but to give itself up "body and soul."

In a second direction—"I, this unknown"—works by Peter Campus, Gerald Minkoff, Peter Weibel and Jean Otth, among others, address the subject of perception. Long ago we were taught to mistrust our senses (the famous optical illusion of the stick that breaks in the water, for example). For a long time philosophy, and for much longer mythology, and more recently science, have taught us to mistrust appearances. It is unfortunate that they have not taught us to do this in the same way, nor on the basis of the same principles, and even less for the same purposes. Hence, the antagonism which results in mutual rejection, each discipline claiming to hold the key to the truth; to the point that with each of them—philosophy, mythology and science—one finds oneself plunged into entirely different realities. Does the stick really break? It cracks, in any case.

Video, for the first time, or at any rate to such an extent, allows us to see and feel how our senses operate, the images we form, how we form (and deform) them, and how we put them together. *Interface*, by Campus, is a case in point. Two images appear on a large glass panel: one is your reflected image, the other your video image. When you move, the two images move in opposite direction. Which is the truest image? Just as television showed us for the first time in history men in a state of weightlessness—which is radically different from an *explanation* for the phenomenon—*so the video reconstructs the experience in which one is oneself participating.*

A third path proposes developing the interface at the *environmental* level. The *natural* environment first; thus Gilbert and George, who as living sculptures remain motionless for minutes on end, with only very rare and restrained movements (*The Nature of our Looking*). An absurd proposition, if it were not for a significant reversal for the viewer. Rather than the human figure detaching from its environment and hence from nature, as is our usual experience, here it is the environment which detaches itself, or rather—the terms changing in their turn—it is a branch, a leaf, a bush, moved by a slight wind, which become the "actors," while Gilbert and George are gradually conventionalized into a double commemorative medal, to end as petrified as fossil prints.

Again at the level of the natural environment: the *act of love* which is almost always soiled in the pornographic species, finds it opposite in the videotape made by Les Levine (I have never seen another like it): a couple, through the video recorder, discover their own caresses, movements and murmurs, without breaking their intimacy, or turning the act into an exhibition. *Only video seems able to respect the special nature of lovemaking*, the other media invariably reproducing desire and, in reproducing it, commercializing and distorting it. Forsaking the "product," spurning the idea of merchandise and discouraging the voyeur, video in this case combines eroticism and modesty.

In this same "loving" vein—I use this term carefully—are included such other experiments as Frank and Laura Cavestani's work with Indians, which we might at first be tempted to call ethnographic studies were it not that video allows them to avoid both the aims and dissertations of ethnography in order to take the approach of the "other-and-I" (I use hyphens intentionally). *Show, see, observe, deduce, and infer*, presuppose a distinction between object and subject. Linguist terms sanctify this vision. What happens when ethnocentrism, even in the name of science, has discretionary powers and gives common currency to such concepts as "primitive," "savage" and "civilized"?

Along the same lines, Shirley Clarke, with the aid of one or more portable monitors, creates a veritable *affective environment*, establishing a relationship with the monitor she holds in her arms like a child. We might be tempted to say: Is the machine, as the object of human tenderness, the portent of electronic maternity? We are all familiar with the joke that Europeans condescendingly tell of American children and their three parents: father, mother and television set. And yet it is true that we have entered a new age, the age of the *media* (300 million television sets, more than one and a half billion viewers, representing a third of the population of the world) and that we must tame these *media* as our ancestors tamed wild animals. In fact, with the advent of community television, for instance, it is no longer absurd to speak of *domestic media*.

Douglas Davis and Ira Schneider have participated in this undertaking in their own way. In a work completed in 1975, Schneider built a video environment in the shape of a map of Manhattan—hence the title *Manhattan is an Island*—comprising twenty-five monitors on which six videotapes show scenes taken from various places in New York, from the excursion boat circling the island, from a helicopter flying over the city, and so on. It is in no way a model, a game, and least of all a gadget. As I "watched" (is this verb still appropriate?) *Manhattan is an Island*, I was struck by the presence of these images and sounds of several young people who, squatting, kneeling or lying down, chatted calmly as if the megalopolis, paradoxically reduced by video to the dimensions of its inhabitants, had regained a human, and most importantly, welcoming dimension. No mock-up, this, but simultaneity, as if the gesture, having too long been adulterated, commercialized and corrupted, sought to purify itself through the video screen in order to become once again fraternal, the alliance which it was originally.

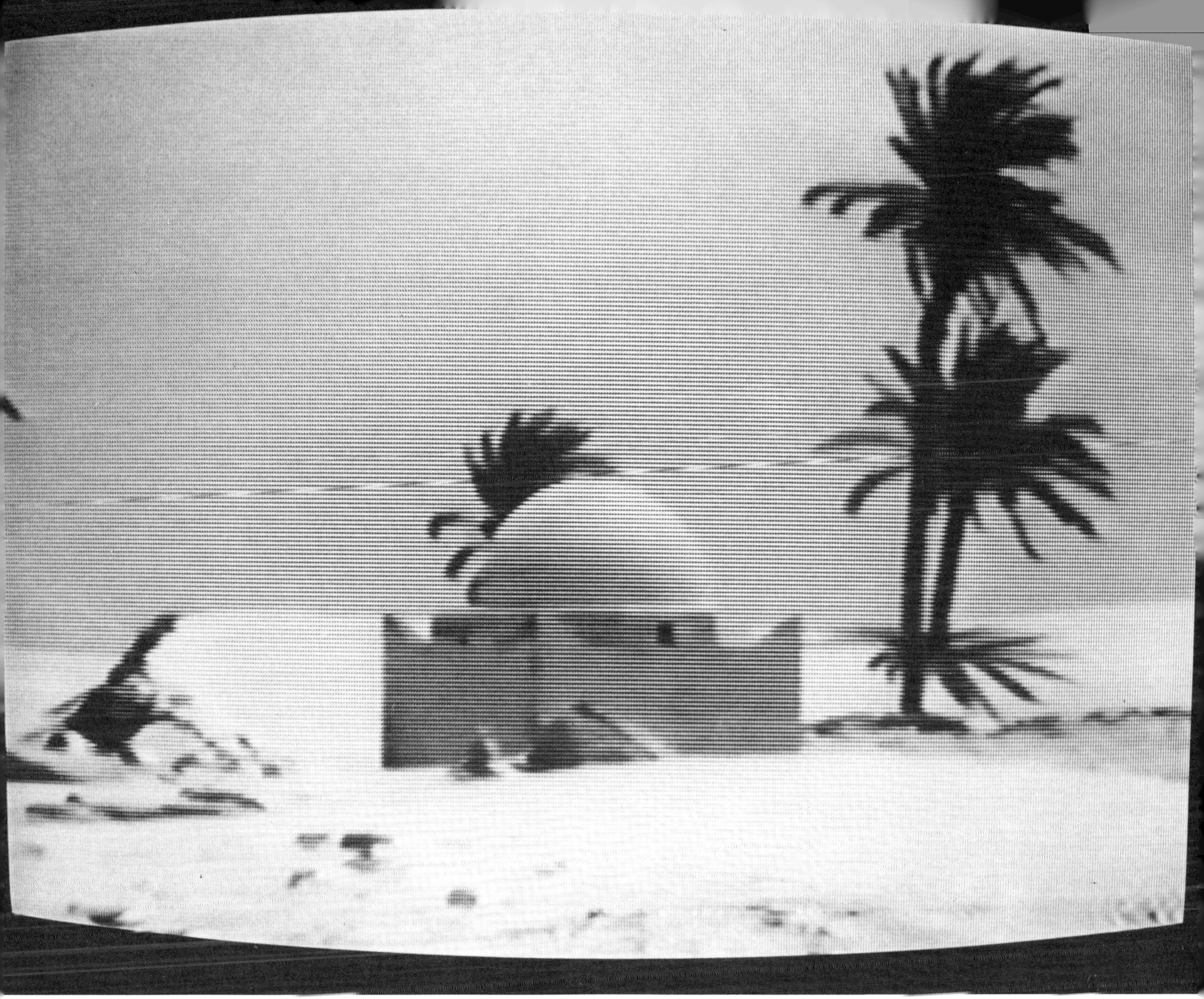

Fig. 8-4 Bill Viola, "Chott el-Djerid (A Portrait in Light and Heat)" 1979 (Photograph by Kira Perov, from The Museum of Modern Art, New York).

Along these lines, which we could perhaps call "The world and us," Frank Gillette, a contributor to *Radical Software*, has developed a philosophical and poetical ecological research which also combines cybernetics and long personal meditation.[24] At his exhibition at the Everson Museum in Syracuse (1973), a series of environments entitled *Subterranean Field, Terraquae, Gestation/Growth*, for example, forced us to participate in the life of plants and animals, and even the live birth of hundreds of baby chickens. This piece challenges our observation and, diametricaly opposed to the documentary, undertakes to create a *sympathetic biotope*.

The directions I have indicated here should be considered as itineraries taken rather than categories. Can I be more specific? The task is particularly difficult because video touches on all possible uses which the artist, in the absence of established aesthetic criteria, may choose to make of VCR for purposes of which he alone is aware (I will return to this subject later). Nevertheless, it might be useful to draw a few distinctions of an instrumental nature.

Monovideo and Multivideo

The videotape that the artist creates and which is intended to be shown either on a monitor or a television screen, I call *monovideo*. It is the most widespread form of production up to the present. However great the diversity of the subject matter, the situation involved in monovideo is relatively simple: on one side the videotape projected onto a screen, on the other side, the viewers. The relationship is thus bipolar, as in the theater, cinema, or in front of the television. Here I speak only of the topography of the situation, not taking the value or diversity of the contents into consideration.[25]

The advantage of monovideo lies in the fact that the cassette is easy to send and that, if the receiver has the technical equipment necessary, the tape can be viewed on the spot. Thus, many museums today have video booths for video viewing. There are either set programs (specific days, hours, artists, tapes) or, in the case of these institutions, the viewer can select the work he would like to see, inserts the cassette into the recorder himself and turns it on and off as he pleases.

Monovideo lends itself to the creation of video libraries which, materially at least, are obviously very much like the traditional library; cassettes are arranged on shelves, and a catalogue or card index is available for public consultation. All things considered, the behavior of the video librarian and that of the reader/viewer is similar. This situation, as we have seen during the last two decades, has been especially conducive to the organization of video festivals, which have increased in number throughout the United States, Europe and Japan. With adequate technical equipment—and technicians on hand, in case of technical failure or difficulties—the contributions of the guest artists can be presented to the public as at an exhibition of paintings.

I have already referred, in speaking of video's early years, to the need almost immediately felt by artists to broaden their experience. These more or less complex *video environments* draw upon a spatial relationship which modifies the viewers' behavior. Breaking with the bipolar relationship of monovideo, multivideo, as I call it, offers a multipolarity which involves audience participation. This is not a multi-screen spectacle, but an arrangement in which the intervals themselves are called upon to contribute, as demonstrated by artists as different as Steina and Woody Vasulka, Muntadas, Les Levine, Peter Campus, Gerald Minkoff, Jean Otth and others, not to mention the time-lapse environments of Frank Gillette and Dan Graham.

Used in this manner, video acquires a power of embodiment which reorganizes our perception of space and time and marks a further step in the break with macrotelevision, with its purely entertainment-oriented function.

Video environments, however, have a serious drawback in that they require a great deal of expensive equipment. Unlike monovideo, which relies on cassettes easily sent anywhere at low cost, environments pose transportation problems and must be set up anew at each presentation site. They are, therefore, almost only found at institutions which have technical and financial means at their disposal, and this tends to restrict their distribution.

As for *installations*, another form of multivideo, these utilize not only a group of screens, but different media objects even, and occasionally the direct participation of the artist. An example of this which comes to mind is the piece by Veronique Mori, presented in Geneva in 1981 in the context of *Video Mix-Media*. Two television sets face each other on the floor. In the thirty centimeters between them there is a houseplant, a pointsettia to be exact, whose flowers spread like leaves, except that they are red. On the two screens the double face winks, turns, or moves

Fig. 8-5 Frank Gillette, "Oracle," a video installation at the Whitney Museum, New York, June 1, 1983 (Photo copyright Francene Keery).

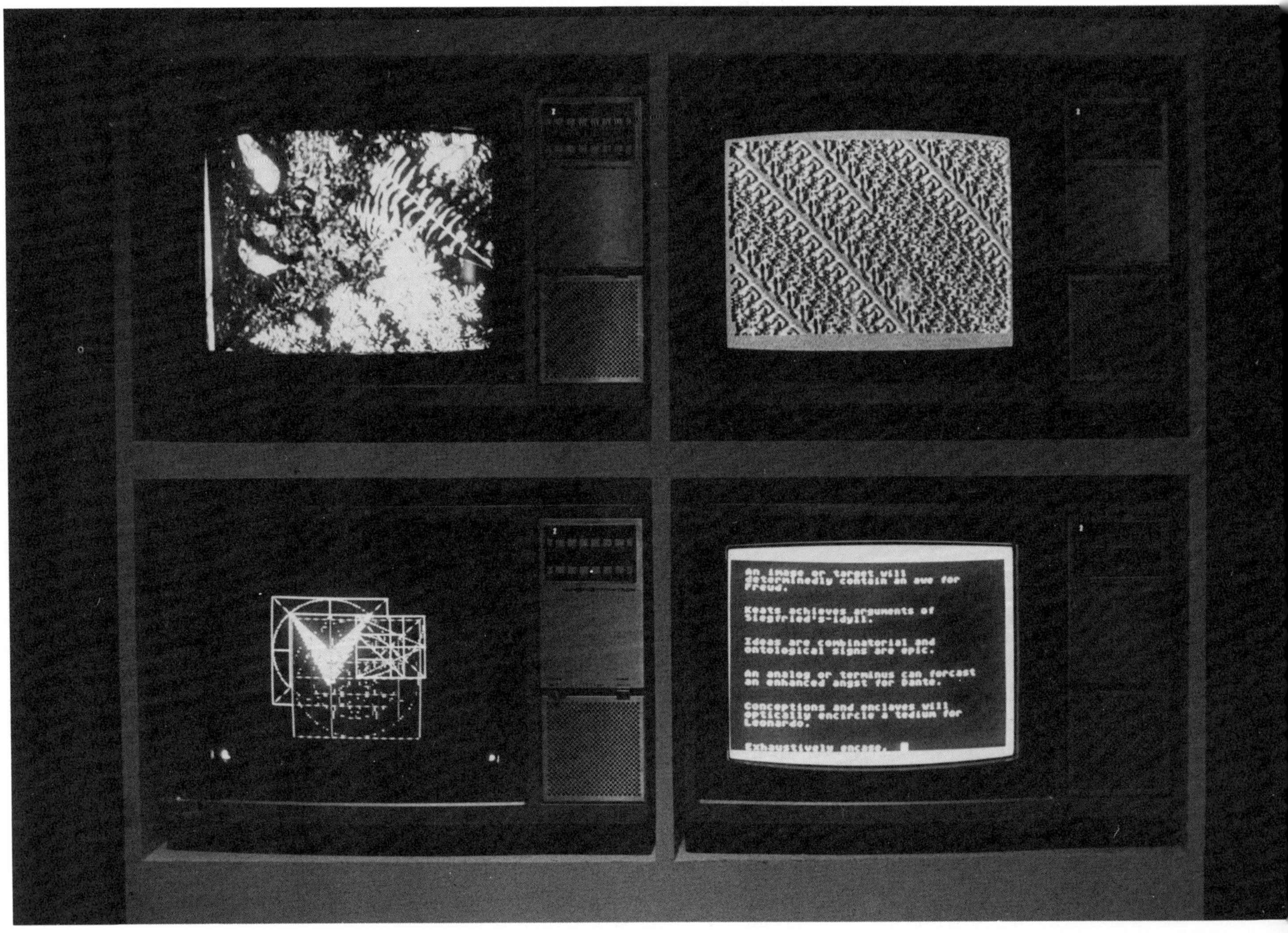

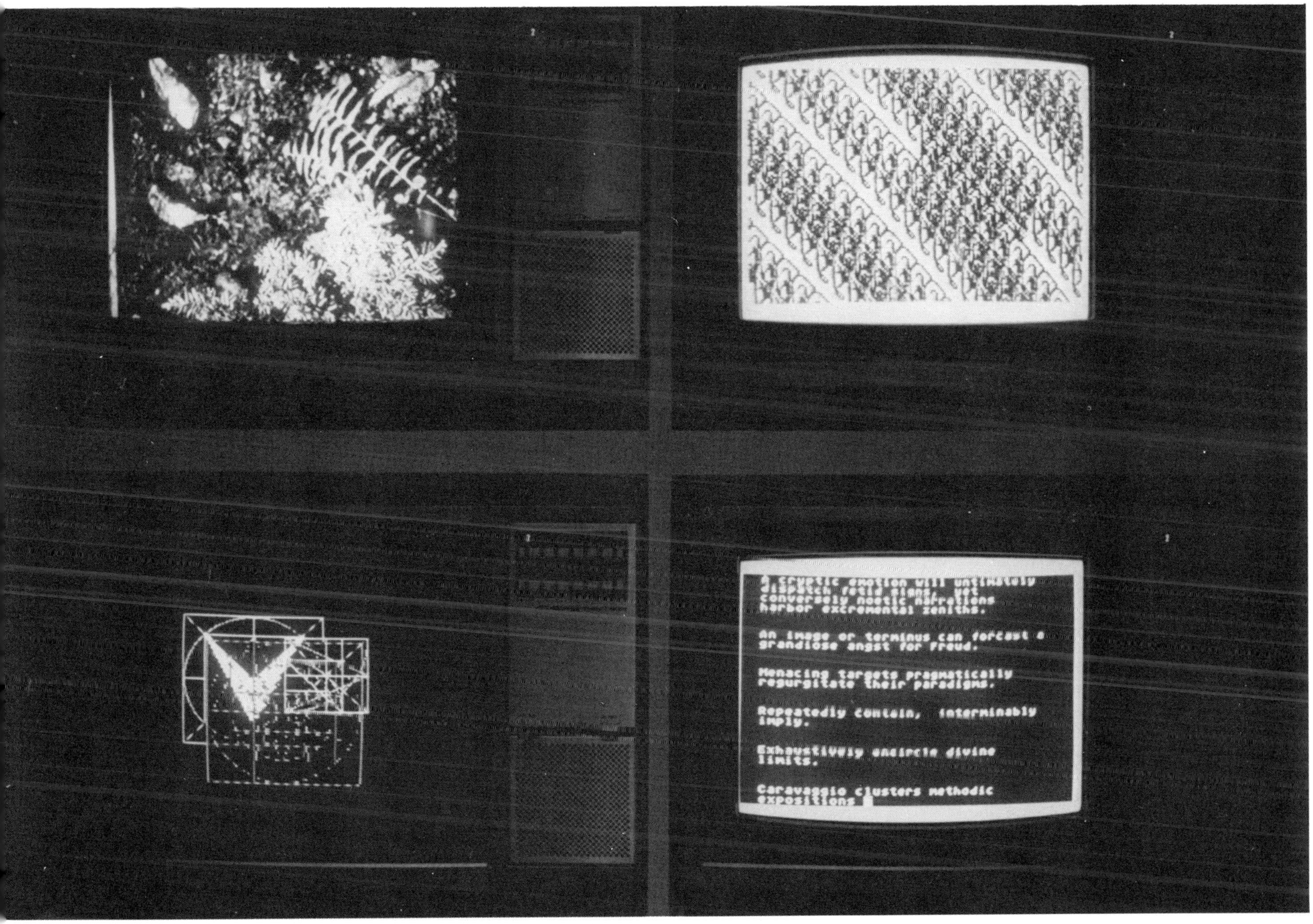
A cryptic emotion will ultimately
conversely noetic narrations
harbor excremental zeniths.
An image or terminus can forcast
grandiose angst for Freud.
Menacing targets pragmatically
regurgitate their paradigms.
Repeatedly contain, interminably
imply.
limits.
Caravaggio clusters methodic
expositions

aside. A guillotine basket? But the head is alive, as if, having rolled to the ground, almost knocking into the plant, it comes back to life, not real life but the ghost-like existence given it by video image. And suddenly, intuitively, it seems that the electronic iconography in which we like to see the very image of reality is perhaps only an ectoplasmic emanation emitted at eye-level by a ceaseless, endless screen. The fact that the television set can come down from its stand on to the floor and, instead of being isolated in the middle of the living room, mix with plants or other objects, challenges our very concepts of the world and our bodies.

At the Crossroads of "Topics"

When examining a painting in a museum, I place myself in a *typical situation* which consists (to reduce it to a few characteristics only) either in identifying the painting, or evaluating its contribution to art, or situating it in an historical context, or simply admiring it. However, in order to examine, study and analyze the work, one must choose an appropriate "place" neither too near nor too far away. This "place" also involves a time element: the length of time needed to examine the work. The situation of the "painting spectator" is determined by spatio-temporal conditions and cultural rules which permit and establish appropriate behavior. This could be illustrated, to remain in the area of painting, by the social ceremony of varnishing day at an art exhibition, when crowds of people invade the gallery or museum, eating, talking and drinking (often all three at once) without bothering to look at the paintings, whereas it would be considered scandalous to turn up the following day, drinks and snacks in hand. Contexts assign roles according to the conditions and purposes which make up the *typical situations and relationships* of our social practices.

Taking a broader view, we might say that all social behavior, stemming as it does from communication, i.e., involving interactions between people and techniques, is based on "topics" which, through repetition, become the norm.

It is precisely this "normality" that all video artists, almost without exception, attack, either as in the early works of Nam June Paik, Wolf Vostell and Jean Otth, by intentionally deforming television images or establishing absurd connections between the television set and other objects,[26] or by introducing, through mono- or multivideo, atopical relationships which "pervert" (in

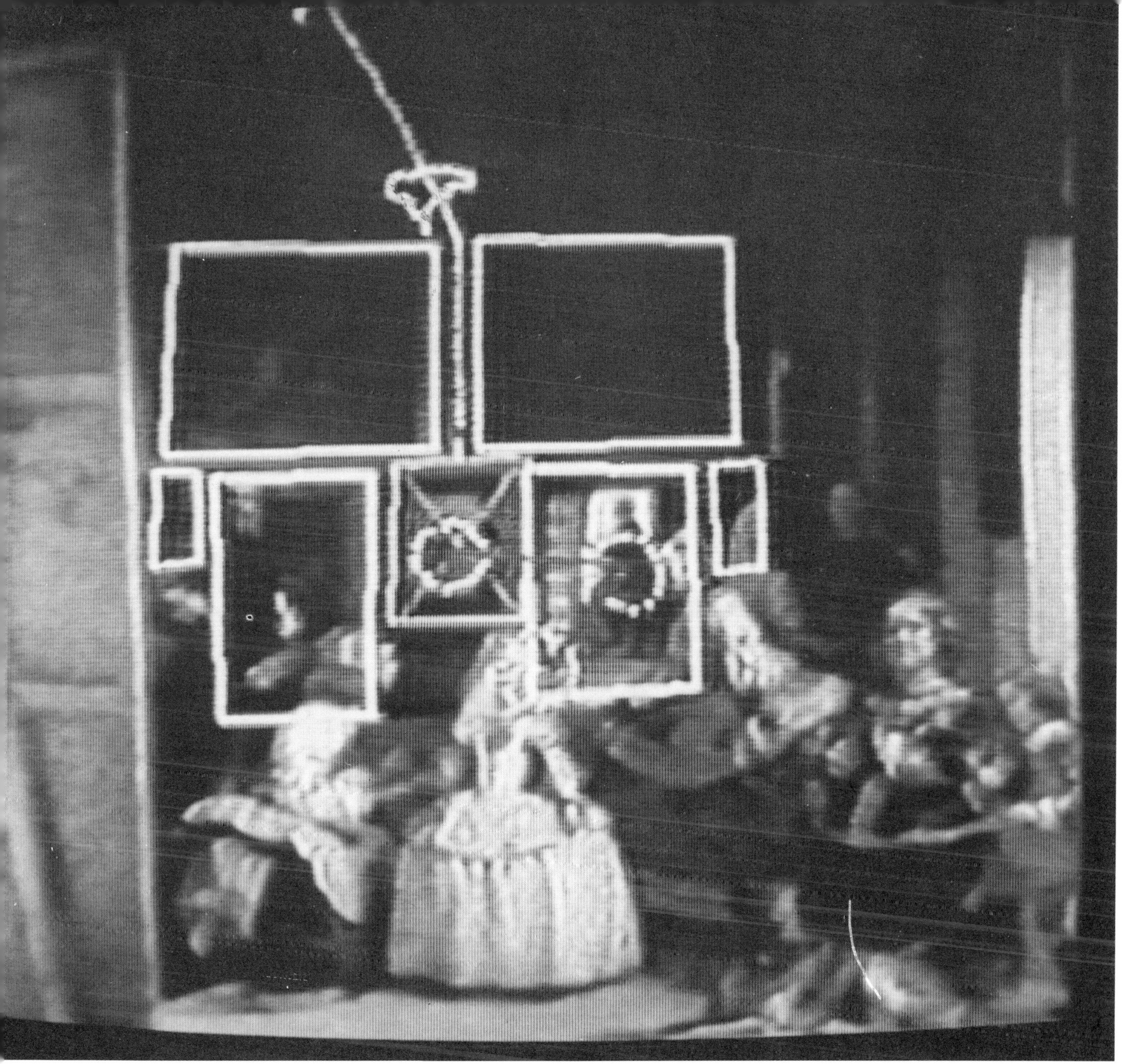

Fig. 8-6 "The Looking Glass" by Juan Downey.

the etymological sense) the normal functioning of the media. This is the first effect of video art, which I call the *dislocation effect.* Nothing is stranger than comparing a television program with a videotape. Willingly or not, it is difficult to avoid a *feeling of strangeness.* Why does video function so differently? In fact, as video experiments multiply, it is surprising to realize that *all videotapes depart from standard television material.* This observation is even more surprising when one considers the number of similarities existing between the two. Video uses the same apparatus as TV, a set which is turned on and off, the manual operations are the same and the electronic signals are identical. Videotapes project on the screen what we are accustomed to seeing on television: images organized in sequences, in black and white or color, accompanied by sounds, speech or noises. In other words, there is nothing *technical* to distinguish video from TV; yet despite the technical similarities, we could not have a stronger impression of being confronted with two fundamentally different phenomena.

This is because in television the various genres, subjects, schedules, programs are regarded as "natural"; more openly, the hostesses, presenters, journalists; more secretly, the administrative, political and technical departments—in short, all that constitutes in the broad sense the *organization of the TV discourse,* which is the result of the "topic" established between the producer on the one hand and the receiver/audience on the other. This correspondence might appear to be a happy coincidence (pre-established?) of interests if it did not stem simply from a progressive assimilation of the medium's logic by the audience.

As long as a medium prevails, its hegemony makes it the privileged model of communication. This has been confirmed for over five centuries by linguistic communication, which has made the book the leading authority (as it still is in the classroom). This has also been confirmed for almost fifty years by audiovisual communication, which has made television the privileged moderator of our everyday lives. In fact, any medium is an arbitrary and complex production which appears less so owing to our familiarity with its operation (witness the use we make of television today).

We can therefore better understand why video artists are so intent on effecting a caesura with television to restore the issue of communication to the production level, which is its proper place.

In attacking the nature of the image, as they often do, these artists disclose one of the strictest imperatives of television and the basis of its very topic: realism. All television pictures must be clear and sharp; their credibility depends on their reliability. Fuzziness and approximation are tolerated only exceptionally, when production conditions are hazardous, as in wartime, for instance. In all other cases it is television's duty to create a feeling of "you are there"; the power of the medium is dependent on its capacity to compete with our powers of perception. Can one imagine an impressionist or cubist-style news program? Viewers would immediately complain or think there had been a breakdown. Realism requires that the image identify with the reality (in linguistic terminology; the sign as the referent). Yet this coincidence is always an illusion, *for the simple reason that an image is never real.* This is revealed by many video artists when they attack the self-assigned authority of the screen, which is always only the expression of an image that, if not manipulated, is at least processed. Thus genres are challenged—news broadcasts, variety and sports programs, dramas, etc.—in other words, the dramaturgy and rhetoric of the most powerful media in existence today. In this *video criticism* there is a considerable blend of irony and humor, but beyond the satire it shows that television, which tends to take itself as the natural order of things, is an order founded on technology and the social relationships which govern it. By emphasizing subjectivity—sometimes to an exasperating extent—video artists show us that the authority invested in television is a matter of power. In so doing, they refuse to permit any media to be "the voice of France" as President Pompidou termed French television; the beginning and end of all communication is the person engaged in it as its subject. Outside the field of polemics, although it occasionally indulges in it, video criticism sheds light on one of our basic problems: the individual is not lost in the mass to which the mass media tend to reduce individuals. The individual's voice, just as his identity, must remain inalienable; the effect of mass reality should not be mistaken for reality.

In addition to the *dislocation* effect which I have briefly described above, video, as used by video artists, tends to *relocate*; that is, it aims at showing us places, or networks of places which, regardless of their apparent bizarreness, fit in with our own profoundest experiences. The latter, whether they be on the conscious or subconscious level, are not *immediately* communicable, but *mediately* communicable. Every medium is therefore the place and the means by which the field of our shared experience and interpersonal relationships is established. Psychoanalysis has revealed much of the subconscious, and has brought its innermost workings to light. However, analysis (as the term would indicate, and as Freud insisted time and time again) proceeds in a rational manner through the organization of rational discourse. That is to say that the discourse, whatever its object may be, applies to it the logic of the concepts on which it is based. As linguists have pointed out, the structure of a language affects the thinking it expresses; it molds it to its own forms. This is true even for philosophy itself, which is dependent upon the mode of linguistic expression in which it is expressed.

When a video artist like Naumann devotes a videotape to the painstaking examination of his face and grimaces, we could draw a seemingly outlandish analogy by saying that he is only *visualizing* what Descartes *conceptualized* when he decided to pretend that all things entering his mind were no more real than the illusions of his dreams. In proceeding in this manner, Descartes did not use discourse to increase or verify his knowledge, but to discover, within the concepts of which it is composed, the confirmation of his own thinking substance. The conviction of "I think therefore I exist" is based on the very process of communication—in this case conceptual—that Descartes uses to formulate his experience.

Similarly, a video artist like Naumann (and this is only one example) does not use mobile images and sounds in order to say something about his state of health, his aesthetic appearance or his mood (as when one looks in the mirror or when shooting a film). The images he records are less informative or scenic than they are heuristic: they constitute the artist's process of communication—in this case audiovisual—through which he discovers his identity during the *experience* he visualizes on the video screen. Just as with Descartes concepts and discourse ceased to be mere auxiliaries designed to establish or preserve conceptual knowledge, so for Naumann mobile images and sounds cease to be mere auxiliaries designed to establish or preserve visual knowledge. Just as the concept for Descartes was no longer used only to serve purpose, but became the experience of his thinking activity, so Naumann's mobile images formulate the possibility of an *audiovisual identity* which disclosed itself in the very course of projecting the videotape. However slender a chance it might offer, video could well become, as the mass media close in on us, a new form of "electronic cogito." At the crossroads of the topics sanctified by our social practices, particularly those relating to communication, video feeds a doubt which, like that of Descartes, has the virtue of posing a question.

Similarly, at the opposite end of the spectrum, video allows us to plumb the depths of the subconscious. Some may object that this is the subject of psycholanalysis and that Freud's entire work, as well as that of all those who followed the master's path with varying degrees of fidelity, has tracked down even its most secret mechanism. However, whatever is involved—impulse, unconscious acts, Freudian slips, plays on words, dreams, everything in short that constitutes the obscurest side of our being—the analysis is always conducted in the light of rational discourse. It seems to me that monovideo, and multivideo even more, often delve as deep into the subject, *but do so using a radically different approach*. In lieu of analysis through rational discourse based on concepts, video proposes configurations of images, sounds, objects and movements which appeal to our emotions more than our intellect. It is as if the artist, through his work (and the expression "artist's work" could be equated with "dream work") visualizes that which analysis limits itself to conceptualizing. Installations, environments and video sculptures take on the appearance—with the help of the third dimension—of *models* of our subconscious, while analysis remains on a linear level. The audiovisual is not merely an extension of linguistic communication; it introduces an entirely new space. But while the audiovisual almost everywhere, because of television, has taken on the image and force of a superego, video art has the capacity to explore, among other things, the *ego* and the *id*, which is a stumbling block for most topics, particularly those of the media.

Such a task is particularly difficult because video art is not widespread and receives only limited attention from the public and critics, so that the artists run the risk of losing themselves in an idiolect unintelligible to the others. One has only to watch visitors at a national or international video art exhibition to notice that very few watch tapes from start to finish, and that most devote only a limited period of time to each piece, or just wander from screen to screen. In the best of cases, through chance circumstances or the obstinacy of an organizer,[27] video art attains the status of a *dialect*, by which I mean the medium of expression common to certain groups. But nowhere has it succeeded in affirming itself as a common, widely used language on a par with painting and sculpture. This situation, can, of course, be interpreted as a weakness; but I am convinced that it can also represent a strength in that, in the face of the hegemony of the mass media, it remains, however precariously and at times inadequately, the source of that "untamed thinking" which is our guarantee of freedom.

The Coming Market: New Products (New Forms of Communication?)

All things considered, and at the risk of playing the prophet with all the suspicion attached to the term, I will say that the same thing is happening and will go on happening with video as happened with the automobile. Who would have dreamed that the automobile, now barely one hundred years old, would so transform our traditional environment as to make it unrecognizable? Yet it is a fact that a product as simple as the automobile, a temporary mobile residence on wheels, has imposed a new physiognomy on our cities in order to meet the demands of traffic and has convulsed our countrysides by replacing the old roads and paths with a straightjacket of motorways; in short, nothing in our styles of living, working or thinking has not been affected by it. Video is very possibly in the process of doing for communication what the automobile has done in the field of transportation and means of locomotion (for better or worse is not to question). Whether in public or private transportation, the automobile has added individual technolocomotion to our original two legs. Similarly, at the communication level video has added to the public communication of TV a private means of communication, which, while not replacing the postal or telephone systems, is in the process of fundamentally changing our communication relationships. In the case of the automobile these changes only came about once the mass market as we know it today had been organized.

The organization of the video market is still in its early days, but all signs point in a similar direction. Publicity, one of the more significant signs, is a case in point. Publicity forges that alliance between producers and distributors, on the one hand, and consumers on the other, which testifies to the existence and

Fig. 8-7 Frank Gillette, "Aransas" 1978 (from the University Art Museum of the University of California at Berkley).

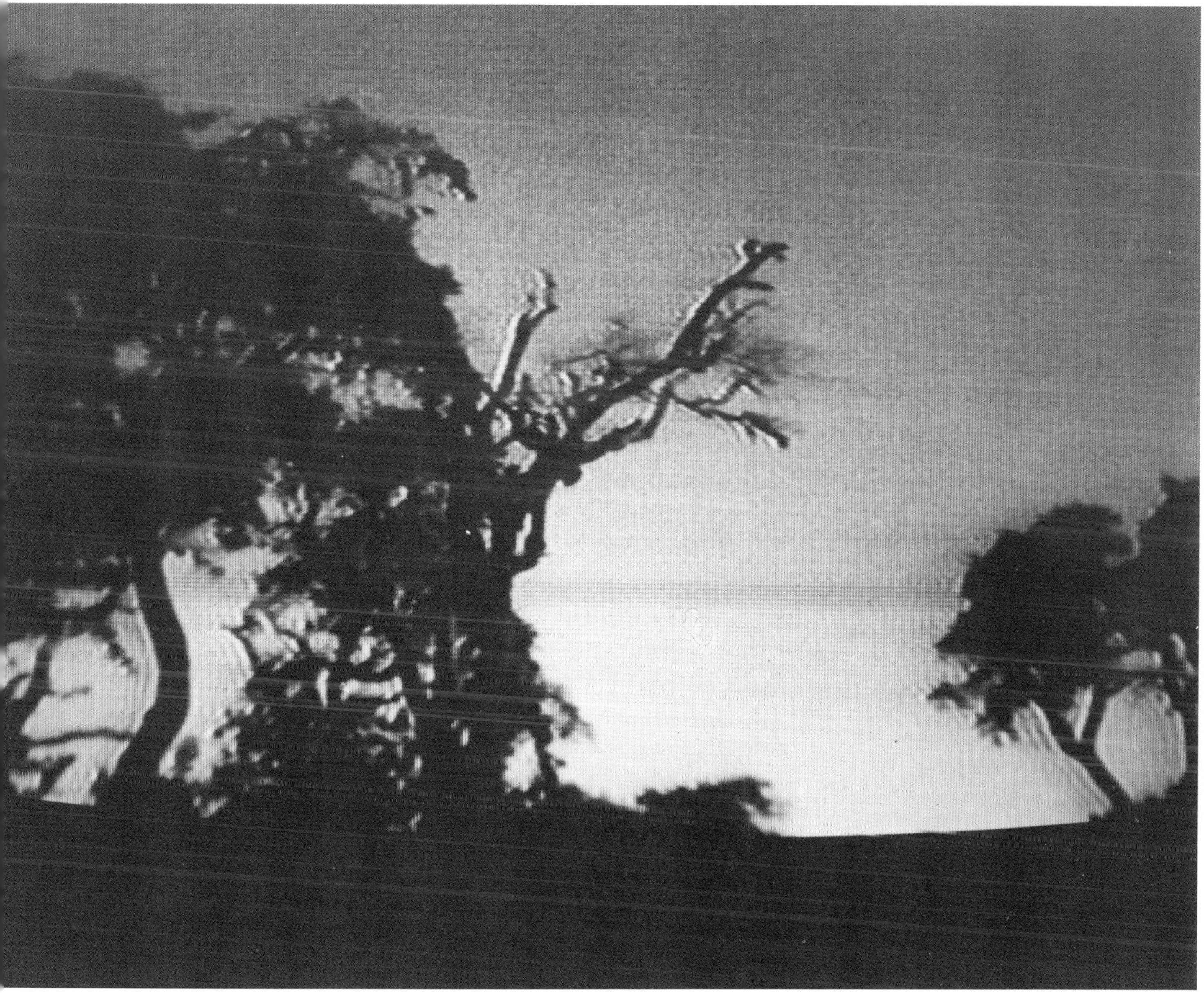

vitality of a market. Today, there is no newspaper, magazine, advertising spot or prospectus which does not give place of honor to the new communications equipment: video.

In the abundant publicity we can discern the outlines of the "magnetic field" of the new product, whose strong points—the selling points—are roughly nine: (1) the specific characteristics of video, (2) its many uses, (3) its performance, (4) reliability, (5) feasibility, (6) economy, (7) adaptability to future needs, (8) price, and (9) the promise of "Videoland" (as one speaks of Disneyland).

I do not have time here to elaborate on each of these points, only a few of which require further explanation in our particular context. It is understandable, for instance, that manufacturers should emphasize the fact that their equipment is becoming increasingly compact and lighter in weight; that it is easy to operate ("a child can use it" is a popular theme); that it can record for hours on end and be programmed at will (each new feature is heralded as a victory); and that technological progress ensures its efficient, economical performance, which means competitive prices that bring these products within the reach of all. Of most interest from the point of view of our study is the emphasis placed on the specificity of video. Video not only adds a new communications machine to those already existing, but introduces a new type of communication—perhaps even, as I will show later, a new type of communication environment. Like the other audiovisual media, it breaks away from the demands of writing and books and moves outside the realm of linguistic communication by developing images, sound and movement. Unlike the fixed photographic image, it captures movement, as indeed the film has already done; but its magnetic recording, doing away with the need for film development and allowing for unlimited erasing and re-recording, gives video a considerable economic advantage. Moreover, the situation has so evolved that new uses are being prepared for the electronic image, some sporadically in the course of development, some still at the planning stage. It is, therefore, not surprising that manufacturers should emphasize something of what this evolution portends: "Enter the *videoworld of Sony*[28] and discover the sixth sense," announces the firm of the same name with solemn assurance. Without a doubt it is "videoland" and "videoworld" which are beginning to appear on the horizon.[29]

Having said this, what are the main uses of the video now on the market or about to be marketed?

1. *The possibility of recording in parallel*: while watching one program, another one, shown at the same time, may be recorded for future viewing on videotape (a solution to family arguments). The VCR can also be programmed a day, or even several days or weeks, in advance to record a program which is to be shown during one's absence. With this feature, much touted by the manufacturers, video is considered to be an instrument which allows one—to a certain extent—to escape the disadvantages and limitations of television. There are also a few clever features in their own right: freeze-frame, speed up and slow-motion controls.[30] *"A la carte" television*, as it is called: everyone makes up the menu he/she wants. We must remember, however, that each item on the "menu" is created, prepared and served up by TV alone!

2. Secondly, the video market has led to the production of *audiovisual cassettes*—like music cassettes—whose field of application is virtually limitless. In principle, videocassettes could (will?) form the equivalent of current day libraries; in principle, too, they can become the depositories and the vehicles of human knowledge, instruments made even more extraordinary by the addition of the visual to the conceptual dimension. Providing, of course, that there are publishers to produce them, distributors to distribute them and customers to buy them.

   What does the videotape market look like today? Four major trends —of varying importance—predominate in video catalogues.

   The first includes cassettes offering *films* (i.e., adventure, burlesque, comedy, musicals, drama, classics, horror, crime thrillers, science fiction, westerns, etc.), which are categories I have taken from Sony's latest catalogue, *Sonyscope*. This selection is in keeping with the desire of many people (a desire which, when lacking, is created or encouraged by the manufacturer/publisher) to have "films at home" instead of having to go out to the cinema.

   The second, and not the least important trend, comprises the burgeoning production of "X-rated" cassettes (in other words, pornography).[31] Libertine or licentious literature has existed throughout the ages; pornography, a more recent phenomenon, owes much to progress in color reproduction. With cassettes, however, it has made a real breakthrough. The market for it is all the greater in that contents, however repetitive and stereotyped, correspond in all countries to the same repetitive and stereotyped drives, and need no translation.[32]

The third trend covers an eclectic selection of cassettes ranging from sports to variety programs and concerts, which can be grouped under the general heading of entertainment.

A fourth trend is made up of "how-to" cassettes (how to learn, to sail, play golf, cook, etc.).[33]

Judging from the early days of the mass video market which, it will be recalled, is barely five years old, it is clear that videocassette recorder usage remains linked to that of television, even if it claims to be liberated to a certain extent from its constraints; and that the cassettes marketed are mainly mass-consumption cultural products, as can be seen from a brief glance at the catalogues distributed by the publishers. As for video art, I can find no mention of it in these catalogues, which shows very well its marginal status. More surprising is the absence of the plastic arts, which one would have thought an important role in video. This is again proof that the producers, with their market imperatives, care little for the specific possibilities of the medium; *they only produce what is likely to sell on a large scale*. This may only be a temporary phenomenon, but as things stand it has to be recognized that "electronic" culture is subject to the law of the market place.

3. *Electronic games*. The popularity of these continues to grow. The market has therefore already introduced, in addition to those sold separately in the trade, cassette video games which can be inserted directly into a video cassette deck. Thus "the Atari Cassette Library today comprises thirty-seven game programs: games of skill (e.g., Space Invaders), of strategy (chess), sports (football, Pele), of chance (casino) and education (basic maths). A total of over 1,500 different games which will provide hours and hours of exciting entertainment, either alone,

or with family and friends. This collection grows each month with the addition of new game programs . . ." The use of the videocassette recorder for games (which I do not have time to analyze here), varies considerably from the traditional games of which Piaget, among others, demonstrated the importance in his essays on *genetic epistemology*. The replacement of manual activities, which activate the child's body as well as his mind, by purely entertainment-oriented activities, which relate only to the sense of sight—itself reduced to binary logic—is bound to have considerable influence on the child's development.

More broadly, it is video in the classroom (or video instead of the classroom?) that is being challenged.[34]

4. *Personal Television*, as it is called in a Nordmende advertisement (the Vision of the Future): ". . . it is more than home television! More than the mere recording, playing back or slowing down of the offerings of official channels or prerecorded cassettes"!

"*Personal television means going out and making your own TV*"! To do so, don't take just any videorecorder—take the two-unit Nordmende 350. Strap the recorder over your shoulder (it weighs under 12 pounds), grab the Nordmende C225 video camera, and you're off"!

"*At home that evening*, simply hook up the recorder, turn it on, and you have *truly* personal television . . ."

This use of the VCR, still embryonic, is, in fact, an extension of family photography and "home movies." Now anyone (provided they possess the necessary equipment, whether Nordmende or another brand!) can record his children at play, his vacation in the mountains or at the seaside, birthdays, and his wife's charms (which can always be erased in the event of divorce, and the cassettes left blank for the future spouse).[35] It matters little that this type of recording becomes stereotyped fairly rapidly, as does most home photography, the important thing being that the sale of equipment and cassettes should follow a rising curve. Alongside Kodak, Polaroid, Yashica, Nikon and Canon, new empires have appeared: Sony, Matsushita (JVC, VHS), Philips, Thomson and Ampex.

The camera has already become an individual "medium" in our industrial society; video is now in the process of taking its place.

5. "*Institutional" Television* (videocommunication for professional use) enables institutions and communities —universities, businesses, unions—to produce videograms devoted primarily to information, training and promotion. Although it may be an in-house service exclusively, its production has been steadily increasing.[36] How to rationalize the job of a service station attendant? How to welcome a new bank customer? How to react in a dangerous situation? How to display a product? How to be a floorwalker? The videogram is much more effective in many situations than a verbal or written explanation. Even dating services use it, since the advertisement can now be accompanied by a videocassette, which converts the anonymous concepts of handsome, tall, nice, affectionate, romantic, rich, etc., into seductive (?)—or at least credible—images addressed to prospective partners.

This is only a brief overview of video applications, since video is still in its early days and the organization of the mass market dates back only a few years. Its development is dependent on an increase in production and sales. While demand was slow at first, it has become so great that the Japanese have scheduled production of some 20 million VCR's for 1982.[37] By all indications, this trend will continue, stimulated by the development of videotex and telematic systems at the personal levels.[38]

Video communication is tending to become an *audiovisual environment*, fed by an ever-increasing number and variety of sources (television, cassettes, satellites, telephone, databanks, computers, institutional and personal videograms). Another no less significant indicator is the advent of *video clubs*,[39] on the line of book clubs. The number of *video magazines* and reviews, which has been growing steadily over the past few years, is yet another indication of the vitality of this phenomenon. Their aim is summed up in the strident slogan of one of them,[40] "video concerns you." Conscious of the importance of this trend, the manufacturers have not stood idly by. For the more clairvoyant of them (and are they not bound to be so, since the world market is at stake), the home videocassette recorder, even equipped with a personal video camera, is only a beachhead. They are, therefore, trying to build up around the videocassette recorder a whole series of related products, on the lines of hi-fi speaker equipment. The *video environment* is, therefore, no metaphor; it is in process of becoming both a technical reality and a fact of civilization. The Thomson-Brandt Group, among others, has observed this, and "to celebrate worthily the inevitable marriage of video and sound . . . offers the first system designed for the home audiovisual environment." In addition to the monitor and electronic selector, the system features "an outlet configuration allowing for the hookup of two videocassette recorders and a camera, as well as a videodisc player and a personal computer." This firm looks boldly into the future in saying: "The audio-amplifier-projector allows *all existing hi-fi system equipment and future innovations*[41] to be interconnected, but it can also take all sound channels from the television-video part of the system. Thomson-Brandt has envisaged the video system in the broadest and most evolutive sense of the word by anticipating our future "audiovisual happiness" (satellite, cable, videotexts, etc.)."

Without going further into this rapidly developing situation, it is fair to predict—or at least imagine—that, just as the automobile revolutionized our environment, so audiovisual systems are in the process of revolutionizing communication.[42] Video mediation is preparing to conquer all fields and all publics. It should be examined with a view both to clarifying the type of relationship it creates and establishing the place of video art which, with all its originality, is a part of this evolution but is not restricted to the dictates of the market.

A New Relationship—Video Mediation

For thousands of years, communication has mainly been effected through linguistic and iconic messages. In the former, linguistic symbols serve as intermediaries; in the latter, images or more broad representations, are used. To be transmitted, linguistic and/or iconic symbols need to be *re*produced, *re*presented orally, in writing, painting, sculpture, or by any other means of *re*production. However, reproduction requires a space which provides the support in appropriate material form, forms to fill the space, rules to order it, and finally a certain stability; in other words, representation is impossible without a certain fixity of the message, which alone ensures the length of life—whether long or short is of little importance—which it needs to serve its message-relaying purpose. These structural elements which make communication possible have taken on a myriad of forms throughout history: tombs, palaces, cathedrals, paintings, frescoes, statues, utensils and tools; in short, the world of images has been traditionally characterized by an infinite variety of forms, techniques, materials, dimensions and functions.

With the new audiovisual media, particularly television and video, a fundamental change has taken place. The key operation is no longer representation, but *transmission*. The viewer, instead of reactivating stabilized forms, is drawn into the movement of the image-messages. The screen, by definition, excludes all representation and even all possibility of representation, as it only functions if one image gives way to the next.[43] Based for thousands of years on the polymorphous spatialization of images, iconic communication tends increasingly to make the video screen the omega of all communication. This radically new mediation, which today extends throughout the world and to which we are all subject without even noticing it, affects the very basis of our behavior[44] by producing effects which video artists are perhaps the first to have perceived.

From before recorded history to the dawn of the twentieth century, man called upon *incandescence* to produce the light he needed, whether it was in the form of a wood fire, torches, candles, or the white heat of a filament enclosed in a vacuum-tight bulb. The electronic image, however, comes from *fluorescence*, in which free electrons generate luminous photons. Unlike incandescence, in which light is linked with heat, fluorescence allows matter to be energized without generating heat. Hence the term "cold light" given to it by the German scientist Wiedemann in 1888. An image is thus produced by the impact of electrons on the luminescent powder on a screen. This endows video with characteristics not possessed by the other media, even the cinema: ". . . the cinematic image is created by transforming variations in the intensity of light at the level of the object perceived (the object-scene filmed by the optical camera) into chemical variations at the level of the filmsupport, which relays through a projector these chemical variations in the form of an image on the screen. The production of a TV image involves more complex relays: light variations are transformed—or 'analyzed'—into electronic variations at the level of the picture-taking tubes (electronic cameras); these produce electromagnetic variations at the level of the recording heads (videotape recorders) and the magnetic tape which has received and recorded these variations sends them back in a symmetrically opposite process in the form of electronic variations and then the light variations which constitute the TV image."[45]

The electronic signal comes from invisible rays which are suddenly made visible by a flow of electrons; the empty screen is at once filled with a moving flow of images which, from nonexistence, come into existence instantaneously. This is a radically new experience, for hitherto we have always perceived images, even in photography and films, in outline or printed form. Today these images *appear* with no indication of their source and with no apparent means of retaining them, combining immediacy and fugitiveness.[46]

Without going into details, we must remember that, unlike the optic nerve, which has hundreds of thousands of tracts, the video screen has only a single one. This means that there is never more than one point on the screen, which may seem as unlikely as it is incomprehensible. In fact, the electronic image is created by the paradoxical course of a single point of light which is perceived in the second degree. It is an imaginary image constructed from the "frozen" path of a bodyless point; hence its ghost-like character, without past or future. Arriving without any perceivable vehicle (electrons escape perception through their size, number and speed), for the electronic image everything is possible; its capacity for metamorphosis infinite.[47] As in the act of creation in the Book of Genesis, the screen creates forms, beings and figures out of electronic chaos, apparitions having neither physical life nor remains. Fingers stretched toward the screen will touch, feel and encounter nothing but the crackling of static electricity. A book, on the other hand—an abstract medium if ever there was one—leaves its presence and the actual weight of its pages in the hands of the reader, even when he interrupts his reading.[48]

Video mediation is thus much more than a means of communication. In the form of the *audiovisual environment* for which it is being prepared, and which Thomson-Brandt assures us will guarantee our future "audiovisual happiness," it tends to become the most popular, and even exclusive, form of mediation. It is as if the Screen, instead of bringing us together in a cosmogony as masks and idols once did, sets itself up as the purveyor of our every need: information, entertainment, education, games, as well as the most domestic of activities (one can already shop for groceries via video, without money changing hands: accounts are debited or credited automatically). Video mediation has moved from the sectoral to the universal and permanent. Our behavior is being affected, and will be increasingly affected by it in the future. Piaget's epistemology was based on the idea that our actions precede our thought patterns and that the progressive interiorization of these actions results in the system of operations which we ultimately use—an irreversible system, unstructured at the outset, having a reversible balance as man grows from childhood to maturity.

Video mediation, through its holistic extensions, has tended increasingly to produce a *referential illusion*. Every language and every medium is characterized by the fact that there is always a gap between the sign and the reality it designates. Thus, the word *ox* is not the ox itself, but refers us back to the animal known by this name. On the other hand, the video image as it appears on television, especially in live broadcasts, does not merely designate an event, but gives the viewer the feeling, the sensation even, of "being there," of participating directly in the event. To a certain extent, video mediation does not reproduce events in the same way as the other media: it creates the very presence of reality, which in fact it produces. Hence, the feeling of "illusion-reality" which is its typical feature and makes it so omnipotent.[49] This is probably the first time in history that a phenomenon of this sort has occurred on such a large scale. The danger is all the greater because the "ghost-like" fascination of the electronic image is open to every use, every treatment[50] and every manipulation.

Where is the "audiovisual happiness" looming on the horizon of the new technology going to take us? A recent UNESCO study gives us one answer where television is concerned.[51] Its most striking conclusion is that: "There are unwritten laws which make for a structural similarity between broadcast schedules." This means that, regardless of the size of the country or its political regime, there is an international programming probability which can be observed in the form of common genres (news broadcasts, sporting events, dramas, news magazines, etc.), in similar scheduling and in the design and treatment of programs.[52]

The unearthing of these implicit rules reveals two things: it first demonstrates that they stem from the transmitters, not from the viewers, and that they are therefore the work of the production apparatus; and, outside the field of television itself, it shows the way the video market is going. Very probably it will adopt unwritten rules of its own which will make both audiovisual equipment and products look alike[53] and, much like the driving code for motorists, will increasingly impose their regulations, including sanctions, on the circulation of image-messages. The social energy of the media in closed circuit!

And Video Art?

After what I have just said, it is to be wondered whether there can still be a question of video art, or even whether I was wrong to give it such importance at the beginning of this study.

Video art is, in fact, in a doubly precarious position: first it uses electronic technology which, as we know, the market aims to monopolize; and second, video art, as we have seen, is not just a simple addition to painting or sculpture, which, in fact, it challenges on several points.

Until recently, art seemed to be both the favored expression of a civilization and, for us, the favored means by which we could learn about that civilization. Palaces, tombs, sculptures, paintings, triumphal arches, churches, cathedrals, stained glass windows were all "monuments" (as etymology attests) which preserved the collective memory by assigning it an identity of its own. Today, the electronic image, of which television remains the main purveyor, tends to replace this "monumental" memory.

This is no mere substitution; the very nature of our memory is in the process of changing. So long as memory was backed by stable materials (even the pages of a book), it proceeded by a slow process of sedimentation during which it retained those elements necessary to establish and maintain a certain unity: hence the gods, and later the kings, princes, and military leaders whose majesty or glory gave meaning to the community. But who today pays any attention to statues scattered throughout our cities, the heritage of only a recent past, which the motorist passes without seeing or merely takes advantage of an empty space around them to park his car? Gone are the days when heroes' features were cast in bronze, carved in stone, or painted. Long-term memory deteriorates when, as is the case today, durable materials are no longer used. The new memory—should we call it "short-term" memory, or present memory (the very terms are contradictory—is the one generated by the audiovisual media, and "the news" has become its supreme form)? The video screen must be "up-to-date," or nothing at all; hence it is voraciousness. But all media are based on signs, and it is no mere word-play to add that they are accustomed to separating the "signal" (worthy of the dignity of being signified) from that which is not (undignified and insignificant becoming synonymous here), *according to its own logic*.

The place and role sought by video artists in the new, yet already worldwide phenomenon of electronic image production is easy to guess. As we have seen, in accepting the challenge of the mass media they attack the code of those media as well as the market. Moreover, they explore what the mass media consider *insignificant*, of no value, and which they ignore, dismiss or neglect. Video artists thus live an interstitial adventure conducted through established filters, aware that it is in the "inter-regions" that will emerge or appear—not truth—but the unveiling of meaning; whereas the code knowing no meaning, merely ordains. In so doing, video artists adopt the fruitful scientific approach taken by Norbert Wiener, father of cybernetics, when he wrote ". . . the most fruitful areas for the growth of the science were those which had been neglected as a no-man's-land between the various established fields. . . . It is these boundary regions of science which offer the richest opportunities to the qualified investigator."[54]

Is this reason enough to devote one's interest to so marginal an activity as that of the video artists? It is as if democracy periodically felt the need to invent its own Socrates to rid itself of the assurance of the Sophists, whether man or media.[55]

There is always the hope, however remote, that if communication changes hands it may not only change its language, but its objectives as well. This seems to be happening in video art today. It is for this reason that I attach so much importance to the artists who—let me repeat one last time—are not content to add yet another means of expression to the arts, but who are moving fearlessly into the electronic environment, despite the manufacturers' claim that they alone can ensure our "visual happiness." Researchers, fighters too, and guides in any case, these artists remind us that happiness, while it can be helped by technology, is first and foremost the affair of men, and that meaning, all too often constricted or misused by the powers that be, can only be revealed by respecting differences. Which is precisely the purpose served by art and artists.

Notes

1. In the Eastern European countries, television is a service provided by the State. In Western countries, it professes to be a public service, but the monopoly system, with which only Italy has broken, makes it clear that the authorities are not yet ready to loosen the reins, however relaxed their actual grip may be. It may thus be said that the monopoly principle operates in various ways throughout Western Europe.
2. Briefly it may be noted that, although in the United States there is no monopoly, all broadcasting licenses are issued by the Federal Communications Commission (FCC).
3. Although radio waves remain the main method of transmission for the time being, cable use is continuing to develop and, with the advent of fiber optics, will most probably become predominant. This technological development offers the advantages of improving picture reception, increasing the number of channels available to the viewer, and controlling broadcasting—which represents a definite advantage for the authorities.
4. At the last Consumer Electronics Show (CES), which is reserved for the entertainment electronics trade, several dozen manufacturers displayed satellite television reception equipment. This equipment is becoming increasingly less expensive and an entire system can cost as little as $500. According to *Video Actualite* No. 19 (April–May 1982): "Already inaugurated, or about to be inaugurated, are the "Penthouse" (12/81) and "Eros" (1/82) channels broadcasting, as their titles might imply, adult programs; family programs (HTN 1/82), women's programs ("Daytime," spring 1982), and jazz with "Black Music Service" (8/82) and "Nashville Service". At the risk of generalizing, this development can only be regarded as revealing. For an overview of the satellite T.V. issue, see *Science & Vie*, No. 776 May, 1982.
5. Extensive documentation is available on this subject, which may seem paradoxical given the relatively limited number of cable installations so far authorized in Europe. On the other hand, the United States and Canada have been experimenting heavily in this field, and their systems are highly diversified (see footnote 6).
6. I do not refer to those cable systems which are content to relay the national programs with the occasional addition of other programs. By mesotelevision I am speaking of cable television at the community level, which only too rarely rejects the "ready-made" or "ready-to-watch" program.
7. The order to these terms may be inverted, especially as the effects of an anesthetizing ideology make us forget that macrotelevision in fact involves not a trilogy but a tetralogy, its advertising mission being as important as the others, if not more so.
8. "Proclamations from the magisterium, even if they be dogma in the strictest sense of the word, are in need of reinterpretation," Cardinal Deopfner, Archbishop of Munich, stated at the 1969 symposium at Coire attended by

some 115 bishops. "And even if, with the aid of the Holy Spirit, they contain a truth which transcends time, in other words, a truth which is objectively valid for all time, they formulate it in the temporal language. They are always proclamations subject to historical conditions, imprisoned in concepts conditioned by a given period of time or a given system and resulting from certain concrete situations or a specific event. This is why they always express the truth inadequately, from a given standpoint, in piecemeal fashion, taking into account certain aspects and with a specific audience in view. Thus, to understand them, one must be well acquainted with those conditions." (*Le Monde*, July 9, 1969).

9. Ludwig Bertalanffy, *Theorie Generale des Sustems physique, biologie, psychologie, sociologie, philosophie*. Dunod. Paris, 1973, pp. 38–39. "General System Theory," G. Braziller, Inc. N.Y., 1968.
10. Erwin Schrodinger, *Qu' est-ce que la vie*? (translation by Leon Keffler), Paris, Club Francais du Livre, 1949, "Sciences" serie No. 3 Leon Brillouin, *Vie, matiere et observation*, Paris, Albin Michel, 1959.
11. Andre Leroi-Gourhan, *L'Homme et la matiere* and *Milieu et technique*, Paris, Albin Michel, Coll. "Sciences d'Aujourd'hui," 1971.
12. This is the same "strangeness" we note in all illustrated film magazines, although in this case it is tempered a great deal by our familiarity with this art form. On the other hand, isolated images of video art not only furnish us no clue as to its nature, but by artificially equating it with painting, often give it—the ultimate absurdity—the appearance of the bastardized version of Monet, Rodin or Picasso.
13. (Editor's note.) Any book or article about video art illustrates this point. Reproductions of video art on the printed page therefore are an inherent distortion or even a misrepresentation because of this difference. This makes video art, at least in its fullness, in principle inaccessible to those people who do not have access to video exhibitions where the art is shown on video screens.
14. See René Berger, *Art(s) and Power(s)*, ICUS 1982.
15. This concept of property deserves to be discussed here. The plastic arts —painting, sculpture, and engraving—are unique in that they are dependent on *material support*. Unlike other arts, such as theater, ballet, concerts or poetry, which take place over time, they are by nature like chattels which can be bought and sold. These articles must, however, be of such size and weight as to be easily transportable. The black market in stolen art is proof of this. As far as I know, operas and ballets have never been stolen in the material sense of the word. Similarly, insofar as the plastic arts are concerned, it is impossible to steal a cathedral or church; on the other hand, there is a thriving market in stolen paintings and sculptures. The theft of art work defines the scope of the art market and the concept of property.
16. A history of video art has yet to be written. Information remains scattered in different locations, and is most often found in video art festival catalogues. However, in the special issue of *Cahiers du Cinema* entitled "Video Art Explorations," hors serie–10, Dominique Belloir offers an extremely interesting typology based on the technical possibilities of the medium. See also: Anne-Marie Duguet, *Video, la Memoire au poing*. Paris, Hachette (coll. L'Echappee Belle), 1981.

17. Nam June Paik, Videology, exhibition catalogue, Syracuse.
18. Howard Wise subsequently closed his gallery to devote his energy entirely to organizing the distribution of videotapes.
19. Cf. *San Francisco*, KQED-TV, Gene Youngblood, p. 281.
20. I attended the first women's video festival there in 1972.
21. An historical overview of video art in the United States is provided by Barbara London in "Independent Video: the First Fifteen Years" in *Art Forum*, vol. XIX, No. 1, September 1980, pp. 38–41.
22. In this connection, I am tempted to mention the course that I presented at the University of Lausanne in 1969 entitled *Esthetique et Mass Media*, at which were presented such innovative works as those of Rene Bauermeister, Gerald Minkoff, Jean Otth, Fred Forest and others. Cf. A. Willener, G. Milliard, A. Ganty, *Video et societe visuelle*, Paris, Tema, 1972. It was Guy Milliard, an assistant at the Lausanne Institute of Sociology, who showed us the first examples of the sociological applications of video.
23. Cf. Dominique Belloir, *Cahiers du Cinema*, "Video Art Explorations," hors-serie–10, *op. cit.*
24. Frank Gillette, *Video: Process and Meta Process*, Syracuse, N.Y. Everson Museum of Art, 1973, exhibition catalogue. Cf. Frank Gillette, *Between Paradigms. The Mood and its Purpose*, New York, Gordon and Breach, 1973, and "Interface" book.
25. The connection with "established" art forms such as the theater, cinema, or television is generally facilitated by the existence of explicit or implicit "genres," save in the area of purely experimental art. Video relates more specifically to the latter.
26. Wolf Vostell: "When I associate a sickle or a pile of shoes with a television set, I am not complying with a formalist principle in order to create a moving plastic object taking up space, but I am seeking a psychological truth which is dependent on the fact that the sickle or the pile of shoes only take on their real significance to the extent that they are situated in the context of a television program. The result is the creation both of a plastic reality (a sculpture-event) and a psychological revelation closely tied to the program being televised." Quoted by Dominique Belloir in "Video Art Explorations" *Cahiers du Cinema*, hor-serie–10, p. 27.
27. Such as Jorge Glusberg, director of the Centro de Arte y Communication (CAYC) in Buenos Aires, who for ten years has been organizing video exhibits regularly in the United States and Europe, as well as Japan.
28. My underling.
29. With the genius characteristic of some Japanese businessmen, the president of Sony, speaking of the Walkman, which was practically ignored by Sony's competitors at the time but whose subsequent rapid worldwide success is known to all, said: "Our competitors are wrong to think that we market new products. What Sony is interested in is the creation of new patterns of behavior." (I quote or paraphrase this from memory, which in no way detracts from the "divinatory" character of the statement, which is applicable to video as well.) Technology is only a production process if it is confined to the marketing of products; it becomes a factor of civilization (I reserve my

judgment on this point for the moment) when, as Mr. Aiko Morita says, it addresses our behavior patterns, and thus our imagination. It is the very problem of the new culture emerging from technology which is raised in essence here.

30. One example among hundreds: "JVC, champion of the video world," is the title. The text reads as follows: "Mundial '82. Behind the technical prowess of the world's greatest soccer stars, it's an important game for JVC video cameras and video-cassette recorders as well. JVC, the official Mundial audio and video equipment supplier, inventor of the VHS standard (accounting for over 80% of all sales in France), is right up there with the champions. In order not to miss a single play, even if you're not at home, you need a JVC video-cassette recorder. With our special "programming feature," it will record the game for you. All of them. And you'll be able to relive the "goal of the century" as often as you like. Or study it by speeding up, slowing or stopping the picture. All of this at the touch of a finger, or by remote control. JVC equipment, from the basic model to the tri-standard Pal, Secam and NTSC international model, are disconcertingly easy to use. A child can easily program the latest front-loading JVC7600S model for a two-week recording period. And behind JVC technology, JVC selection and JVC dynamics, lies JVC maintenance: over 700 JVC video specialists in France alone. Long live video entertainment! From the Mundial to Moliere: that's tele-liberation!"
31. To give the reader an idea of one of the more distinguished (!) advertisements of this sort, I quote: "THE STARS. The three most extraordinary X-rated video cassettes. Barbara Streisand and Jane Mansfield in HOLLYWOOD SPECIAL. PUBLIC TRANSPORTATION and *2 Plus 2*, with Miss France 1980." Under the suggestive photo is marked: "mail order: 320 francs apiece": and the address, which I do not feel obliged to repeat here.
32. There are many publishers specializing in the production of X-rated cassettes. Other firms, such as Sony, whose catalogues are more varied, endeavor to distinguish between "eroticism" and "X-rated films." (Does this distinction exist outside of these catalogues headings?)
33. In this broadly cultural area are to be noted the laudible efforts of some publishers, such as Video Encyclopedias, whose video magazine *Globe* aims, with its ethnological and other serious documentaries, to provide a high quality encyclopedia. On the other hand, how should we classify tapes like that of the Princess of Wales, alias Lady Di? For a few pounds a video cassette offers the English view of a "slice of the life" of the fiancee and then the wife. An hour of pictures, interviews, discreet emotion and even a few tears, and comments by her friends and even suppliers. The next cassette will be devoted to the baby. (Cf. 24 heures 15–16 May 1982).
34. This problem will undoubtedly be the subject of extensive studies, which have only just begun. I will limit myself to one example here; a fictional scenario appearing in *Video International* M2899-11 September '82, whose tone is apparent in these few lines: ". . . Remi walks over to his television and video set, which are sitting on his desk. He selects a cassette. This morning's program is to be "French History: The Hundred Years War" featuring his favorite actors Mickey, Donald, Goofy and Pluto who set out on a crusade and are pursued by the mean Wolf Brothers . . ."

35. "Your son is well behaved for the first time in his life: It's his first communion.
Your baby gleefully pokes his chubby fingers into your husband's eyes.
Your enemy is being buried.
So many unforgettable moments that you would like to keep forever.
There's only one solution: the video camera.
You manage to stay on your windsurf board for two minutes.
Your picnic is a huge success.
You give your grandfather a pair of roller skates for his seventieth birthday.
So that your friends will believe you, and to remember Grandpa's face as he opens the package, there's only one solution: Your portable video equipment.
Finally, if Nostradamus was wrong and you're celebrating your twentieth wedding anniversary, there's only one solution: video games."
Cf. VIDEOTECH, vol. 1, "Cameras et derives."

36. "In 1969 alone, ITVA International Television Association's members produced 75,000 industrial videograms which represented 25,000 continuous viewing hours, or more than all broadcast television in the United States!" (Charles A. Siebenaler, Medias electroniques).

37. An estimated 10 percent of all households in Japan and 5 percent in the United States already owned VCRs in 1981. In France there were only half a million of these machines (the Secam standard and 33.33 percent VAT constitutes serious handicaps).

38. Two French examples from among the hundred-odd experiments being carried out throughout the world: "Today, 2,200 individuals equipped with terminals share the 157 services offered by Teletel 3 V (from the names of the three communes involved in the experiment: Vilizy, Versailles, Villacoublay). The press, publishing, radio, TV, business and administrative and public organizations each offer roughly twenty different services (practical information, programs, prices, ordering services, and guide to legal rights and procedures)." *Le Monde*, 27 April 1982, "La Nouvelle Communication I. —Trois visages due videotex," by Richard Clavaud. In Nantes, "Telem has four thousand pages of information classified under nine headings, which range from municipal proceedings to senior citizens' services, and include social services, education, transportation, postal and telephone services, housing, leisure time activities and culture. As a start, fourteen terminals have been installed in various public places throughout the city; this number will soon be increased to thirty. Citizens of Nantes have free access to these terminals." "Parallel to interactive videotex, France has developed a videotext system broadcast on microwaves. Broadcasting authorities prefer to use the term "teletext" (not to be confused with Teletex, which is a transmission system using the telephone to transmit pages recorded on a word processor). Teletext and videotext are both considered to be part of what is called 'videography.' The French teletext system has been christened ANTIOPE (acquisition numerique de televisualisation d'images organizees en page d'ecriture)." (*Le Monde*, 27 April 1982, p. 13, "Les magazines ANTIOPE").

39. "Enter the Sony world of prerecorded cassettes" urges the Sony Video Club, which already has close to 100 outlets in France. The Video Charm Club, on the other hand, proclaims itself—unafraid of the bad pun—the first "videocassex club" and offers a magazine, *Video Prive*, in addition to its catalogue.

40. *VIDEOREVEUE*, a Swiss magazine, repeats: ". . . at a time when everything happens on the screen, video has become a modern necessity." The text of this advertisement should surely put the "cultural" community on the alert!
41. My underling.
42. "The Festival of Sound and High Fidelity became this year (Paris 1982) the Audio and Video Image Festival." *Le Monde* commented on this as follows: "The sound festival is dead: long live the peritelevision festival. To open an article like this is surely to be told by the organizers that I am exaggerating. They are probably right. But perhaps it is only running the risk of being right too soon."
    "In any case, a characteristic feature of the advertising, is that this year's Festival has not been given a number. Last year, the Palais des Congres welcomed the twenty-third International Festival of High Fidelity Sound. This year it welcomes, from 7 to 14 March, the International Audio and Video Image Festival. The eye has entered the kingdom of the ear in force, an entry accepted with poor grace by the latter. French television's Channel 1 (TF1) has taken the spot traditionally reserved for Radio France. This evolution was no doubt inevitable." *Le Monde Dimanche*, 7 March 1982.
43. The still-frame feature is a convenience which in no way changes the nature of the phenomenon. Moreover, all images, whatever their source and whatever form they take, become "standardized" into electronic images to fit the screen.
44. To quote *Videorevue* publicity once more: ". . . at a time when everything, or virtually everything, happens on the screen, video has become a modern-day necessity." Americans who spend an average of five or six hours in front of their screens each day, have for long been aware of this. Indeed, today, most people cull their world news and entertainment from electronic images. Neil Armstrong's landing on the moon, which was seen live by almost one billion viewers, is perhaps the first "monument" of this new reality, which is echoed by the "divinatory" advertising for the 1982 World Cup: "Our goal is to bring all World Cup goals into your home." *Video Magazine* has already taken over from the news programs: for example, "1981, L'ANNEE CHOC," the first video news review, in fifty-six minutes, covered the main stories of that eventful year, month after month, combining short sequences of world news with brief human interest features. Among the 120 events covered were the El-Asnam earthquake, the attempted assassinations of John Paul II and Reagan, the IRA suicides, the release of the American hostages, Borg at Roland Garros, etc.
45. Roger Dadoun, *Le Plus-que-parfait due cinema et l'impartait de la television*, in Revue L'ARC No. 50 Aix-en-Provence, Gutenberg, pp. 32-40.
46. Unlike the other media, and what is traditionally called art, videograms and TV broadcasts only exist when viewed on the screen; materially, the tape or cassette furnishes no idea of the image, perception is indissociable from the technical operation.
47. Computers are already being used to generate images which have nothing to do with reality. This activity has been further developed through synthesizers, and will find even more applications with the advent of the digital image.

48. At most, one could compare video with the cinema, whose images have no more physical presence on the screen than do video images. But the cinema and the video "situations" are quite different. The former implies an occasional decision which involves leaving the house, while the latter is a part of our domestic scene and a familiar element in our daily lives.
49. It has often been observed that American children, by constantly watching television, are becoming increasingly unable to distinguish between information and fiction, even accepting commercials as reality.
50. For example, it was impossible to distinguish during the moon landing broadcasts between what was actually happening and the explanatory sequences which were indicated as such by the word "simulation" printed on the screen.
51. *Three weeks of television*, UNESCO (Cultural co-operation series), Paris, 1983, p. 47. A team of researchers studied the television programs broadcast during a three-week period in seven developed countries as varied as Belgium, Bulgaria, Canada, France, Hungary, Italy and Japan.
52. I stated this myself in *LA TELEFISSION: Alerte a la Television*, Paris, Casterman, 1976. The UNESCO study confirms my statement. (Cf. Roland Cayrol, Le Monde, 13 May, 1982).
53. As already demonstrated by the prerecorded video-cassette market and what is known of the program being prepared for television satellites.
54. Norbert Weiner. *Cybernetics, or Control and Communication in the Animal and the Machine*. Paris, Hermann, 1958, pp. 8-9.
55. This aspect has been studied at the political level in particular by Anne-Marie Duguet in *Video, la Memoire au Poing*, Paris, Hachette, 1981, L'Echappee Belle series. Therein, the author describes a number of experiences, so there is no need for me to do so here. It is surprising, however, that this work ends with an "unconditional" condemnation of video, which it accuses of having let itself be taken over by what we have seen of the market, but I still refuse to give in to it (a profession of faith?). I might add that although Socrates was condemned, Plato's *Dialogues* live on.

Addendum. Dany Bloch, *L'art Video 1960–1980/82*. Ed. Flaviana, Locarno.

ART VIDEO, Retrospectives et Perspectives. Charleroi, Palais des Beaux-Arts, 1983.

René Berger, *Videokunst oder Die kunsterlerische Herausforderung der Elektronik*. in Bettina.

Gruber–Maria Vedder, *Kunst und Video*, DuMont Buchverlag, Koln, 1983, pp. 55-61.

The Cultural Impact of Video Art: A Discussion of René Berger's "The Challenge of Video Art"

Andree Paradis

Introduction

If, as the *happy few* believe, we are presently completing the last phase of a long process identified as the convergence of all cultures towards a planetary culture, the contribution that art is making, and more specifically video art, is fundamental to this unifying process. It has often been said that art, like science, is a universal transcultural value. It might deal with the imaginary world, and be less rational than science; it still emerges from a basic need of every human being to grasp the reality of his environment, to try to give meaning to it and to communicate his beliefs through words or specific art forms. That unique experience of art is thoroughly analyzed in the philosophical text of René Berger's *The Challenge of Video Art*. It is philosophical in the sense given to philosophy by Vladimir Jankélévitch,[1] which consists in giving all possible thought to a question, and doing this thoroughly from all angles, no matter what the cost. When it is time to sort out from the entangled, to free from intricacy, Berger relies on his scientific knowledge to build a fresh and pertinent approach to the meaning of art, its evolution through history and the challenging new fields offered to it by modern technology.

He has clearly established that communication has always been a vital function of any civilization or society which confirms the ability to *transmit*, *disseminate* and *control*. States were literally built around networks of communication, such as China, Rome and the Inca Empire, to mention a few. As a result, those famous roads gave birth to exchange of all sorts—goods, objects, ideas —and art from the beginning was part of these trading values. Then came the revolution created by Gutenberg's discovery[2], which we still benefit from; and the era of the railroad which opened a new field of communication; followed by all the great discoveries of the twentieth century, giving us unlimited powers of production. But as Ortéga y Gasset notes in *Révolte des masses* (1926), our era offers those fabulous powers, but does it know what it really wants? Apparently everything seems under control, but it is without direction. To be the master of oneself is, and will forever be, a crucial problem. The artist, I believe, is always close to that searching process and as the goal, stated at the beginning, is the working towards a universal culture, he is one of the most active instruments in the process, with or without his awareness. He will, at the same time, continue to assure the necessary diversity of all cultures by highlighting their universal traits.

The Electronic Image in Our Daily Life

It is acknowledged by all that no other system of communication has had so much impact — whether social, political or cultural — as television. Radio, like telephone, relies on the voice. We seldom hear of radio art or telephone art, although I feel strongly that oral transmission is a form of art and that radio, through its powers of evocation, has a potential of environmental presence comparable to none. Walking through the bush or skating in the city plugged into the transistor is a familiar sight these days. On the other hand, according to René Blouin, "A Canadian video artist, Tom Sherman, who questions our attitude towards television, demonstrated well in the videotape *East on the 401*, that it is impossible to drive a car while looking at television. Two cases of passivity caused by the domination of technology, two cases where the level of attention required provokes hypnotism."[3]

It is a fact that television relies on the visual image and that its influence grows continually. Images are being transmitted at a rate that cannot be measured and already the dread of being caught with a shortage of images is building up. Television series, soap operas, variety shows, game shows, and films, which occupy about two-thirds of programming all over the world, whatever the political regime or economic condition, bring home "a permanent public performance." It is difficult to measure the impact of the constant bombardment of images, but it is evident that it has an influence on the "collective imagination."[4] Strong poles of diffusion like the United States are world conveyors of occult messages and standardized formula, leaving the cultural diversity that is much needed to fertilize this coming universal civilization in danger of having less and less meaning because its powers are relatively smaller. Television, patented in the twenties, reached the public in the early fifties. It has become a strong institutionalized instrument of communication and its monopoly of what is to be shown and when went unchallenged until the recent development of new technologies (cable, satellite and portable video). Very little space was given during that time to art experiment with television. Of course, good series were produced on art history and programs of information on artistic events; also, interviews with artists, but creative experiences with the cathode tube were limited, no doubt because they would have proven too costly in view of the relatively small market. One name, however, has to be mentioned. Jean-Christophe Averty, in France, has succeeded in creating sheer poetry with images, and he is recognized as having done for television what Meliès did for film.

It might be a bit strong to state that the revenge of artists, frustrated in not being more involved with television, came with the birth of the video. But the intensive development of video as an artistic medium has occurred over a relatively short period of time, a little more than two decades. Not all practices of video, however, concern art. Again, only a small portion of this enormous new field of communication, with many practical aspects, is reserved for art, but it remains a space of possible conquest and already artists like Nam June Paik, Andy Warhol, Robert Rauschenberg, Bob Wilson and Les Levine from the United States; René Bauermeister, Gerald Minkoff and Jean Ott from Switzerland; Michel Jaffrenou from France; and Eric Cameron, Tom Sherman, Marshalore, Robert Morin, Colin Campbell and General Idea from Canada, to mention only a few, have mastered it.

A confusion existed from the beginning as sociological artists like Fred Forest (France) and Pierre Falardeau (Canada) took hold of the medium and were able to reach a large audience of their followers but, (alas!) it must be added, bored many others. It became quickly evident that true video art does not thrive solely on political and social content. Often it specifically criticized television itself, especially its commercial formats, and sometimes it dealt critically with more traditional art forms. At the beginning of the sixties some wide experimentation with images was popular, but attracted serious criticism for being "tedious and self indulgent." The artist's use of the medium, small format, portable closed-circuit video, to seek a reality not evident on the surface of things, was described by Peggy Gale,[5] "as a kind of personal journalism that probes the psyche through associations, memories, juxtapositions of facts and intuitions" or, as Berger also underlined in reference to Norbert Weiner, father of cybernetics, a probe into a world of things considered of less importance, "the neglected no man's land between the various established fields" with hidden values to discover.

The Ills and Hopes of Video Art

It is clear that video art has not attempted to replace any form of existing art. First of all, it has a very short history, and the freedom of the practicing artist is quasi-total. Not being subjected yet to established "diktats," he is experimenting to find a path through a new visual language that will expand knowledge and modes of perception.

In *Independent Video: The First Fifteen Years*,[6] Barbara London explains that in the late seventies, video in the United States declined slightly and that it was only partly due to the recession. Some artists were frustrated, showing their works repeatedly to the same local, closed audiences, while others became discouraged when their videotapes or larger installation pieces did not sell. On the other hand, the poor image of portable equipment drove some artists to want to produce better images; but the prohibitive cost of the best cameras and most sophisticated editing facilities would not allow them to do so. In the eighties video equipment is being considerably refined. Sony Corp. has developed what it claims is the world smallest and lightest home color video camera. The CCD-G5 weighs only 2.24 pounds and uses a device which transforms light into electric signals, replacing the camera tube. Producing videotapes will then become easier and artist videoproducers will develop quality images inexpensively with their own equipment. The old problem of distribution remains more serious. Up to now only a few museums, galleries, and artist-run centers have been able to show videoworks to the public and establish comprehensive video archives. Some of them have already taken on the responsibility of protecting and maintaining video material for future audiences.

The problem of the individual viewer is of another nature. The valid criticism of some of the artist's video works of the sixties has not helped to build a large audience for the recent excellent productions available. A more comprehensive interpretation of the seventies' art activity has yet to be articulated. Little consideration has been given to the viewer who feels completely lost most of the time in front of a static ambiguous image that forces him to his last entrenchment. Very few critics have found the simple, correct approach and vocabulary that would help to bring the public and the artist-videomaker closer. When we admire how far back in time André Malraux takes his readers to situate an artist like Piero della Francesca; when we appreciate all given material and information that, in museums, usually accompanies even the smallest exhibition, it is difficult to understand why practically no introduction and no adequate information is given on video art. The art discourse so far concerned with theoretical aspects of video art has been of little help in understanding this new creative medium. Some artists like Tom Sherman have taken the problems into their hands: they produce the written part of the experience very successfully in the form of catalogues, which become a link between the literary world and the visual one.

In Canada, my own country, video art has received considerable support from the Canada Council. In the early seventies the Council adopted a policy of funding artist-run centers in many regions of the country with a basic level of equipment that would permit diversity, flexibility and accessibility[7]. The primary function of these centers has been to allow artists to work directly with video as a medium of artistic expression. These facilities have also permitted the integration of video with other art disciplines. The accomplishments of the video program within a few years of operation have been dramatic, and directly responsible for the evolution of Canadian video art to a standard of international acclaim. In Canada, since the early seventies, video art has enjoyed a very high profile within the arts, with consistent representation at major international art exhibitions.

The Canadian Pavilion at the 1980 Venice Biennale, organized by the National Gallery of Canada, consisted entirely of works by Canadian video artists. Canadian video artists also made a strong showing at the San Paolo Biennale (1978), and at the Paris Biennale (1980), and took an important part in the Canadian exhibition at the Berlin Academy in 1982. All the artists represented at these prestigious exhibitions were supported in their professional development by the Canada Council, and their works were largely produced through facilities subsidized by the Council. The funding level of the Canada Council has grown from $53,000 in 1973–74 to $428,000 in 1979–80. It covers aid to the individual to undertake video productions, aid to organizations for video productions and installations, and also provides funds for national and international distribution. A modest level of funding is available for non-profit centers which present exhibitions of video works. Special projects, communication funds and equipment funds are also subsidized. Sales of Canadian video tapes are increasing, and one acclaimed event was the recent purchase by City-TV of *The Television by Artist* series produced by A Space Gallery in Toronto.

The hopes for tomorrow's video art can be far reaching. In the next decade, videodisc and video installations will provide a wide range of choices and the videotape done by artists will have more and more impact in our cultural environment.

Notes

1. Vladimir Jankelevitch, Béatrice Berlowitz, *Quelque part dans l'inachevé*, Paris, Ed. Gallimard, 1978.
2. Marshall McLluhan, *Gutenberg Galaxy: The making of typographical man*, NAL, 1969.
3. René Blouin, *Ingenieries culturelles*,, text to be published in *Vie des Arts*, Vol. XXVIII, winter 1983–1984.
4. Philippe Wade, *L'audiovisuel, faux débats et vrais enjeux*, Paris, Ed. Fayard, 1983.
5. Peggy Gale, *Vidéo, regard introspectif*, in *Vie des Arts*, Vol. XXI, Spring 1977, p. 17.
6. Barbara London, *Independent Video: The First Fifteen Years*, in *Artforum*, September 1980, p. 39.
7. *The Canada Council's Programs in Film and Video*, Ottawa, May 1981.

The Children of the Tube

Frederick Pohl

René Berger's admirable paper says nearly all that can be said about the present state of video art. It is a very impressive work, not least because it appears to me that it is a very courageous man who will venture to predict the future forms of any art.

Of course, the shape of any kind of art is, to some extent, limited by the available technology. Palestrina or Mozart could not have written for the Moog synthesizer. In fact, they could not have written for the large modern symphony orchestra of one hundred instruments or more, since the technology of the musical eras before—let us say—approximately the time of Beethoven had not mastered the problem of tuning so large an array of instruments to play together.

Nevertheless, the nature of artistic genius is such that he who possesses it finds a new way to exploit the available technology. The time when that occurs is not necessarily coincident with, or even fairly soon after, the time when the technologies become available, either to use or to inspire. Picasso's painting could have been anticipated centuries earlier, and certainly no later than Leeuwenhoek's lenses or Newton's prisms. But it was not, because Picasso himself did not then exist. The artist must obey certain rules, but the genius among artists is he who discovers new rules arising from opportunities no one else has perceived. I am sure it will be so with video art.

So the great revolutionary developments in the art of video cannot, I think, be foreseen—at least by me. I suspect that any person skilled and wise enough to predict them would also be skilled and wise enough to practice them, and so no "prediction" would be involved. What can be foreseen, I suggest, is only the more obvious and even trivial prospects— but nevertheless prospects which have very significant implications for both the forms of art which ennoble our lives and the more mundane aspects which shape the ways we live them.

The recently released Woody Allen film, *Zelig*, seems to me to foreshadow some interesting developments. We have all seen so many wonderful special effects in cinema and video emanating from the workshops of Steven Spielberg, George Lucas and a dozen other masters, that we expect marvels. But *Zelig*, I think, demonstrates a capacity I had not hitherto expected, of putting the face and form of any human being into any actual record of any event. So far the process has dealt only with black and white film, but surely color is only a step away. In *Zelig* Allen appears taking his turn at bat as a member of a great historic Yankees baseball team, as an intimate of President Hoover, and in many other roles as he is electronically inserted into contemporary news films of half a century ago. If it is possible to tell that the film has been tampered with, that possibility was not evident to me.

What a way to rewrite history! Now that we have passed George Orwell's year of 1984 we can see that his vision did not go nearly far enough; Big Brother need not bother to reprint newspapers with the "correct" stories in them when he could so easily electronically revise actual news films to show whatever he wished. Those faked Soviet films of past May Day parades in which officials in disgrace are simply blotted out, leaving a visible gap, could be replaced with wholly authentic-seeming newsreels with anyone they like in the position of honor. What a potential for propaganda: Imagine Adolf Hitler with such resources at his command—perhaps staging some vicious anti-Semitic scene, and electronically placing Winston Churchill or Franklin D. Roosevelt in the middle of it. Nor need such uses be only political; the potential for blackmail is immense. I am reminded of the film *Caligula* in which, it is said, once the featured actors had said their lines and departed, stand-ins were hired to perform pornographic acts. Their faces were never shown, and they were intercut in the existing film to make it appear that the stars themselves were engaged in the acts. Yet, with the technology used in *Zelig*, the very faces of the stars could have been shown, as indeed, the same sort of thing could be done for any citizen confronted with a divorce or custody hearing.

I do not, of course, suppose that the future prospects inherent in electronic manipulation of images need all be so worrisome. There is surely an immense potential for good, especially in the field of education. A combination of techniques—perhaps time-lapse films with computer fill-ins, maybe with some animation—could give us an opportunity I cherish, namely, the chance to see things that are real but which no one has ever seen: the drift and collision of tectonic plates as the continents are formed and move; the development of a child, its relative proportions changing as it grows; the evolution of primitive life forms to present ones—perhaps even extrapolations beyond the present so we can see what the future holds for our descendants.

I think there are also great possibilities in the video representation of non-visual perceptions. We have become used to seeing radar screens which penetrate fog or darkness to show us something our unaided senses could never see, Tank commanders and machine-gunners see by infra-red, submarines display their environment by sonar. The human eye is blind to all of these, but video makes them all visible.

Even more interesting is the false color representation of things we can actually see, but which our optic systems cannot entirely decipher. Thermal photography reveals cancers and foci of infection in what, to the unaided eye, is healthy flesh. Satellites photograph a hemisphere, and by displaying the results in false colors let us identify mineral deposits or insect invasions or drought-threatened areas. Here are great opportunities for both creative art and illustration. Art, in that the bringing out of hidden meanings by means of non-visual perceptions and false color may offer as revolutionary a new graphic style as those of Picasso or Dali. Illustration, in that we can display "sights"—perceptions, rather—that by definition cannot be "seen" in any other way. What does the inside of Mammoth Cave look like to a bat? Sonar video can show us. What are the sources of the sounds that frighten a small child alone in his bed in a darkened room? We can make a picture of those sounds to show him, and perhaps to ease his fears.

Yet I must say that, exciting and informative as Berger's paper on video art surely is, the sections I found most challenging were those which dealt with the nature of video itself. I will forever be indebted to Professor Berger for the fine and insightful phrase "ballistic information," which so clearly describes the nature of almost all of present "communication" by means of television —and, indeed, of a major fraction of current communication by any medium. As he says, "Ballistic information is more concerned with hitting its target than with establishing communication, which implies exchange."

"Exchange" is a need which must be met in order to produce true participatory government. We see very clearly that most exchanges between the governors of any modern society and their governed are actually exchanges of "ballistic information." Presidents and congresses and other rulers hurl laws and decrees and acts at their people. The people respond by returning ballistic ballots—in a democracy like the United States they do; in other parts of the world the ballistic information may take the form of real projectiles from real guns in a revolution. There are many intermediate forms of "communication" from governed to governors—strikes, demonstrations, letters to the president, petitions and a thousand others. But they, too, are ballistic in nature because the element of interaction is missing.

And yet there are developments which promise to restore that interaction, and perhaps return the complex series of mediations, negotiations and compromises we call "governance" to the ideal of the Greek agora or the New England town meeting. It is in the medium range of video—that area Berger has defined for us as "mesotelevision"—that we can find the first signs of this. Cable television has provided access to individuals and groups on an almost censor-free basis. The access is not complete, but it is a beginning. Taken along with that other experiment in interactive television, Qube, in which the viewers have an opportunity to express opinions and reactions in real time by transmitting signals from the broadcasting studio, it seems that the technology is already in place for a non-ballistic form of television. I have written a science fiction novel in which I depicted a city where all major policy decisions are made through an interactive video system, combing features of Qube and community access TV to produce a forum for mass participation in the affairs of the state. As Berger says, "the debate . . . becomes a *discussion*. A new type of creativity appears. . . . Rhetoric gives way to exchange." And the hurling of "ballistic information" is replaced by an attempt to achieve consensus.

As you can see, I find Berger's paper extraordinarily stimulating in many ways. I feel that art of any kind, video or otherwise, is unlikely to flourish unless we are able to find means of improving communication and, in particular, improving the interactive communication between governments and the governed—and here I believe "mesotelevision" can make an immense contribution.

Chapter Nine

Art and Technology in the Future

Art and Technology in the Future

Magoroh Maruyama

The task assigned to me by the committee chairman has the following specifications: (1) to discuss art in human environments which result from technological advances, such as outer space communities, subterranean communities and underwater communities, including the role and the use of art in such environments; (2) to focus on a relatively near future, i.e., on the first generation of such communities, and to exclude discussions of more distant future when the inhabitants of such communities will be those who are born there and consequently feel more at home there than on the surface of the earth, as well as a still more distant future when the human body itself will have undergone some evolutionary changes in nonterrestrial environments; (3) to put an emphasis on material details, such as color, light and other sensory characteristics of such environments, and other material conditions which affect the nature and the use of art.

For the past decade or two, there have been numerous technological, physiological, psychological and sociological studies of such existing and projected communities, much more than can be discussed even within the limited specifications of this article. Among the various types of extraterrestrial, subterranean and underwater communities, there are many similarities and differences. For the sake of conciseness, I have chosen the following format in writing this paper: to take one type of extraterrestrial community as an example and discuss its many aspects; then to point out the differences between this type of community and the subterranean and underwater communities. Consequently the section on extraterrestrial communities is much longer than the sections on subterranean and underwater communities. But the reader can realize that the discussions on many of the aspects of the extraterrestrial community apply to subterranean and underwater communities as well. Some of the differences between the three types of communities are: extraterrestrial communities will be much more spacious than subterranean and underwater communities; the subterranean communities will have the smallest possibilities for landscaping; the underwater communities will have the largest visual effects from the seasons and the movement of the moon.

Outer Space Communities

The population of the first generation extraterrestrial communities may vary from a few persons to 10,000 persons per unit. The communities of later generations can be much larger. The design of several first generation communities has been carefully studied in the NASA/Stanford project since 1975.

Let us take the Stanford Torus Model as an example (NASA, 1976), in the design of which I participated. Its overall shape is like a bicycle wheel with a diameter of approximately 1600 meters. The diameter of the "tire" is approximately 130 meters. It accommodates residential areas for 10,000 inhabitants, agricultural areas to feed them and to oxidize the air, and industrial areas. The "wheel" rotates like a bicycle wheel, one rotation per minute, to create a centrifugal force equivalent to the earth surface gravity at the outer periphery of the tube. Areas closer to the center of the wheel will have weaker "gravity." The center of the wheel is a weightless area. Other non-rotating weightless areas for special industrial use can be attached to the wheel. Design of extraterrestrial communities which have a spherical form, a cylindrical form, and other forms have also been studied in various projects. In all cases, the communities rotate to produce artificial gravity.

Extraterrestrial communities of these various forms will have the following characteristics: (1) Abundant sunlight and solar energy; (2) Spaciousness: extraterrestrial communities are more spacious than subterranean communities and underwater communities for the following reasons: (2a) The costs of construction

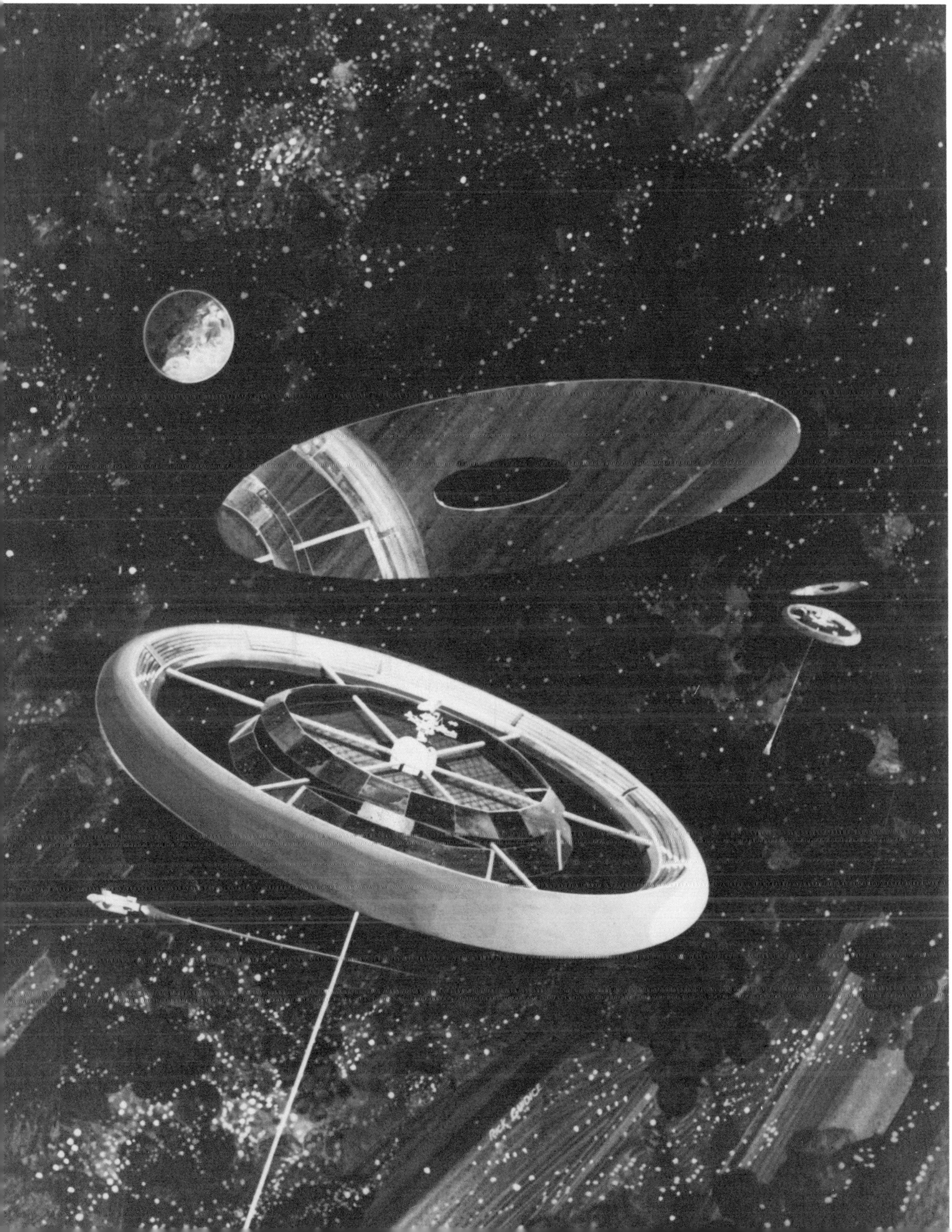

Fig. 9-1 Drawing of proposed space colony, showing the colony and its reflecting mirror (NASA).

of extraterrestrial communities are affected more by the mass (weight) of the materials which compose the physical structure than the volume enclosed in it. For the first generation communities, the materials which go into the structure must be lifted up from the moon surface and the earth surface against the lunar and earth gravity. The transportation costs are enormous. Later communities will use more asteroid materials which involve less transportation costs. On the other hand, there is plenty of empty room in outer space. Physically larger communities are more efficient in terms of the volume-per-weight ration. Furthermore, spherical, cylindrical and torus forms have a greater volume-per-weight ratio than high density forms used in the subterranean communities and the sheet form used in the underwater communities. Subterranean communities must economize the volume because of the excavation costs, and the underwater communities must be vertically thin and horizontally spread in order to avoid high water pressure which rapidly increases with depth. (2b) Each extraterrestrial community includes agricultural areas for wheat, vegetables, fruit, cattle, rabbit and fish, not only for food production but also for oxydation of air. Humans and animals convert oxygen into carbon dioxide, which must be re-converted back to oxygen by plants. On the other hand, subterranean communities and underwater communities can obtain food from food-growing localities, and take fresh air in from the earth's atmosphere. (3) Diurnal and seasonal cycles can be artificially created; (4) there is a lack of weather (5) and a lack of sense of movement, both inside and outside the communities; (6) there is a lack of environmental noise; (7) and there is a scarcity of certain types of materials which cannot be produced within the enclosed ecosystem of each community.

In order to discuss aesthetic considerations in more detail, let us return to the Stanford Torus Model.

From the aesthetic point of view, it is important to note the following:

(1) The longest line of sight inside the tube is about 600 meters. (2) Three residential areas are separated by agricultural areas. (3) The ground is the side of the tube most distant from the center of the wheel. (4) The "daylight" (sunlight) reaches the ground through the "ceiling" (the side of the tube closest to the center of the wheel) via several sequences of mirrors, and the direction of

the sun as seen in the mirrors is stationary with respect to the ground. Nighttime darkness is achieved by closing the blinds. (5) The volume of the air inside the torus is too small to achieve the "blue sky" effect, and light diffusion is achieved from the surfaces of objects. Consequently, shadows will be very sharp, shades will be dark, and color hues inside the shades are strongly influenced by the colors of the objects from which the sunlight is bounced off. One way to counteract this tendency is to place glazed glass (cloudy glass) in the ceiling. This will diffuse the light. But the color effect will be very different from the blue sky effect. In the blue sky effect, colors in the shades are tinted toward blue, while colors in the sunlight are tinted toward red. In order to produce the same effect slightly blue glazed glass should be interspersed with slightly orange clear glass. (6) The agricultural areas appear mostly green. However, the agricultural areas are likely to consist of multilayer shelves rather than one flat land surface, and include ponds and pools to raise fish. (7) The volume of the air is not sufficient to produce cloud and rain. Irrigation systems must be used. (8) The torus is efficiently air conditioned (without air conditioning, the entire torus will become a hot greenhouse). (9) Since there are no rains, strong winds or extreme temperatures, there is no need for roofs or walls, and the architecture can have a very open style. Buildings will not resemble gothic cathedrals or skyscrapers, and may tend toward Polynesian huts without rooves, but multi-tiered. (10) The first generation extraterrestrial communities will use mainly lunar materials for its construction. Aluminum and silicon are abundant on the moon.

Therefore, aluminum and glass will be extensively used. Wood will be scarce. On the other hand, bamboos grow faster than trees, and bamboo materials will be much used. Petroleum is non-existent. Therefore, plastic and synthetic fibers will be scarce, unless vegetable oils can be converted into synthetic materials. There may be a revival of the use of cotton, silk and wool, along with the use of fibers from bamboo, straw and agricultural plants.

First generation residents of extraterrestrial communities will want to make their life as similar as possible to the life on earth, just as first generation immigrants on earth tend to preserve their home cultures. Chinese immigrants in San Francisco preserved the pre-communist Chinese culture long after the Communist Revolution. For example, Communist China simplified the Chi-

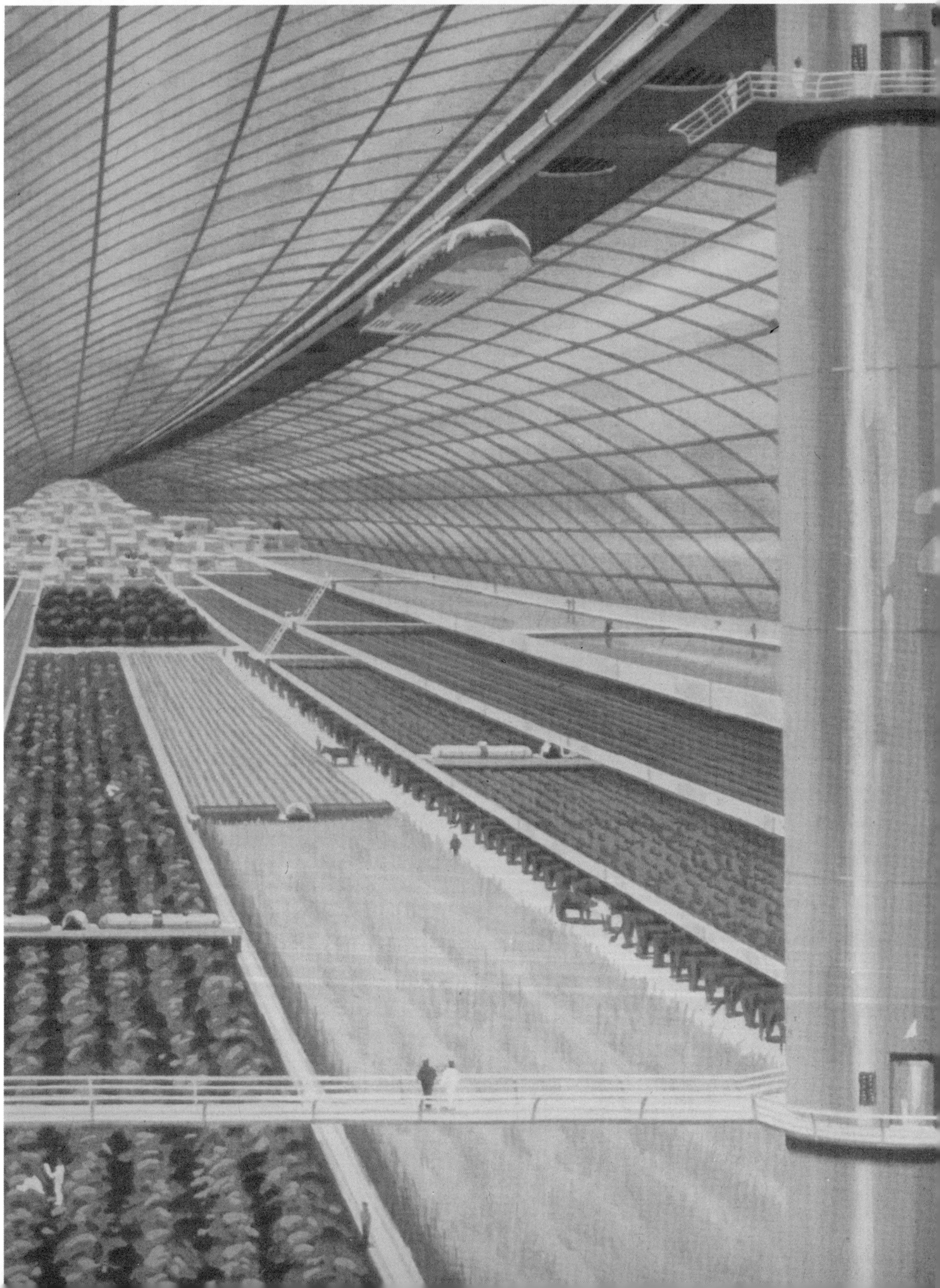

Fig. 9-2 Drawing of the interior of a proposed space colony, showing the agricultural area (NASA).

nese writing system. Today the printing types of old Chinese letters are available in San Francisco, but not in China. Similarly, the French-Canadians in Quebec speak old French, not modern French. It may also be true that first generation residents of extraterrestrial communities will remain culturally more traditional than those who stay on earth and develop new art forms.

Much of the technology will be used to preserve or re-create earth-like environments and art forms. One of the things the first generation extraterrestrial residents will miss will be an earth-like landscape. There are several ways to deal with this problem. One is to create miniature landscapes and gardens. On earth there is a specialized field in the art of landscape simulation, developed mostly by museums of natural history. The number of expert artists in this field is very small, but this art can be revived in extraterrestrial communities. There are also landscape architects, especially the Japanese garden designers who are experts at creating miniature landscapes. The other solution is to use new technology, holography, for example, to create three-dimensional landscapes without solid material. Since outer space has much empty space, very large holographic images can be projected between structures within the enclosed community.

Even though many of the first generation extraterrestrial residents may tend to preserve terrestrial art forms, there will be others who will experiment with novel art forms. One new opportunity is in the weightless environment. This will open new avenues for at least two fields of art: sculpture and dance. Sculptors will be able to build gigantic sculptures. Dance in the weightless environment will be completely different from the terrestrial dances. Floors cannot be used to jump from or to stabilize the body on. Dancers must use one another's bodies to pull to or push from. Residential areas are provided with artificial gravity.

The living environment in extraterrestrial communities will be characterized by: (a) lack of noise; (b) lack of movement.

Terrestrial life is full of all sorts of noises. In urban areas there are noises from automobile traffic, airplanes and road repair. In rural areas winds may rattle tree leaves, or birds and insects may sing. In contrast, in extraterrestrial communities there tends to be only a small volume of vehicular traffic, no wind or rain, and very little need for road repairs. Birds and insects may sing, providing pleasant sensory input. (One researcher discovered that the brain of persons who grew up in a Japanese culture processes bird and insect songs as well as sounds of wind and waves in the dominant cerebral hemisphere as if these sounds were human voices; while the Europeans' brain processes these sounds in the non-dominant hemisphere.) (Tsunoda 1978). In the relatively noiseless extraterrestrial communities, auditory aesthetics will become important not only to provide pleasant experiences but also to prevent sensory deprivation.

The environment in extraterrestrial communities will have relatively few moving objects. Even though there is air, there will be no strong wind. Tree branches will remain motionless. There will be no fast-passing clouds, even though in larger communities of later generations with a large volume of air, cloud formation may occur. The first generation communities are likely to be orbiting around the earth. Their orbital velocity is fast. But because of the lack of a background landscape against which the movement can be directly perceived, the community will appear to be hanging motionless in a vast, empty space.

Each of the extraterrestrial communities will be physically isolated from one another. However, travel between communities will be very convenient. It takes very little energy to move a vehicle from an orbiting community to another since there is no need to fight gravity. A slight push from a community will enable the vehicle to continue its flight to another community. However, the use of any propulsion device on the vehicle itself, whether it is for trajectory correction or for landing deceleration, is to be avoided as much as possible because, in outer space, there is nothing to push against, and the propulsion device must eject something, whether it is a gas or solid material. It is undesirable to waste materials, because all materials must be lifted up from the moon or from the earth at high transportation costs. This means that the vehicle will travel at a low velocity, and the travel time will be long. For busy persons, teleholovisits (televised holographic communication) can be more convenient than physical visits which involve travel.

There are several psychological problems inherent in an artificial, controllable environment such as extraterrestrial communities. One is that of sensory deprivation, solipsism, low tolerance for something going out of control, and autism. Another serious problem is narcissistic, hierarchical rationalism. Art can contribute toward reducing some of these problems.

Solipsism is a state of mind in which everything appears to be in a dream with nothing existing outside your brain. Even your friends are a part of your dream. They don't exist. Since everything is your dream there is no need to make any effort. Nothing is real. You become very apathetic, and very lonely.

The child learns to deal with reality, precisely because reality is different from imagination. When you are hungry, you cannot eat a toy. You have to learn to find food, buy food or cook food. In a highly controllable environment where temperature, humidity and other conditions can be set at will, and where everything is quite predictable, you may get a feeling that there is no difference between what you think and what actually happens. Under these conditions some individuals may fall into solipsism.

Another danger is that persons who are used to an easy life may collapse under the weight of a very small difficulty or frustration. They may succumb, explode, turn aggression toward others or toward themself, or completely withdraw to autism. Art, if designed well, can contribute toward reduction of these problems. The important point is that *art can bring more reality into life*. Professor Daniel Charles has already discussed the point that art is an integral part of reality, not an escape. I concur by saying that in many African and Asian cultures, art forms such as folklore, garden designs and flower arrangements are expressions of, and tools of, the articulation of the principles of interpersonal relations and ethical conduct. (Maruyama 1981, 1982)

One of the basic human needs—which Maslow did not realize—is the need for self-heterogenization and contextualization. This need is very strong among some Africans and Asians, but it also exists among "Westerners" in varying degrees (particularly among artists). In the highly controllable extraterrestrial community environment, opportunities for self-heterogenization and contextualization tend to be limited. Art can alleviate the situation.

But let me first explain the process of self-heterogenization and contextualization, which are concepts alien to the Aristotelean logic. In the Mandenka culture of West Africa (Camera 1975), each individual is continually changing due to new experiences, new contacts, social duty rotations, and other events. Through these changes the person learns to understand the same situation from different points of view, and develops a poly-ocular vision. In Japanese factories workers are rotated between jobs, sometimes as many as 40 different jobs for each worker (Koike 1978). This enables the workers to think in one another's mind and feel mentally connected. This need to feel mentally connected is very strong among the Japanese. This connectedness is achieved through self-heterogenization in job rotation. In contrast, job rotation in Sweden fulfills the following needs which are very European: (a) to prepare human "spare parts," i.e., workers who are capable of replacing sick workers; (b) to relieve the fatigue of specific muscles; (c) to reduce boredom. Compared to these European reasons for job rotation, the African reason (poly-ocular vision) and the Japanese reason (feeling of mental connectedness as well as poly-ocular vision) for self-heterogenization are interpersonalistic, even though some Westerners do have a similar need. The Mandenka has another systematic way to cultivate their poly-ocular vision. Formally, the Mandenka society has a hierarchical structure. For example, a man must obey his father, who in turn must obey the former's grandfather. But the man is in a joking relationship with his grandfather, and this gives him an effective feedback channel over his father's head. There are many other feedback channels as well. A man's joking relations are his older brothers' wives, his maternal uncles, etc. A woman also has several sets of joking relations. And for the entire society there are official clowns who can criticize and ridicule anyone with impunity, including the kings. Obviously the joking relations are important as feedback channels. However, a less visible but more crucial function of joking relations is that they help one's capacity for a poly-ocular vision. Persons in joking relations can criticize each other and must take the criticisms without getting angry. They are forced to see themselves from others' points of view. During one of the adolescent initiation phases a person has taken the role of a clown, enjoyed the role, and thus later can identify with his or her critics and internalize the criticism.

How can art help the process of self-heterogenization in outer space? The answer is by bringing more varied reality into life: If a person can control his environment he/she tends to fall into his/her own rut. From the point of view of psychology, constant repetition of the same stimulus is almost the same as lack of stimulus. The person must be reminded that there is more in the reality than his/her own rut. Art is an expression of principles of interpersonal interactions and ethical conduct (Maruyama 1980, 1981). Art can remind the person that there are principles other than his/her own. Thus, art products which reflect different principles are particularly indispensable in highly controllable environments.

"Contextualization" means "to define oneself in terms of the context of the relations with others." This, by itself, is not an entirely alien concept to Westerners. For example, Martin Heidegger (1927) has discussed various types of "to-be" such as Dasein, Zuhandensein, Vorhandensein, and Mitsein. However, as Kimura (1972) and Hamaguchi (1977) have thoroughly analyzed, there are enormous epistemological differences between the Japanese sense of contextualization and the European/white American (EWA) sense. The EWA concept of the individual is ontological and based on the Aristotelean concept of identity, negation and oppositional logic. But in the Japanese concept, the "human being" is the same as "the relations between persons." It is based on a relational, heterogenistic, harmonistic, non-negational logic instead of the ontological and oppositional logic of mutual exclusion, negation, separation and categorization. But this verbal explanation is almost always misinterpreted and reduced to EWA concepts by EWA readers. A more effective way is to illustrate it by means of art, particularly through architectural designs as follows:

In the European conceptualization, space is defined as something between masses; for example, between walls or between buildings. Space itself is a transparent mass, and has boundary, shape, volume, identity, specialized function and immutability. Within a house rooms are specialized into bedroom, dining room, living room, etc., and are filled with specialized furniture. The house itself is intended to separate the humans from their environment.

In contrast, some of the basic principles in the traditional Japanese architecture are: (1) boundarylessness; (2) the outside-in principle and black hole principle; (3) convertability and connectability of rooms. A room in a traditional Japanese house can be a dining room at meal times, a bedroom at night, a living room for the rest of the day. Furniture is stored away and brought out for a specific function. Partitions between rooms can be removed. Moreover, the entire house is often designed to be connected to the outdoors or to be a part of the outdoors. The outer shell of the house can be removed. When the partitions between rooms are opened, one can sit in the innermost room of the house and see and smell the garden. Moreover, the garden may penetrate into the house like a deep recess and as a sort of room. Eaves may extend over the stepping stones in the garden. A moon-viewing platform may protrude from the house floor beyond the cover of the roof. All these are expressions of the outside-in principle. Another principle is the black hole principle. A house may look dark and inconspicuous from outside, as if camouflaged. Yet the inside is very bright, because all the daylight is sucked into the building.

The same principles of connectability and convertibility apply to the concept of the individual, the aesthetic design, and the relational, heterogenistic, harmonistic, non-negational logic. Kumura and Hamaguchi say that "ningen" human being—the word consists of "nin" (person) and "gen" between—is actually "gennin" (relation between persons). Tetsuroo Watsuji (1935) characterized the Japanese person as "kiwamete henka ni tom shitsuteki tayoosei ni oite onore o miidashite iru nipponjin." (The Japanese discovers himself through the qualitative heterogeneity which is rich in extreme changes). I have discussed the archaeological, prehistoric and historical development of these Japanese principles elsewhere (Maruyama 1980, 1981).

Even though contextualization cannot be accommodated in the official Aristotelean logic, many Westerners, particularly artists, practice it. In outer space living inhabitants can easily lose the sense of the context with other communities, both terrestrial and nonterrestrial. Art products based on contextual principles are particularly valuable in extraterrestrial communities ("art products based on contextual principles" is a topic I would love to discuss in depth, but the committee chairman advised me that it is outside the scope of this symposium.).

Let me move onto the problem of narcissistic hierarchical rationalism. Fifty years ago Japanese philosopher Tetsuroo Watsuji was inspired by Martin Heidegger's book *Sein und Zeit*, and produced a book *Fuudo*, which might as well be called *Sein und Umwelt*. He had many interesting and useful insights and observations in the book on the relations between design principles and environment. Relevant to our discussion is the following consideration:

Watsuji proposed that people who developed their cultures in monsoon regions, where the climatic events such as rain and storms brought both abundance of harvest and disaster, were dependent on the weather which had both regularity and unpredictability. They developed an attitude of receptivity and high adaptability to changing conditions. This had a great influence on their social interaction patterns, philosophy, aesthetics and religion. In human relations they tended to be interactive and mutually accommodating. They saw animals and plants as interactive partners of humans. They tended to develop heterogenistic, harmonistic logic which were reflected in their aesthetic designs and religion. For example, the traditional Japanese gardens and flower arrangements were based on the principles of harmony of heterogeneous elements and avoidance of repetition (Maruyama 1980). Often art was a tool for articulating the principles of human relations. Furthermore, many of the Japanese art designs are based on the principle of continuous change and evolution, in contrast to the European principle of discontinuity and stepwise development. As an example Watsuji takes *makie*, horizontally long, scroll paintings on which the sequential parts of a story are fused together to make a design whole.

On the other hand, people who lived in Europe had a much more benign, docile climate. Climatic disasters such as storms, hurricanes and floods were much milder in degree and much less frequent compared to those in the monsoon regions. Added to this was the fact that weeds and crop-damaging insects were rarer in Europe than in monsoon regions. Japanese farmers had to spend 80% to 90% of their labor fighting weeds and insects during the crop growing season. But the European farmers had no comparable problems: plants grew more or less automatically until harvest time. The tempo of harvesting was also much easier for European farmers. In Japan the harvesting had to be accomplished within a very narrow time scale, between the end of the growing season in August and the typhoon season which began in September. A short delay could mean the destruction of the crops by typhoons. Therefore, harvesting required very intensive work and a high degree of coordination and cooperation among the farmers. On the other hand, the European farmers could accomplish their harvesting at a much less hectic pace. The Japanese climatic environment required the farmers to work several times harder than the European farmers.

To the Europeans, nature not only was benign, but also had a high degree of regularity and predictability. In fact, the European sewage and river systems would be unable to handle a fraction of the volume of precipitation discharged regularly on the lands in the monsoon regions. Animals and plants were also more easily under human control: very little human effort was needed to keep them under control. Trees have regular forms, due to the lack of strong winds in most parts of Europe. Even where there are strong winds, such as on the Normandy coast in France, all trees are bent in the same direction, resulting in another type of regularity. In contrast, trees in Japan tend to have irregular forms due to stress and damage by storms and avalanches.

In Europe, humans were masters over animals and plants, not interactive partners. Together with the regularity in the nature, this fostered a hierarchical, rational view of the universe. Regular forms were considered beautiful, European designs are often based on principles of geometric regularity and repetition. Furthermore, the European hierarchical logic required that there should be something which was the highest. This "something," a produce of European logic, became the European concept of God. In European theology, God was often defined as "that, to which there is nothing superior." The rationality and regularity of the universe were attributed to this "god" since the time of the early Greek philosophers. In the sixth century B.C., the scientist Anaximandros conceived the notion of infinity as the inexhaustible protosubstance of the universe. As no materials knowable by human senses fulfilled his requirements for the infinity-substance, he thought that the substance must be beyond all human experience. Xenophanes applied this concept to religion. Proud of his "discovery" and scornful of the human-like gods of Greek mythology, Xenophanes declared that his infinite god was incomparable to humans and was eternal. In the fifth century B.C., Anaxagoras invented the notion of a power-substance that penetrated into things and caused them to move. He thought that the power-substance must be a soul, and must have order and purpose. He ascribed rationality to the power-substance. In aesthetics, the Greek philosophers' notion of beauty often consisted in regularity and arithmetic proportions. Much of European art, particularly European architecture and music, still retains the same concept.

A little later, sophists taught the principle of identity and the law of contradiction, and originated the tendency toward the formation of the ontological logic of mutual exclusion, negation, opposition, separation and categorization. Plato gave higher reality to abstract ideas than to concrete things, and advocated that true reality had no materiality. Here was the beginning of the dichotomy between the mind and the body. The fallacious notion that art is an "escape" from the material world, which is held by some European intellectuals, had its origins in those views of Plato.

Actually, the European image of god was an idealized human. Europeans created god in the image of man. This was logical because Europeans considered humans to be the highest category of living creatures. Even human exploitation of animals, plants and the rest of the environment was sanctioned by this god. I would consider this to be a narcissistic, human-centric religion, even though Watsuji did not use such a term.

The relevance of Watsuji's theory to the art in extraterrestrial communities can be appreciated in the following questions:

(1) Will the highly controllable environment in the extraterrestrial communities tend to foster hierarchical design principles?

(2) Will it encourage narcissistic art?

(3) Will it favor geometric or mathematical forms of art?

(4) Will it encourage monotheistic art?

(5) On the other side of the same coin, will the cultures in extraterrestrial communities evolve toward hierarchism and homogenism to the extent of excluding heterogenistic, interactive harmonism? If so, will the heterogenistic, harmonistic designs become a tool for "escape" from "reality" instead of being a part of the reality?

We can also ask questions in reverse: Can the art forms influence the choice of philosophical types in extraterrestrial communities? We can ask a still more basic question: Does the technology determine our environment, or can we first choose the types of culture we want, then design environments conducive to such cultures, and mobilize technology to construct such environments?

As I stated earlier (Maruyama 1976), with the advent of the era of extraterrestrial communities, we have reached the stage of our civilization where we have to think in terms of hitherto unknown

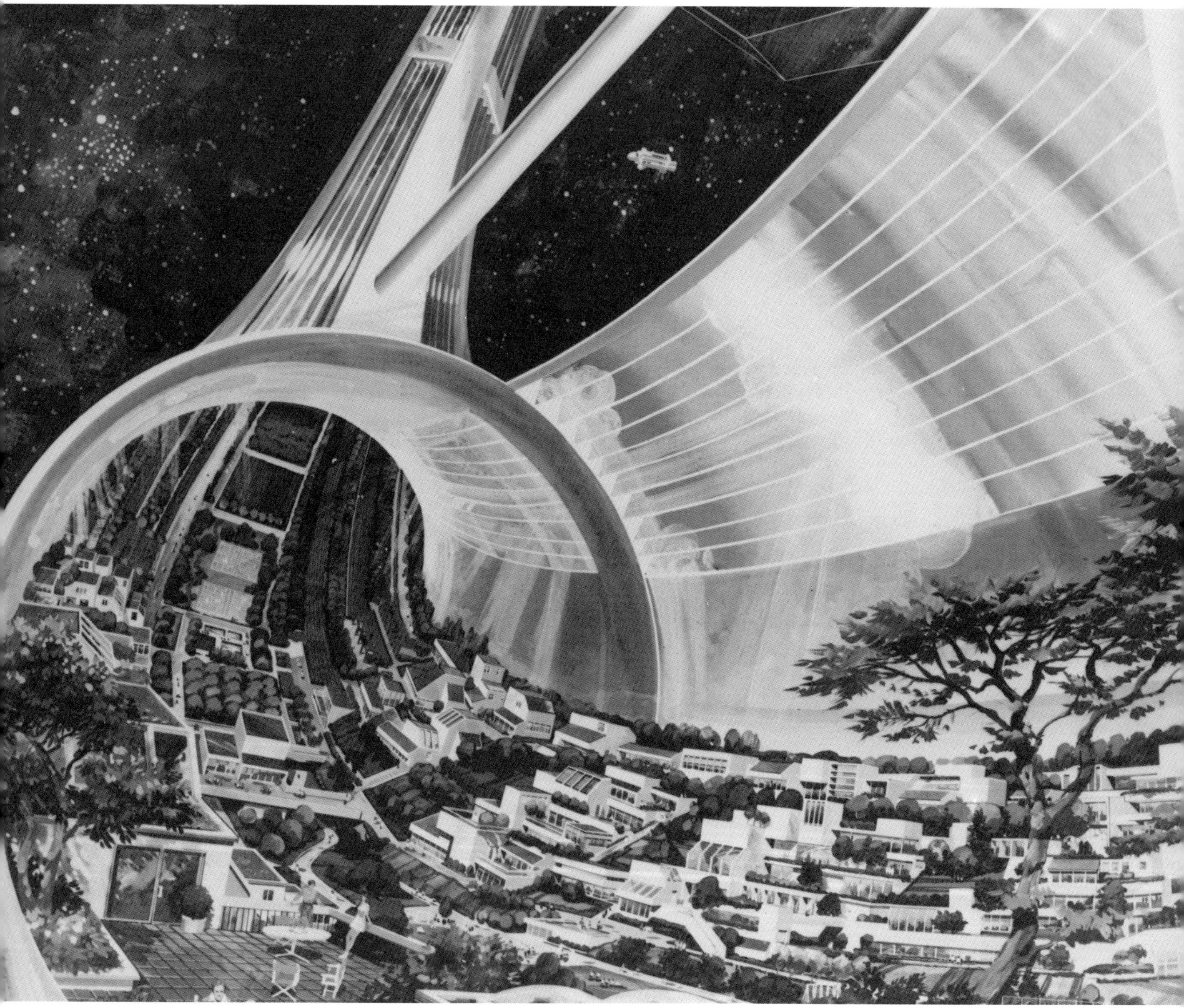

Fig. 9-2 Drawing of both the interior and exterior of a proposed space colony, showing the agricultural area (NASA).

cultural options. In the extraterrestrial communities, many of the constraints which restrict life on earth are removed. Temperature, humidity, seasons, length of day, weather, artificial gravity and atmospheric pressure can be set at will, and new types of social organizations and social philosophies become possible. The thinking required is far more than technological and economic. More basically, it is cultural, aesthetical and philosophical. We will be in a position to invent new cultural patterns and then choose environmental designs to fit the desired cultural, aesthetical and philosophical goals.

Instead of worrying about the environmental effects on art, let us now consider the choice of the design of the environment, i.e., choice of art in the wider sense of an environmental design.

One important consideration which was clear from the beginning of the first NASA/Stanford project on extraterrestrial community design, which took place in 1975 and in which I participated, was that we need heterogeneity both within each community and between communities. While Americans tend to believe that differences ferment conflicts and sameness leads to peace, the Mandenka of West Africa think that diversity enables mutually beneficial cooperation while uniformity generates competition and conflict. The Mandenka view is ecologically correct. The basic principle of biological and social processes is the increase of heterogeneity and mutually beneficial combinations among heterogeneous elements. The richness of life in a tropical rain forest or on a coral reef is due to heterogeneity of species. If all animals ate the same food there would be a food shortage, and if some species remain uneaten by some other species there would be an intolerable accumulation of corpses. Even solar energy is used maximally if different organisms use different means to absorb it: tall trees, short trees, vines, algae. But it is a mistake to consider the advantage of heterogeneity only in terms of resource diversification and disaster risk dispersion. The argument for resource diversification is based on the zero-sum game assumption. Disaster risk dispersion can be supported even by those whose models are non-interactive. Both are true and valid. But there is a more important advantage in heterogenization: mutually beneficial positive-sum interactions. For example, there are birds who eat food debris between the teeth of alligators. The alligators get their teeth cleaned, the birds get food, and the potential predators of the birds are scared away by the alligators.

It is advantageous to have different types of cultures in outer space for two reasons. One is the positive sum interactions between diverse cultures. The other is the increased possibility for optimal matching between individual types and community types. Individuals are born different, even from the same parents. There must be many different types of communities to choose from for each individual.

Having accepted the necessity to make heterogeneous communities, we must invent designs and art forms for these different communities. Many designs can be tried out, experimented with, weeded out or further developed. Here are some different types of designs briefly summarized from more detailed discussions (Maruyama 1980, 1981):

Type H	***Type I***	***Type S***	***Type G***
Homogenistic Hierarchical Classificational	Heterogenistic Isolationistic Random	Heterogenistic Interactive Homeostatic	Heterogenistic Interactive Morphogenetic
Unity by repetition, similarity and symmetry. Geometric regularity is beautiful. *Examples:* Gothic, Islamic architecture. French, Italian gardens. Chinese and Japanese Yamato architecture.	Each building or element is unrelated to others. Random, capricious, haphazard. *Examples:* Manhattan, Los Angeles downtown.	Avoid repetition of similar elements. Static harmony of diverse elements. Asymmetry. Completed equilibrium which cannot be disturbed. *Examples:* Some of Japanese gardens and floral art. Japanese Yayoi culture.	Changing harmony of diverse elements. Avoid repetition and symmetry. Designed for simultaneously multiple as well as changing interpretations. Deliberate incompleteness to enable addition of new elements and changes. *Examples:* Some of Japanese gardens and floral art. Japanese Jomon culture.
Defined as between masses, between points, along shafts. Space is a transparent mass, has shape, volume, boundary, identity, specialized function. Space opposes mass and other spaces.	Unrelated units.	Miniature universe containing all phenomena. Internal interaction of heterogeneous elements. Space represents interrelations and harmony, which may be steady or cyclic.	Locality with its own quality and feelings. Connected to other localities. Often thought of as around something (rock, grove) which is considered to be condensation of the quality which permeates the locality.

Type H	Type I	Type S	Type G
Volume and mass opposing space, and opposing one another. Often mass represents human; space represents nature.	Each is its own expression.	Represent phenomena and elements of the universe. Objects support one another, and do not oppose space. Object is quality, not quantity or mass.	Represent growing process, vitality, vicissitude. Objects stimulate one another. Interplay of space and objects. Object is agility, not substance. Space is movement.
Buildings are designed to persist in time. Permanence. Buildings embody eternal principles.	Permanence is unimportant.	The form persists while the building materials may be renewed from time to time. Include seasonally changing elements to emphasize each season and annual cycle.	Increasing heterogeneity and symbiotization. Design permits addition of new elements, pattern change. *Example:* Katsura Palace
Tension and extension. Straight lines spanned between points. Lines and surfaces of mass extended to create shafts of space. Lines radiating from or converging to a point. Curves created by physical contour of mass.	Caprice. Random.	Stability. Triangles of unequal sides (not physically present but conceptually underlie between elements). *Example:* Some of Japanese floral art.	Flow. Spiral or curve suggested by relative position of objects, but not by objects themselves. Curves extend and do not return. *Example:* Some of Japanese floral art.
Pre-planned and sequential. Perceiver maintains his identity as a point which moves in space.	Random.	Perceiver permeates simultaneously in all parts, and feels invisible parts and behind objects. The entire design is perceived simultaneously, not sequentially.	Generated by perceptual experience which has multi-branching effect, not single-tracked.
Separate the inside from the outside.	Separate one household from another.	Continuation of outside into inside. Removable outer shells. Garden continues into house. Floor extends to outdoors. River flows under floor. Passive appreciation of nature and environment. Absorb the outside into the inside.	A base for activity and interaction with environment, such as a Japanese farm house.

Type H	Type I	Type S	Type G
Specialized rooms (bed room, dining room etc.) occupied by specialized furniture. Rooms are separated by walls and can be locked.	Specialized to individual user's needs rather than by generally categorized functions.	Continuity, connectability, convertibility and self-heterogenization. Partitions between rooms can be removed. The same room can be used for different functions. Furniture is stored away and brought out for varying functions. *Example:* traditional Japanese houses.	Continuity, connectability, convertibility and self-heterogenization. Partitions between rooms can be removed. The same room can be used for different functions. Furniture is stored away and brought out for varying functions.
Superior can monitor subordinates, but subordinates may not see superiors' intention or activities. Concern with patent, copyright etc. based on zero-sum game assumption.	Visual, acoustic and other insulation for each individual. Insulation is considered to increase efficiency. Negative-sum game assumption.	Maximum sharing of intimate concerns. Group solidarity. Positive-sum game assumption.	Less concern with privacy, patent and copyright, because interaction is seen as positive-sum game which generates new inventions.
Socialize within homogeneous group. Hierarchy of groups, subgroups and supergroups.	Self-containment. Freedom from social obligations and freedom for individual caprice.	Mutual dependency. Concern over not disturbing the existing harmony. Perpetuation of familiar relations.	Make new contacts, new network. Explore new patterns. Go beyond own group.
Neat categories, clear without context.	Individual meaning.	Meaning is contextual and interdependent.	Ambiguity is basic to further development and change.
There is "the best design" for all persons.	Give each person a cubicle where he/she can make own design.	Different designs preferred by different individuals can be esthetically combined in an interesting whole.	Individuals stimulate one another by their varying designs. Experiment with new combinations.
Majority rule, consensus or "educating" to make people appreciate the "best" design.	Do you own thing. Let others alone.	Compensation for hardship on any single individual regardless of which design is chosen.	Compensation for hardship on any single individual regardless of which design is chosen.

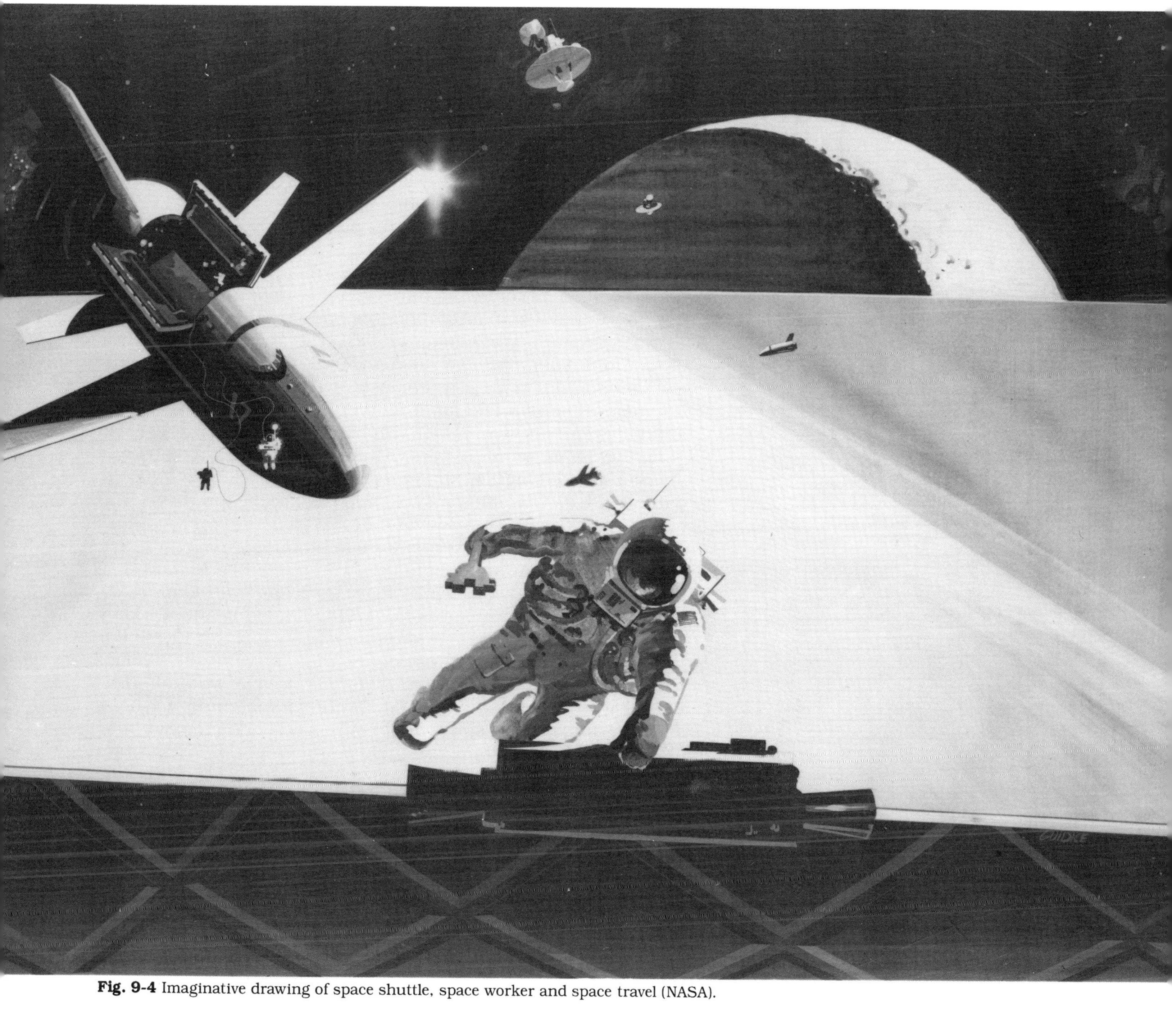

Fig. 9-4 Imaginative drawing of space shuttle, space worker and space travel (NASA).

Subterranean Communities

Subterranean communities will become much more practical than earth surface communities in some parts of the world: (1) regions with large diurnal fluctuations of temperature, such as in some high altitude regions; (2) very cold regions; (3) very hot regions; (4) regions with very strong winds, sand storms, tornadoes, hurricanes, hail; (5) regions with frequent freezing rains at temperatures around the freezing point; (6) regions with very heavy snowfall.

Even in very cold or hot regions the subterranean temperature at the depth of one or two hundred meters can be around 20 to 25 degrees centigrade. In most other regions a comfortable temperature can be attained at a much shallower level, often at the depth of five to ten meters. Even in temperate regions underground storage facilities remain at a constant temperature throughout the year. Kansas City, for example, utilizes extensive underground storage facilities. Subterranean communities may consist of any combination of residential areas, work areas, industrial areas and even agricultural areas. Some vegetables grow in darkness, such as mushrooms and soy bean sprouts, while others can grow under artificial light.

Environmental characteristics of subterranean communities are: (1) lack of natural daylight; (2) lack of weather; (3) lack of a large landscape; (4) shortage of space associated with high excavation costs, and consequent low ceilings in living and working areas.

The lack of daylight and use of artificial light will affect color-related art, such as painting; while the lack of weather may induce weather-related art to flourish as a substitute, such as auditory and visual displays which give a sensory stimulation of weather.

A lack of a large landscape is a problem shared with the extraterrestrial communities. Solutions are partly similar to these discussed in the section on extraterrestrial communities: miniature landscapes, and transmitted or recorded and replayed landscapes. However, there is one difference. In the case of extraterrestrial communities, there is a vast, empty space outside the enclosed communities, and immense, holographic sceneries are possible. This is unfeasible in subterranean communities. A substitute is the use of wallscapes. In most of the existing architecture, walls are flat. But walls can be irregularly surfaced and indented to create an illusion of a landscape effect, especially if combined with the art of miniature gardens. In the 1950's I saw subterranean factories in Sweden which utilized very effectively the irregularly shaped walls for landscape effects. In Sweden much of the ground consists of granite, and excavated surfaces produce irregular shapes.

The low ceiling creates another challenge to art: ceilingscapes. Until now, the ceiling has been insufficiently utilized by interior decorators. Most of the time the ceilings in residential and working areas are left blank, even though in cathedrals and auditoriums the ceilings may be decorated with murals, patterns and chandeliers. (Chandeliers were originally invented to support candles). In conventional architectural common sense, a low ceiling is undesirable. For one thing, a low ceiling gives the impression that the space is "cheap." But this is based on the assumption that the ceiling is a visual barrier. However, if we invent the art of ceilingscape, a low ceiling can become a great advantage: the lower the ceiling, the more sense of movement and change when one walks around under the ceilingscape.

The lack of noise is also a problem shared with extraterrestrial communities, and the solutions are also the same.

Underwater Communities

Underwater communities will be used along the coasts of Arctic and Antarctic lands and islands and along the coasts of some seasonally hot lands. Below the ice of the frozen ocean, the temperature does not fall much below the freezing point. Underwater communities may consist of residential areas, working areas, industrial areas, agricultural areas and fish culture areas.

Environmental characteristics of underwater communities are: (1) They must spread horizontally rather than vertically because water pressure increases by one atmospheric pressure for every ten meters of depth increase; (2) They are built as near the water surface as possible below the ice layer, and consequently receive daylight which is filtered toward blue; (3) Since they are in polar regions, the daytime is very long during the summer and very short during the winter; (4) Some weather changes are visible through windows; (5) There is a lack of weather inside the communities; (6) Natural seascapes may be available, and artificial seascapes may be constructed.

The horizontal spread means that the surface-to-volume ratio is high. Particularly the ceiling-to-volume ratio is high. The use of "skylight" windows will be practical. Since both the length of daylight and the temperature vary seasonally, the skylight quality will have four distinct seasons. In the underwater communities, spring can be defined as a season when the daytime is longer than the nighttime, and the water surface is still covered with ice. Ice keeps the water surface waveless, and the skylight will be almost motionless. I do not know whether the ice filters colors toward blue as the water does. If it does not, the light quality will be less blue during the spring than during the summer. Moreover, the blueness changes with the tide: more blue at high tide, and less blue at low tide because of the change in the thickness of the water layer. This adds a semi-diurnal rhythm. The timetable of the tides is related to the position of the moon. Therefore there will be a monthly cycle as well. Ice has two other characteristics related to light, especially when there is some snow on it. It reflects the sunlight and the moonlight back to the sky more than water surface does. Consequently, the skylight windows receive less light. Secondly, it diffuses light. Summer can be defined as a season when there is no ice on the water surface and the daytime is longer than the nighttime. During this season waves act like lenses, and there is a rhythmic change in the light pattern. The colors are filtered toward blue. Autumn can be defined as a season when the daytime is shorter than the nighttime, and there is still no ice on the water surface. Winter can be defined as a season when the nighttime is longer than the daytime, and the water surface is covered with ice. In both autumn and winter daytime is short or non-existent. Therefore the skylight effect is minimal. However, a full moon will light up the water to a surprising extent. In both hemispheres the moon at the full moon position rises higher in winter than in summer because it is 180 degrees away from the sun.

For most people the blue-colored environment has a depressive effect, and this must be counteracted by artificial lights inside the underwater communities to offset the effects of the very blue skylight. The blue filter effect will be more pronounced in the waterscapes observed through the windows on the side and bottom surfaces of the building. In sceneries on the earth or on hazy days distant objects are more faded than closer objects. Similarly, in the waterscapes, distant objects get bluer than those close by. This adds an uncanny feeling to waterscapes. To counteract this, some reddish objects can be placed in the waterscapes: the more distant, the redder and brighter.

Another characteristic of the waterscapes is that the wave motions during the ice-free seasons create rhythmic movements of seaweeds. These movements are quite desirable aesthetically and psychologically. Where there are no seaweeds, artificial objects can be placed which move with wave motions. Design of waterscapes is an art yet to be developed.

Conclusion

The art in extraterrestrial, subterranean and underwater communities will have very important roles to play, not only for aesthetics, but also for many psychological, social and even epistemological reasons.

References

Camara, Sory (1975), The Concept of Heterogeneity and Change among the Mandenka, *Technological Forecasting and Social Change*, 7: 273-284.

Hamaguchi, Esyun (1977), *Nipon Rashisa no Saihakken*, Tokyo: Nihon Keizai Shinbunsha.

Heidegger, Martin (1927), *Sein und Zeit.*

Kimura, Bin (1972), *Nito to Hito tono Aida*, Tokyo: Koobundoo.

Koike, Kazuo (1978), Japanese Workers in Large Firms, *Keizai Kagaku*, 26:1-37.

Maruyama, Magoroh (1976), Designing a Space Community, *Futurist*, 273-281.

Maruyama, Magoroh (1980), Mindscapes and Science Theories, *Current Anthropology* 21: 589-608.

Maruyama, Magoroh (1981), Denkmuster: Metaprinzipien der Umweltgestaltung, *Garten und Landschaft*, 806-815.

Maruyama, Magoroh (1982), Mindscapes, Workers and Management, *Japanese Management*, edited by Sang M. Lee and Gary Schwendiman, New York: Praeger.

NASA, (1976), *Space Settlement*: a Design Study, NASA publication SP-413.

Tsunoda, T. (1978) *Nipponjin no Noo*, Tokyo: Taishuukan.

Watsuji, Tetsuroo (1935), *Fuudo*, Tokyo: Iwanami Shoten.

Art and Space-Age Communities

Arnold Berleant

Professor Maruyama's purpose here is to consider the place and nature of art in future human environments that are the product of an increased range of technological innovation. This leads him to focus on communities that are wholly created by means of space age technologies as almost totally self-sufficient social and physical units that encompass their entire environments. These, he finds, may be in the form of extraterrestrial, subterranean, and underwater communities, or what I shall call space age communities.

It is true that communities have always originated at some historical point and have developed as the product of human design, whether conscious or not. And with increasing frequency, especially in the past century, communities have been designed as totalities before they were built. Yet the communities of which Professor Maruyama speaks are different for several reasons. One is that they are largely or entirely self-contained and self-sufficient. Another is that their construction and operation rely heavily on a highly complex and sophisticated level of scientifically developed materials and technology. The design of these space age communities, then, represents a challenge that could easily assume the character of science fiction but for the fact that not only have such communities been the subject of serious planning efforts by NASA and other such organizations, but that the technical means are near at hand by which they can actually be constructed. Thus, such design decisions lie not in the realm of fictional imagination but become concrete and necessary parts of the planning process.

To speak of art as a factor in such discussions may seem superfluous but for two considerations. One is the recognition that every social group in whatever form of organization and development displays artistic activity that is an integral and essential aspect of its life. There is no reason to suppose or to desire that a space age community will be any different. What may be different is the kind of contribution art can make in such wholly fabricated environments. This is a matter to which I shall return later.

Another reason for regarding art as more than superfluous in such a community is that a creative aesthetic factor is inherent in and inseparable from the very act of fashioning an extraterrestrial or other such human habitat. In a profound sense, to talk of art in such communities is a redundancy, for the elaboration of a community as a total, self-sufficient social and physical entity is itself a fundamentally artistic undertaking. Viewed in its essential nature, the planning of such a community, in all its material, perceptual, social, and institutional dimensions, is a basic act of artistic creation—the shaping of an entire realm of experiential being. Such a task represents a challenge of the highest order, and the fact that it must be concerned with practical considerations detracts not at all from its aesthetic character.

As Professor Maruyama observes, the spatial conditions of such a community will be decided by the design of its physical structure. These determine the limitations, the needs, and the opportunities with which the designer will work. Now, since the shaping of such experiential conditions rests with those who make the design decisions, the ultimate question becomes this: apart from broad technological limitations, are there any normative requirements, including aesthetic ones, that intrude on the designer's actions and compel, urge, or at least make preferable certain choices over other ones? When the designer is free from the constraints of gravity constraints that have exercised the greatest influence on architectural design up to the present time, can that person follow his or her stylistic proclivities, private intuition, or theoretical allegiance in fashioning the forms and qualities of such a design? Will space age communities be equally satisfactory when shaped by the imagination of a Gropius or a Gaudi, a Léger or a Monet? The answer to this question is, I suspect, more than a technical or even an aesthetic one. It raises the moral issue of what constitutes a *good* (or just or humane or proper) community and this forces us to recognize that in shaping the human environment lies the ultimate and necessary fusion of the aesthetic and the ethical. Let us see how this is so.

Maruyama develops at length several basic contrasts between European and non-European, especially Japanese, patterns of thinking, of conceptualization, of ways of viewing the world. There is self-heterogenization, a process in some cultures (e.g., the West African Mandenkas and the Japanese) whereby various social roles are tried, practiced, and shifted in such a way as to encourage the development of a wide range of social perspectives (poly-ocular vision) or a sense of relation with others (mental connectedness). A related process is contextualization, by which a person is defined through his or her relations with others. Self-heterogenization can be contrasted with the Western penchant for rigid social roles that define sharply the sphere of permissible action. Contextualization is seen to differ from the pervasive individualism of the West, which values discreetness, separation, self-sufficiency, and even isolation. Another contrasting pattern of thinking is the Western narcissistic hierarchical rationalism, which idealizes the regular order of nature, assumes its anthropocentric character, its presumed order of dominance and submission, and its assumption of discontinuity and stepwise development. This is unlike the Japanese, for instance, who emphasize continuous change and development, harmony among heterogeneous elements, and the avoidance of repetition.

Now, the questions Maruyama raises about such alternatives suggest certain choices between essentially Western and non-Western cultural values and forms. This brings him to the basic design question: does technology determine the nature of our environment or can we use technology to construct whatever environment we choose? His answer lies with the latter option. With the technological ability to determine most of our environmental conditions in such wholly planned communities, setting at will their spatial characteristics, temperature, humidity, weather, atmospheric pressure, length of day, and gravity, the opportunity seems to be available to develop forms of social organization and conditions of human experience hitherto impossible and perhaps inconceivable. This raises fundamental questions about social orders, social values, and cultural patterns, and it creates the possibility of setting in advance cultural, aesthetic, and ethical goals.

With such a range of choices technologically available, do the same limitless possibilities for modes of life apply to the people who will inhabit them? Are human beings infinitely malleable or are there biological, social, philosophical, and even aesthetic demands that lead us to make certain determinations here? The answers to these questions are not entirely speculative ones; they require the cooperation of scholars and scientists in many disciplines to offer well-grounded proposals — environmental psychologists, designers and planners, anthropologists, and especially, philosophers and artists. Many human needs are identifiable, and while their satisfaction is possible under a variety of different cultural and environmental forms, these may not always be equally desirable. Empirically grounded judgments can be made about the likelihood of certain conditions over others to produce human fulfillment, or happiness, in Aristotle's sense of the term. While it is foolish to set certain, fixed goals, open-ended ones like the satisfaction of needs and interests, the development of productive and creative capacities, health, the expansion of awareness, and the enlargement of moral consciousness are among the central values for a fully realized human life. There are many alternative conditions for achieving such generalized ends, to be sure, but the conditions for impeding them are far more numerous.[1] Thus, design decisions that are governed by such considerations are ultimately a philosophical activity and this activity requires philosophical awareness.

For our purposes here it is important to learn whether aesthetic needs stand among the needs whose satisfaction is part of a fulfilled human life. If there are, as I and others believe,[2] then the design of space age environments must meet aesthetic requirements along with other ones. What do we mean by aesthetic needs? These are needs of a primarily perceptual sort, the human necessity for actual or imaginative stimulation along sensory channels — visual, auditory, kinesthetic, tactile, etc. Some of the most dramatic findings in the psychology of perception have come from the large number of experimental studies on sensory deprivation that have been conducted in recent years. These experiments show that people have an extremely low tolerance for environments that reduce or virtually eliminate perceptual stimulation, so that such conditions quickly lead to hallucinatory and psychotic states. Granting the absolute necessity for sensory stimulation, it is possible that whatever else it may be, aesthetic perception in art is a development and elaboration of this essential human need. It is well known that every social group of which we have knowledge, at every period in human history, develops elaborate artistic forms of aesthetic activity, including decorative arts, rituals, entertainments, and objects and events fashioned to be enjoyed for their own sakes — as we now typically describe works of art. I suspect that a case can be made for art as the satisfaction of aesthetic needs and an extension of just such a perceptual necessity.

When this requirement is implemented by the conscious purpose of the designer, it implies the development of perceptual consciousness, of perceptual variety and interest. Community design, especially the design of space age communities in which every feature is a design decision, is an act of totally shaping the perceptual awareness of its inhabitants. Hence, the forms of spaces and of masses, the distribution of colors and textures, the nature and presence of sound, the allocation of the intensity, quality, and direction of light, the determination of the diurnal and other cycles that give shape and regularity to time — all these determine the conditions and the character of the human world, which is a resolutely moral act. And in determining such conditions they create a total realm of human experience that bears a remarkable similarity to the imaginative realms of film, of the novel, and I would add, of the other arts, each in its own distinctive way; the main difference being that there is no discernible limit to the experiential realm of a space age community. This makes the design of a human world at the same time a primary artistic act.

We face directly, then, the matter of art, both art as community design — the shaping of a perceptual, an aesthetic environment, and art as the fashioning of specific objects and events — works of art, in the usual sense. Maruyama argues correctly for increasing heterogeneity both within and between communities and suggests that we should encourage different designs and art forms for different communities, offering four contrasting typologies as some basic alternatives. These he derives from cultural patterns, present and past, and I believe that they can be adjudged by their conduciveness to human fulfillment. I suspect that for Professor Maruyama, as for me, they are logically, but not morally, equal. A type that encourages heterogeneity is preferable to one that demands homogeneity (eliminating Type H); one that stresses contextualization is more desirable than one that leads to isolation (eliminating Type I); and a type that promotes change and growth is better than one that supports equilibrium (eliminating Type S). This leaves us with the type (Type G) that favors heterogeneity, interaction, and morphogenetic development. Of course, elements from the rejected types can enter into a prevailing morphogenetic order as long as they do not become dominant. Repetition (from Type H), for example, may be used as an element in pattern or texture; randomly placed buildings, objects, and events (from Type I) may be useful for variety and surprise; and equilibrium (from Type S) may be desirable in limited contexts, such as private spaces, or religious or contemplative places. The moral factor in design, however, is clearly a difficult one that requires its own full examination.

Moral questions aside, it may be asked in what ways these general features should be given concrete and precise form. Will mother cultures tend to replicate themselves and reflect their own conception of the order of things in the design and art of the communities they spawn? Will space age communities simulate selected earth environments, duplicating the customary materials, spaces, qualities, and temporal patterns of the cultures that produced them? Or are there other, more desirable directions in which the designer, now freed from the stringencies of the earth's environment, can move, fulfilling our fantasies with a spirit unencumbered by the exigencies of gravity, weather, climate, and time? Are there primordial needs which must somehow be supplied, such as accessibility to bodies of water, to fire, to open spaces, and to organic materials, or are these merely acquired tastes that can be replaced by substitutes that are more technologically convenient? Here the need for research becomes essential in the planning process.

Finally, what will be the place of art in the conventional sense in such communities? Will art assume a socializing force, helping the inhabitants of such places adapt to the particular conditions of individual communities? Further, will some forms of art develop to complement the design process? Maruyama imagines the emergence of new forms of genres here, such as wallscapes and ceilingscapes (and, I might add, light sculptures) in subterranean communities, holographic landscapes in extraterrestrial communities, and waterscapes in underwater communities.

Independently of these uses of art, will the forms and objects of art break loose from the constraints imposed by the earth's environment and develop in ways unimaginable at present? It is possible to imagine an artistic renaissance developing out of the rich resources of the new technologies and materials, the expanded range and color of light, reduced gravity, and gravity-free modes of human locomotion. Architecture, sculpture, painting, and dance will certainly be transformed, and it is unlikely that the other arts will continue unaffected. The artistic imagination will be set free and, predictably, will work in unpredictable ways. Here, indeed, the art of community design and traditional object-oriented art may combine when the artist begins to suggest the forms and conditions that space age communities may assume. Out of this fusion may well emerge the embodiment of the ideal of many *avant garde* artists today, the full joining of art and life. Perhaps then, in speaking of art and space age communities we must talk in new terms of "lived art."

Notes

1. *Cf.* Arnold Berleant, "Ethics and Science: Some Normative Facts and a Conclusion," *Journal of Value Inquiry*, XI, Winter 1977, 244-258.
2. *Cf.* M. Rader and B. Jessup, *Art and Human Values* (Englewood-Cliffs, N.J.: Prentice-Hall, 1976), esp. ch. 5.

Epilogue

Epilogue[1]

René Berger

Man has always had an alliance with the imagination. Humans owe their living in society to this fact. If salmon travel thousands of kilometers, ascending rivers at the cost of their lives, it is the need to find and return to their spawning grounds that goads them. The armies which followed Alexander as far as the Orient, or those which Napoleon led to the gates of Moscow, were obeying their imagination, which, in our time, is still incarnated in the flag, a "magical" cloth for which we kill and for which we die. Even the audacity of Gothic cathedrals, which is magnified by the stained glass windows of Chartres, has no other source than this other form of the imagination incarnated by the Christ.

By *imagination* I mean that faculty which a society employs to represent and to embody what it calls "reality," and which takes the form of common symbols such as the flag and the face of Christ. On the one hand, the property of symbols is to mediate by impregnating all the activities (political, religious, cultural, economic) of a given society and, on the other hand, the characteristic of symbols is to articulate and to organize into a total system. Thus, each of the civilizations of the past has its own particular physiognomy, which is attested to by the vestiges that those civilizations have left behind: documents, monuments and tools from such places as Lascaux, Karnak, Athens, Rome, Konju and Uxmal — as many remote locations are preserved as are memorialized in the evidence for them in countless exhibits in museums.

This shows the important role which the *image*, (in the general sense of the word, i.e., that which is visible) assumes in the manifestation of the imagination. Palaces, tombs, temples, paintings, sculptures, cultic or everyday objects, *make present* the symbols: abstract signs, images of gods, of kings, of myths, of history — *representation* is, therefore, one of the fundamental dimensions in which the imagination and the real merge. The imagination also manifests itself, in effect, in dimensions which are outside of the visible realm; thus we have the symbols of language, of music, of belief (which are animated by rites and ceremonies), or even those of knowledge, including science. In effect, we can say that the imagination is the aggregate of activities in a given society (from cooking to the style of dressing, from the manner of rendering justice to the manner of measuring time and space) — an aggregate that is *enformed*[2] by symbols. Whether the forms depend on aesthetics, on authority, on tradition or on laws, they have in common their obedience to rules, flexible or rigid, which constitute the *models* to which we conform from one generation to the next. Representation is one of these enforming activities, like custom and duty, except that representation has the advantage of being more "evident." It is by this that we give it a privileged role, especially as there are hardly any other enforming activities which are accompanied in one way or another by visual aids. Justice, politics, and industry have resorted and continued to resort to an iconography, examples of which are busts of emperors or kings, and nowadays placard campaigns for elections; likewise there is hardly a product that is not illustrated with a trademark, with colors, or with a logo. Every system, however complex or subtle it may be, can hardly dispense with having recourse to the image whose fundamental role is to make the often abstract nature of the relations that it employs, if not always visible, then at least tangible.[3]

Now, confining our discussion to representation, we can say that for millennia, until the last century, the role which it has been vested with has always been the business of human hands, whether we are dealing with the pyramids of Egypt, built with the help of thousands of laborers, or whether we are concerned with miniatures, frescoes, castles, parks or sculptures. Tools or machinery, even if they did undergo certain developments, played only a helping role. Indeed, the technique of engraving, which was a development out of the process of printing, modified the production of images; but it is with the advent of industrial civilization that this mutation with which we are concerned has arisen.

Broadly speaking, this mutation consists in substituting technology for human labor. The machine becomes the instrument or factor which motivates all forms of production, production which is absorbed by markets that are ever-expanding, due to an economy that has become planetary. We can infer from this — something which an attentive observer should be permitted, moreover, to mention — that, if past civilizations have expressed their imagination by an enforming of their activities on a material basis, then concerning "representation" and human labor, particularly manual labor, our age is in the process of elaborating, by an ever-encroaching intervention of the machine and technology into all of our activities, a new kind of imagination to which I have given the name *techno-imagination.*[4]

One of the most ostensible manifestations is the change of the city. Yesterday it was still a town, home of a population whose particular physiognomy it reflected while binding it to the collective imagination of the society of which it was a part (to each city its cathedral or its church, to each city its town hall or belfry). Today the modern city, even when not created out of fragments, does not escape remodeling (including of its past), which the techno-imagination nowadays proceeds to do. Whether it wants to or not, and despite the expedient of pedestrian zones, no city escapes the imperative of traffic which configures buildings, squares and streets, parking lots and standing zones. No city escapes the enforming activity of traffic signs and street-markings, by which all come to look alike. Therefore the automobile is not only the means of transportation as we have defined it, neither is it the instrument of prestige with which we have given it credit, nor even the prodigious market over which General Motors and Nissan reign. The automobile has become, in the strict sense, that part of our being which depends on a *motorized imagination*, a network of trajectories and of intersections alien to the history of the city. Thus the enforming activity of modern life structures our behavior by substituting or by adding to the symbols of the past those which are shaped by the practice of the *technical rite*, which today is universal.

And ubiquitous *advertising* is the high priest of this technical rite, which is concealed by the definition that we are giving, that is: the aggregate of means, writings and sounds that are represented and designed to sell a product. It is a functional definition, and therefore a partial one. As a matter of fact, advertising has become the motivating part of our imagination in urban practice. It is advertising which activates the system of production and consumption by endowing products, goods and services (beyond their usage value and even their exchange value) with a semi-transcendental virtue, the "best world," which is expressly validated and guaranteed by the packaging or the trademark. This makes (as once were the symbols within a community) a pre-established harmony, except that, in our case, it is prefabricated in, and by, the market which replaces the community.

Now, just as production exists only to the extent that it always tends to amplify itself, we see that the images which it engenders are carried along in the same exponential movement. It is an uninterrupted flux which unfurls, disperses; it is resplendent with marvels that at once become junk, but the flow always resumes better than ever thanks to the ductility and the speed of modern techniques of reproduction and diffusion. It is impossible to escape the images which industry propels right into (what it was just recently expedient to call) our own "backyard." Catalogs, prospectuses, journals and magazines overflow from our mailboxes, to end up at the foot of the television screen, the latest image-box. Compared to the civilizations of the past which inscribed their imagination in stable representations, our civilization tends irresistibly towards the volatile. Everything dissolves and is replaced. "Traffic," that of automobiles and that of images, carries everything in its wake. Is the world now a nebulous situation?

It is without doubt that volatilization reaches its zenith with television. This term "volatilization" may seem unusual; I must briefly specify it. What is as a matter of fact striking in the development of the image is that it has less and less recourse to traditional supports, and more and more recourse to electronics, the power of which is limitless. Television stations are not at fault for exploiting this power. Nothing escapes them of world events, which are relayed several times a day by the televised press: neither the sports competitions, nor the universe of fiction (film, theatre, science fiction), nor the documentaries (human, animal, plant, land and sea), nor scientific discoveries. Everything can, and does, take form on the screen — fleetingly. Cathedrals or detergents, seaports, horses or heads of state, are nothing but collections of pixels,[5] immaterial puzzles which the numerical image can henceforth treat in its own fashion. The limitless power of electronics is not only a matter of extension; it touches the very nature of things. The notions of objects, space, and time — notions which we laboriously incorporate from infancy on — are defined by a technology that is fond of all ventures, of all transformations.[6]

Proteanism is as permanent as it is limitless. Until the eruption of electronics, images referred to "verifiable" objects, that is, they had a referent: the natural environment which we perceive (city, countryside, room), or the cultural or artificial environment which the world of art (painting, sculpture) represented, as well as objects created by the collective imagination (gods, allegories, symbols, chimeras). All these images were situated, and are still situated to some extent, in frames of reference which allow us to explore them and thus to explore ourselves. But what happens when our representations are in the grip of the cathode screen, which has the power not only to transmit, but also to metamorphose? Objects, categories, criteria, principles, codes, norms, disciplines — in short all that had taken form — is destabilized to the point of losing duration and identity. Does volatilization denote a change of state to which we are unconsciously aspiring or, in any case, one which we are undergoing? Just yesterday, the collective memory relied upon frames of reference which history filled up, according to those matters of weight, real and imaginary, which every society and every civilization selected. Nowadays, it is on the elaboration of a "reverse memory," so to speak, that we are focusing technology. It is to technology (summed up by the phrase "everything is possible," which is its real nature) that the initiative belongs.

It is a volatilization which begins with a process of unbalancing. On the screen, nothing has more weight, neither the Pope, who is followed by a football game (when he is not followed by advertisements), nor the heads of state, all of whom the directors and cameramen arrange as they like.[7] Gone are the great political principles ("liberty, equality, fraternity") for which we would have had a hard time creating images! If the collective memory is made of "weighty objects" which society preserves in order to assure its identity, then the "reverse memory" engendered by the media has nothing to do with these. For the media all that counts is what we might call the "froth of the epiphany." The billions of magi who are the televiewers are no longer allowed to see Jesus, but the apparition of the Apparition, which is prosaically obtained by turning on the set or by changing the channel. The imaginary flows directly over broadcasting — forming an emulsion which we expect to flow without end.[8]

Paradoxically, it is the electronic apparition, of immaterial nature, which gives the televiewers the strongest impression of material reality. Nothing — not press reports, not photographs, not even eyewitness reports — equals the TV images when a racing car careens off the track and explodes at the 24-hour race at Le Mans, or those which showed us the attacks on Reagan and the Pope or, to speak of more peaceful events, the images of the weightless astronauts. All media, writing and speaking included, imply a time of transmission, whereas the video image can be present everywhere simultaneously. But also the nature of the presence is changing. As opposed to discourse or writing or even photography, the TV image directly engages our senses. How can one doubt an image which is moving and in color, which each one knows is seen at the same time by millions of televiewers?

This evolution from stable representation to volatile image seems to me to prefigure what is being produced, on a different scale, in the evolution of means of transportation. For millennia man has hardly known any other means of transportation than his legs, to which have been added, according to the season or climate, the horse, the camel, the boat. But the advent of the railroad, the automobile, and especially the airplane, marks a radical rupture.

For several decades our skies have been woven with a network of airway lines, beside which the network of maritime and river routes seem prehistoric, just as the network of roads and highways already seem dated. The evolution of messages, of which images are one manifestation, and the evolution of travel (in the broad sense of the term) present a troublesome analogy. Pedestrian, equestrian, railroad traveler, driver, airplane passenger — the *human passenger* has left his original state, aboard ever more rapid engines, in order to span ever greater distances.

Telemics is the term that I have proposed for designating this new dimension in which those who are acting and moving are ever more numerous, and to whom I have given the name *telanthropes*. Here also, everything begins with a process of unbalancing. The weight of bodies tends to fade into the background, as does (in the electronic image) the importance of traditional supports. The speed of travel never ceases to increase, and, even if it still takes 24 hours to circumnavigate the globe, the space shuttle promises to reduce this time to a few hours. Our bodies "take off" like a plane. Even if passengers remain unhurt, it is already noticeable that time change, which is commonplace today, affects our physical functioning. For the first time in history, man makes jumps in time which provoke organic short-circuits, to which he adapts. But is he also adapting to psychological short-circuits which he has never experienced until now?

The cultural field which defines us in a sedentary state brusquely enters into collision with other cultural fields which we cross or which we collide with in travelling. The *enforming* which configures our respective identities to the inside of the milieus to which we belong becomes secondary to the ordeal of the *enmotioning*[9] which provokes a *process of accelerated transformation*. The ever more advanced practice of aviation (which has become a mass practice in our time) is one of the most ostensible manifestations of our new telemic dimension. We are not yet at the stage of "telegraphing" our bodies from one end of the planet to the other, although the idea has been conceived since 1950 by Norbert Wiener, the inventor of cybernetics.[10] Thirty years later, the project already seems less chimerical.[11]

Thus it is good that, as much in the transport of information in the form of images as in the transport of organisms in the form of travel, a new type of imagination is in progress. Symbols have not disappeared; they are changing their nature. Whereas myths used them to organize reality in an "absolute" time that preceded history, technology invents those which it needs for making our everyday history out of its realizations. In a world ceaselessly swept by innovations, it is inconceivable that forms of the imagination that were established so long ago (based on stable modes of existence such as those reflected by past civilizations, all of which acted within a limited area in which each had implanted its particular system of symbols) would be maintained. The imagination in progress in our society is linked *nolens volens* to change, which is the very condition of technology and which recognizes no borders, neither political, historical, cultural, nor ethnic.

Is such constraint irreversible? Common sense, with which politicians pride themselves, or humanism, which intellectuals (at least some of them) call for, lean comfortably towards the negative reply. But already the informationization of society slips as much toward common sense as toward humanism. It is an act of society which gives to the computer the power, not only to use information, but for the first time to stock the collective memory outside of the community, under the anonymous and technical form of memory bank. So already we have become, by the billions, "screenothropes" for whom reality, or at least its image (but electronics makes them one and the same) passes through the screen. So here we are in the process of becoming, soon by the millions, "telanthropes" whose habitat is the planet. Velocity, acceleration, extension, innovation, volatilization — such very well seems to be the orientation in which technology is engaging us, and which we have already to a great extent interiorized.

These lightning developments in all domains have however not, as we might expect, condemned the arts that are rooted in tradition: painting, sculpture, engraving, all of which we find in abundance in the expositions organized by the museums and galleries. Even if representation has been profoundly modified, artists, dealers, critics and collectors would have to believe in its demise. Still, judging from the popular success of certain retrospectives—the Impressionists, Picasso, Monet—or the periodic presentation of "Treasures of Art"—Tutankhamen, the Scythian Gold, Ramses II — one would be tempted to think the contrary. Nevertheless, it is important to recognize that contemporary artistic production is reserved to a relatively reduced class of the population, and that when it concerns itself with historical expositions, the public which attends these, whether or not it is numerous, belongs to the wealthy nations, and more precisely to the capitals and a few large cities. There is no comparison with the innumerable consumers of illustrated magazines, comic strips, romantic picture stories, animated cartoons, cinema, songs, variety shows, radio, televised serials, video cassettes — i.e., mass art, the product of the technological revolution, whose distribution is quasi-universal.

Under these circumstances, do the artists who continue to make art still have a future, when the industrial image — infinitely reproducible, cheap, enviably transformable and manipulable, more often immaterial, in short eminently volatile — is becoming mixed with the very air we breathe?[12]

I am not posing this question lightly — a question to which I would like to answer in the affirmative. Sometimes I am inclined to believe it and to write it. Sometimes I doubt it, especially as their artworks are already caught up in a restless movement. Indeed, what is the market in which, first and foremost, their destiny (that is to say their notoriety and their price) is playing, if not the "machine" for converting aesthetic value into market value? Commercial volatilization has the best of everything: products, goods, services, labor, prestige, skill and, finally, nothing that cannot be commercialized. To which the action of the media adds: and nothing which cannot be vaporized into information.

Is there any possible recourse against total technologization? Is it even desirable? The answer is more difficult, especially as the innovations always seem to be improvements which, moreover, in some cases are often liable to prove themselves noxious as a consequence (thus the famous DDT!). Again, an answer is difficult because we do not hesitate, seduced or petrified by the prowess of technology, to endow the machine with its own imagination. However, neither the automobile nor the TV set nor the airplane are anything but appliances. But when they are functioning, and thus when they are existing, their functioning is never reducible (as we so often believe) to the use that we make of them. The machine (to use the generic term), whether we want to or not, makes us enter into behaviors that correspond to its mode of functioning. The driver is not simply a pedestrian who relocates himself in a vehicle. As soon as he is moving he submits to the rules of traffic, just as much as to the technical capacities of his vehicle. Thus the solidarity of drivers is of another order than that practiced by those who are not driving. In the same way the computer is not simply an instrument capable of dealing with a bunch of data in a minimum amount of time. As soon as we use it, it engages us in thinking and solving problems in the perspectives and modalities which are peculiar to it.

By extension, it is correct to affirm that all new technology is pregnant with cultural revolution, something which history has moreover amply demonstrated, from the invention of the bow, the stirrup and the wheel, to the invention of printing and television, all the way to the invention of intercontinental missiles. It is, therefore, the nature and power of our affinity for technology that endows the machine with the power to which we surrender, to such a degree that our imagination subscribes to it. That is why China, for example, while being at the source of numerous scientific discoveries, has never developed science as autonomous scholarship as it has been established in the Occident. In the Orient, man has always been careful not to cut the tie with nature, while in Europe the rupture, which has become a condition of the spirit, has permitted the establishment of nature simply as an object of observation — and of manipulation — with the view of becoming, technically, the master. These are two types of imagination or of attitude which have led to two types of approach, the consequences of which are at the same time both incalculable and irreconcilable.

Technologism, which has become the prevalent mode of thinking of our age, fundamentally depends upon our ever more ingrained support for the increasing power of technology. Just as rationalism is the system which bases reality on the principle (considered unimpeachable) of reason — a principle by which logic orders and conducts operations in order to explain empirical facts by the processes of observation, proof and demonstration— so, too, technologism is a system which, beginning with reason and science, is based on techniques which are developed by industry and the economy with the view of changing and, if possible, of improving our conditions of life and the conditions of our environment. But while rationalism resorts to key ideas, whose virtue is evidence, technologism resorts to machine-ideas, whose virtue is efficacy.[13] Technologism is the first philosophy for integrating objects and machines into its system. Doing this, it aims less at understanding the world than at transforming it. Such is its principle, and technological development is nothing but its unfolding. Thus vacuum tube computers are replaced by those with transistors, which are replaced in turn with computers with integral circuitry (while waiting for the leap which is being prepared by the Japanese who, beyond the fourth-generation computer, are already working on the fifth-generation computer!).[14] Industrial or postindustrial society has made technologism its credo, the power of which strengthens with every innovation. Until it becomes supreme?

This is the question that presents itself at this point. It asks whether the reality of our existence reduces to only efficacious action, which tends to make us machine-beings, or whether it might overlap (particularly in our condition as individuals) the functioning of the system assumed by the society. Judging from the distortions which are produced — nuisances, pollution, menaces — and the alarming sentiments to which we are subjected — fears, anguish, resignation — it is hard not to accept as fact the general malaise which has taken hold of humanity in a world to which, paradoxically, science and technology have brought more discoveries and inventions than the history of man, from his origin, can count.

It also appears to me none the less certain that only the attachment to value (as opposed to efficacy) is able to, if not replace, technologism, then at least to contain it and maybe even reduce it. Whether it has an aesthetic or an ethical nature, value comes from proposing that action is not the only modality of our existence. Contemplation, dreaming, meditation, inner research, the need to love, to believe, to sacrifice oneself—many dispositions subsist, but do not have much weight in the balance of our society, which is smitten with competition and efficiency.[15] But value is not only concerned with certain activities that are considered "marginal" by technology, or with those activities which technology unites with as soon as it can make a profit. Value qualifies the aggregate of our choices, not according to their operational character, but beginning with, and in view of, ends which give form to a conception of a life freely chosen. Efficacy, the basis of technologism, is strategic in nature; it aims at reaching the fixed objective, with the perspective of gaining victory. Value, which is ethical in nature, aspires to get beyond the field of operations. It points toward a horizon that is not reducible to an objective. It interpolates us as acting and responsible subjects of our actions. It invites us to an enforming, rather than to a struggle for power.

Finally, the problem does not consist in opposing value to technologism. It is not a question of combat or of competition (although our modern imagination is almost entirely occupied with one or the other). The attachment to value, of which art is one of the manifestations, can help us, I feel, to put technology in its place and to keep technologism at a distance. Value implies rules that are agreed upon by common consent. It is distinct from constraints which, once our freedom has decayed, impose themselves like fate. At a time when humanity and the planet are threatened with nuclear volatilization, the ultimate "triumph" of technology, nothing is more urgent than saving our imagination from total surrender, by inspiring it with a sense of respect and marvel. Still, we have to learn to have a clear vision.

Notes

1. (Translated from the French by Pauline Pilote Eby)
2. (Translator's note) The neologism "enformed" is used here to express the special meaning of the French 'mise en forme.'
3. This is the case, among others, of mathematical symbols or of musical notation.
4. As with every neologism, that of the "techno-imagination" is not entirely satisfactory; it nevertheless points to the fact that practically all our relations today are mediated by a machine. One has only to think of the series of operations which compose our daily existence, from the electronic alarm clock, travel by car and the use of the telephone, to evenings spent in front of the television screen, and to wait for this new form of happiness to be distributed to everyone—this new happiness which is the "electronic all," toward which already, after the TV, informatics, telematics, video games, the home computer, etc., are tending.
5. Every image is composed of a multitude of points. Each one of these, a pixel, is defined by its dimension, color and light.
6. Among others, via electronic gadgets which never stop multiplying: thus the newscaster who emerges from the depth of the screen to announce the top story, and who slowly fades into the distance with the final event. But it is already possible to unroll a landscape like a napkin, to make it swing on all its axes and to turn it over like a crepe, to expand or contract it, or even, with the help of the computer, to propose "unreal" totalities. What applies to the landscape applies equally well to all that can be represented.
7. Art Buchwald, *Versailles TV Dinner.* International Herald Tribune, June 12–13, 1982.
8. Perhaps this is what explains the success of American serials that are universally distributed, like *Dallas* or *Dynasty*.
9. (Translator's note) This neologism "enmotioning" has been adopted to express the French 'mise en mouvement'; the obvious parallelism between the English "enforming" and "enmotioning" corresponds to the parallelism of the corresponding French terms.
10. Norbert Wiener, *The Human Use of Human Beings: Cybernetics and Society.* First published, Boston: Houghton Mifflin Co., 1950; Rewritten and republished, Garden City, New York: Doubleday Anchor Books, Doubleday & Co., Inc. Quotation here from the Doubleday Anchor Edition, pp. 103, 104:

    In other words, the fact that we cannot telegraph the pattern of a man from one place to another seems to be due to technical difficulties, and in particular, to the difficulty of keeping an organism in being during such a radical reconstruction. The idea itself is highly plausible. As for the problem of the radical reconstruction of the living organism, it would be hard to find any such reconstruction more radical than that of a butterfly during its period as a pupa.
11. Need we recall — but the reminder is almost banal today — that the idea of putting a man on the moon seemed crazy until one night in July, 1969 . . .

    One of the intermediary operations consists in installing extra-terrestrial colonies, a project on which numerous institutes (as many in Japan as in the

United States) worked for several years. Cf. Magoroh Maruyama, *Art and Technology in the Future*, ch. 9, this volume.

12. Need we recall that the artistic expressions connected with the new media, such as video art and computer art, have known until now a very precarious fate. Is it susceptible to improvement? Or must we believe that, barring important transformations (and thus of their utilization in the new audiovisual genre of "clips"), these expressions show themselves to be incompatible with the nature of the new media?
13. Which does not work without raising unexpected questions. It is thus that rationalism can breed systems that multiply ideas; but the property of ideas is to occupy little space: a few library shelves suffice. In return, the idea-machines, the idea-products, induce a congestion, of which the large cities, crucibles of technologism, are the example — a congestion of inhabitants, of housing, of surplus, of vehicles, of waste. . . . This is, no doubt, one of the reasons that the image was technically developed in the sense of "minimum congestion." The painted portrait, framed and hung on the wall, has given place to photographs that we put in an album or drawer, and then to motion picture film, nowadays to magnetic tape, and finally to the videodisc (54,000 images!).
14. All technical activity has followed and follows the same evolution: need we be reminded of the replacement of traditional agriculture by agricultural machines and fertilizer, or the replacement of stone by concrete, or that of the horse by the automobile?
15. Permit me to recall my note in "Art(s) and Power(s) in Contemporary Society," in *Absolute Values and the Creation of the New World: Proceedings of the Eleventh International Conference on the Unity of the Sciences, Philadelphia, 1982* (New York: The International Cultural Foundation Press, 1983), Vol. 1, p. 109, note 34:

    ". . . Symbolic appropriation is no less important than material appropriation; but in today's world everything is organized so that the second be most sought, and to such an extent that the 'symbolic' in industry has become current practice: to sell a car is also to sell the prestige of its brand or model. But all imaginary needs have not (yet) been industrialized. Appropriation has not 'yet' been reduced to the sole purchasing of a product. Then we should also consider the industry which allows for the appropriation of imaginary values by way of photography, the replacing of the original by a rendering which itself becomes a product."

    The movement generated by technologism consists in interposing between the need (however private or secret it may be) and its satisfaction (even if it is spiritual) a commercial product that can be a marketable object. The art of St. Sulpice had already anticipated this phenomenon, which today takes universal proportions. It is thus that playing, which has for so long been the occupation among children, that has been practically confiscated by video games, of which there is no question of less merit or interest, but all of which have in common the fact of being industrial products sold to the general public,

Contributors

RENE BERGER: Doctor of Aesthetics at the University of Paris (Sorbonne); teacher at the School of Fine Arts and at the University of Lausanne; Honorary President of the International Association of Art Critics and President of the International Association for Video in the Arts and Culture; Past Director and Curator of the Museum of Fine Arts, Lausanne, Switzerland (1962–1981); Consultant to UNESCO and the Council of Europe; leading exponent of video art; author of numerous works on art, television, and art and technology; producer of films on art for television.

ARNOLD BERLEANT: Professor of Philosophy at C.W. Post Center of Long Island University, Greenvale, New York; Holder of B. Mus. and M.A. from the Eastman School of Music and Ph.D. from the State University of New York at Buffalo; Has taught and lectured at numerous American universities and published widely; Author of *The Aesthetic Field, A Phenomenology of Aesthetic Experience;* Since 1979 Secretary-Treasurer for the American Society for Aesthetics.

DANIEL CHARLES: Professor of Philosophy and Dean of the Faculty of Arts, University of Paris VIII Vincennes at St-Denis; Graduate of the Conservatoire National Superieur de Musique, Paris; Former Director (1970–1980) of the Department of Aesthetics at the Sorbonne (University of Paris IV) and of the Department of Music at Vincennes (University of Paris VIII); Author of many articles on modern and postmodern music and on general aesthetics, and of works about John Cage; Co-author with John Cage of *Pour les oiseaux* (Paris, 1977), English translation: *For the Birds* (London and New York, 1982).

LLOYD EBY: Instructor in Philosophy at the Unification Theological Seminary, Barrytown, New York; Taught formerly at SUNY,

Albany; Studied philosophy at Washington University (St. Louis, Missouri) and currently Ph. D. candidate in Philosophy, Fordham University, New York; Author of numerous papers on philosophy, theology, and art and religion; Filmmaker, cinematographer and photographer, including a series of photographs of personal filmmakers taken at Millennium Film Workshop, New York (1970 –1974).

ERVIN Y. GALANTAY: Professor of Urban Design and Planning, Swiss Federal Institute of Technology, Lausanne, Switzerland; Graduate of the Swiss Technical University, Zurich; Has taught at Harvard and Columbia Universities; Worked at the Philadelphia City Planning Commission, and in the New York office of I.M. Pei; contributed to the planning of Ciudad Guayana in Venezuela (1961–1964) and of Owerri, the new capital of Imo State in Nigeria (1975); Represented the Swiss government at the U.N. Conference on Human Settlements (Vancouver, 1978); Designer of major buildings for the State University of New York at Stony Brook, Long Island; Author of numerous articles and books.

TAO HO: Architect, Planner and Designer with degrees from Williams College and Harvard; Principal of TAOHO Design Architects, and Core Member of Asian Planning and Architectural Consultants (APAC); Assistant to the late Walter Gropius; Lecturer in Architecture and Fine Art at the University of Hong Kong; Advisor on design to Hong Kong Polytechnic Institute and Advisor to the Hong Kong Museum of Art; Recipient of numerous awards and honors, and author of many papers on art, architecture, development and planning.

FRIEDRICH KNILLI: Native of Austria; Doctorate in Psychology; Former freelance writer and radio journalist (1959–1962); Professor at the Technical University of Berlin; Director of the Institute for Media Studies and Sociology of Literature and member of the Faculty Board, "Science of Communications and History" (1976 –1980); Guest lecturer in the USA, Netherlands, Australia, Mexico, and Israel; Researcher on anti–Semitism and on the Holocaust; Member of international project on Social Conflicts in Television News; Author of many papers and books on literature, television, film, communications, and the comics.

DIETRICH KUNCKEL: Born in Berlin, Germany; Studied at the Technical University in Aachen, Germany, and at Princeton University; Architect and urban planner; Former Professor of

Urban Design, Central University of Venezuela (1973–1977) and Professor of Design at Simon Bolivar University (since 1977); His private architectural and planning firm in Caracas, Venezuela, is Inteplanconsult S.A.

NICHOLAS KURTI: CBE 1973; FRS 1956; Emeritus Professor of Physics, University of Oxford; Educated in Budapest, Paris and Berlin; Vice-President of the Royal Society, 1965–1967; Taught at Breslau, Oxford, City College of New York, University of California at Berkeley, and Amherst College, and has given numerous honorary lectures including the James Scott Prize and Lecture, the May Lecture, the Kelvin Lecture, the Cherwell-Simon Lecture, the Larmor Lecture and the Tyndall Lecture; Honorary Member of American, Hungarian, French, Finnish and German academies of science; Recipient of numerous awards and medals; Author of many books and papers on physics, magnetism, energy and culinary physics.

JOSE MACEDA: Professor of Music and Chairman of the Department of Music Research, College of Music, University of the Philippines, Quezon City; His articles in scholarly journals and his music compositions performed in Asia and Germany are based on a long experience of field music research in Southeast Asia, and on visits to Africa and Brazil.

MAGOROH MARUYAMA: Born in Japan, moved to the USA in 1950; Graduated from the University of California at Berkeley in Mathematics; Studied at the universities of Munich, Heidelberg, Copenhagen and Lund; Ph. D. in Cultural Epistemology; Author of more than 118 publications; consultant to government agencies and NASA; Fellow of the American Association for the Advancement of Science; Has worked in Africa, the Middle East, Canada, USA and Japan in urban planning, community design and technopolis development; Visiting Professor of International Business at the National University of Singapore (1984); Visiting Professor at the University of Hawaii (1984–1985).

ABRAHAM MOLES: Distinguished social psychologist; Director of the Institute of Social Psychology of the University of Strasbourg, France; Writer of 25 books; Former Professor at the University of Washington, Seattle.

WALTER NUTZ: Studied sociology and ethnology at the Universities of Munich, Mainz, Cologne, Amsterdam and Paris, University lecturer in private capacity at the Universities of Cologne and

Gottingen; Author of publications on problems of popular culture and light reading, as well as on the sociology of mass communication; Head of Deutsche Welle's department for audience research.

ANDREE PARADIS: Born in Montreal, studied at Villa Maria, Notre-Dame College, Ottawa, and the Sorbonne; Former member of the Canada Council and many arts councils and arts advisory boards in Canada; Director and Editor-in-Chief of *Vie des Arts*, Montreal (since 1964); Vice-president of the International Association of Arts Critics (Paris); Member of the Universities Art Association of Canada; member of the International PEN Club; Centennial medal and Officer of the Order of Canada.

NICHOLAS J. PHILLIPS: Born in England and educated at Imperial College with research in theoretical physics and fusion physics; Senior Research Fellow, A.W.R.E. Aldermaston (1959–1962); Research Scientist, Sperry Rand Research Center, Sudbury Massachusetts (1962–1963); Theoretical Physicist at English Electric, Whetstone, Leicester (1963–1965); Joined the Department of Physics, University of Technology, Loughborough, Leicestershire, England in 1965 and since 1982 Reader in Physics there; For past 12 years a leading researcher on holography, with more than 18 papers, hundreds of lectures and numerous radio and TV programs on that topic.

FREDERICK POHL: Has been called "the finest writer science fiction has yet produced" by Kingsley Amis, and "the shrewdest thinker on the future, both in fact and fiction, that I have ever met" by Isaac Asimov; Recipient of nearly every award available to a science fiction writer, as well as the Popular Culture Association Award, the American Book Award and many others; His most recent novels are *The Years of the City* (Simon & Schuster/Timescape) and *The Merchants War* (St. Martins Press).

ALPHONS SILBERMANN: Born in Cologne; Studied law and sociology at the Universities of Cologne, Freiburg and Grenoble; With the coming of the Nazis, emigrated to Australia where he became Lecturer in Aesthetics at the University of Sydney; After the war held Professorships at the Universities of Cologne, Lausanne and Bordeaux; Specialist in Sociology of Mass Communication and Sociology of the Arts; Director of the Cologne Institute of Mass Communications; Editor of the international journal *Communications;* President of the German Association of Mass

Communications Research; Author of numerous books and articles.

ROBERT STEPHANE: Born in Liège; Studied political and social science at the University of Brussels; In 1960 an Eisenhower Fellow at Columbia University (USA); Has worked since 1954 in radio and TV in Belgium; From 1963 Chief of the Center of Production of Radio and TV, RTB Liège; In 1972 co-founder of the Coopérative Internationale de Recherche et d'Action Matière de Communication (C.I.R.C.O.M.), and in 1973 Secretary-General of C.I.R.C.O.M.; In 1977 co-founder and administrator of INPUT (International Public Television); 1980 Regional director of the Center of Production of RTBF Liege; 1983 Vice-president of TV 5 (French satellite TV); 1984 Adminstrator-general of RTBF; Producer of numerous TV shows and programs; Has participated in seminars and taught at the Universities of Wisconsin and Berkeley, and the University of Liège.

EDDIE WOLFRAM: Born in Germany, now a British subject; Leading painter with more than 55 exhibitions since 1957, as well as producer, director, set designer and writer for television, a director of TV film commercials, and a producer of pop records; From 1968–1974 a Contributing Editor for *Art & Artists*, and now writes about art for numerous magazines including *Art & Artists*, *Studio International*, *Cimaise*, *D'Ars Agency*, *Arts Review*, *Ink* and *Harpers & Queen;* Since 1971 Head of Fine and Theater Arts, School of Art and Design, Croydon College, Croydon, Fairfield, England.

WOO HYUN WON: Graduate of Seoul National University, College of Law (LL. B.), and of Boston University (Ph.D.); Professor of Mass Communication and Head of Educational Media Institute, Korea University, Seoul, Korea.

Sources

The content of this volume was derived from papers presented in Committee III ("Art and Technology") at ICUS XII ("Absolute Values and the New Cultural Revolution"). The symposium, one of six sponsored by ICUS, of the International Cultural Foundation, Inc., was held at the Chicago Marriott Hotel, November 24–27, 1983, and was chaired by Professor René Berger.

Index

Page numbers in **bold face** refer to sections written by the subject. Numbers in parenthesis refer to footnotes which appear on the indicated page.

A

B

I

J

K

L

M

N

O

P

Q

R

S

T

U

V